Reliability, Maintainability, and Availability Assessment

Second Edition

Also available from ASQC Quality Press

Reliability Statistics
Robert A. Dovich

Volume 6: How to Analyze Reliability Data
Wayne Nelson

Reliability Methods for Engineers
K. S. Krishnamoorthi

Failure Mode and Effect Analysis: FMEA from Theory to Execution
D. H. Stamatis

Weibull Analysis (with software)
Bryan W. Dodson

To request a complimentary catalog of publications, call 800-248-1946.

Reliability, Maintainability, and Availability Assessment

Second Edition

Mitchell O. Locks

ASQC Quality Press
Milwaukee, Wisconsin

Reliability, Maintainability, and Availability Assessment, Second Edition
Mitchell O. Locks

Library of Congress Cataloging-in-Publication Data

Locks, Mitchell O.
  Reliability, maintainability, and availability assessment / Mitchell O. Locks. — 2nd ed.
    p. cm.
  Includes bibliographical references and index.
  ISBN 0-87389-293-3
  1. Reliability (Engineering) 2. Plant maintenance. I. Title.
  TS173.L6 1995
  620'.00452—dc20   94-40648
                                                                     CIP

© 1995 by ASQC
All rights reserved. No part of this book may be reproduced in any form or by any means, electronic, mechanical, photocopying, recording, or otherwise, without the prior written permission of the publisher. No permission is required for limited reproduction of blank probability graph paper, as follows: Figure 4.3: Normal probability graph paper; Figure 5.1: Semilog graph for exponential plotting; Figure 5.4: Exponential plot for larger sample sizes; Figure 9.1: Weibull probability graph paper; and Figure 9.4: Weibull hazard graph paper.

10 9 8 7 6 5 4 3 2 1

ISBN 0-87389-293-3

Acquisitions Editor: Susan Westergard
Project Editor: Jeanne W. Bohn

ASQC Mission: To facilitate continuous improvement and increase customer satisfaction by identifying, communicating, and promoting the use of quality principles, concepts, and technologies; and thereby be recognized throughout the world as the leading authority on, and champion for, quality.

Attention: Schools and Corporations
ASQC Quality Press books, audio, video, and software are available at quantity discounts with bulk purchases for business, educational, or instructional use. For information, please contact ASQC Quality Press at 800-248-1946, or write to ASQC Quality Press, P.O. Box 3005, Milwaukee, WI 53201-3005.

For a free copy of the ASQC Quality Press Publications Catalog, including ASQC membership information, call 800-248-1946.

Printed in the United States of America

   ∞ Printed on acid-free recycled paper.

ASQC
Quality Press
611 East Wisconsin Avenue
Milwaukee, Wisconsin 53202

CONTENTS

Preface — xv
Acknowledgments — xvii

Introduction — xxi
Components vs. Systems — xxi
Reliability and Confidence — xxi
   Parametric Reliability Assessment — xxi
   Families of Parametric Distributions — xxii
   Maximum Likelihood and Bayesian Methods — xxiii
   Normal Theory Assessment — xxiv
   Weibull Distributions — xxiv
Maintainability and Availability — xxv
System Reliability — xxvi

Part I: Basic Probability Theory — 1
Chapter 1: Basic Concepts of Probability — 3

Attributes and Outcomes — 3
The Universal Set — 4
   Subsets and Events — 4
   Unit Events — 5
   Counting Events — 6
Compound Events — 6
   Unions and Intersections — 6
   Disjoint (Mutually Exclusive) Events — 7
   Partitions — 7
Probability — 9
   Acceptable Assignment of Probabilities — 9
   Joint Events and Dependence — 10
   Independence — 10
   Additive Probabilities of Disjoint Events — 10
   Conditional Probability — 10
Bayes Theorem — 12
Problems — 15
References — 17

Chapter 2: Probability Distributions — 19

Introduction — 19
Probability Functions — 19
   Distribution Functions — 20
   Continuous-Valued Random Variables — 20
Families of Related Distributions — 21
   Uniform Distributions — 22
   Binary Random Variables: The Binomial Distribution — 23
   Exponential Distributions — 24
Measures Related to Parameters of Distributions — 25
   Mean: The Expected Value — 25
   Variance and Standard Deviation — 25
   Percentage Points — 27
   Reliability and Percentage Points — 27
   The Mode — 27
Random Sample Summary Statistics — 28
   The Sample Mean — 28
   The Variance of the Sample Mean — 29
   The Sample Variance: Correction for Bias — 30
   Degrees of Freedom — 31
Standardized Random Variables — 31
Problems — 32
References — 34

Part II: Families of Probability Distributions — 37

Chapter 3: Attributes, Go–No Go, or Zero-One Processes — 39

Introduction — 39
Binary Trials — 39
   Orderings or Permutations — 40
   Indistinguishable Orderings of n Binary Trials — 40
   Combinations: Sets of Outcomes with the Same Result — 40
Binomial Probability — 42
   Binomial Distribution Function — 42
   Tables of the Binomial Distribution — 43
   Mean and Variance of the Binomial Distribution — 43

Negative Binomial (Pascal) Distribution	45
Dual Binomial and Negative Binomial Relationship	45
Symmetry of the Negative Binomial Distribution Function	46
Average Sample Size: The Negative Binomial Mean	46
Lower Confidence Limit on the Reliability: Discrete Distributions	47
Beta Distributions	48
Confidence Limits for the Reliability	50
Beta Distribution Percentage Points for Reliability Assessment	50
Problems	51
References	53

Chapter 4: Normal Distributions   55

Introduction	55
Standardized Normal Distribution	56
The p.d.f. and the d.f.	56
Shape, Shift, and Scale	57
Standard Probability Table: Symmetry	57
Probability Plotting and Graphing	59
Normal Probability Graph Paper	59
Plotting Points	61
Probability Plotting With Grouped Data	63
Goodness of Fit: The Lilliefors K–S Statistic	63
General Procedure of a Goodness-of-Fit Test	63
The Lilliefors K–S Test for an Unspecified Normal Population	65
Approximate Confidence Limits for Attributes Data	67
The Standardized Form	67
Approximate Reliability-Confidence Limits: Yates Correction Factor	67
The Lognormal Distribution	69
Lognormal Probability Plotting	70
Interpreting a Lognormal Plot and g.o.f. Test	70
Goodness of Fit	72
Appendix to Chapter 4: Advanced Topics and Related Distributions	73
Some Properties of Related Distributions	73
Normalizing the Probability	73
Change-of-Variables Technique	73

Mean and Variance of a Normal Distribution 74
Chi-Square Distributions 74
   Distribution of the Normal Sum of Squares 74
   Percentage Points of Chi-Square Distributions 76
Student-t Distributions 77
Problems 77
References 80

Chapter 5: Exponential Times to Failure 83

Introduction 83
Failure of a Single Component 84
   The Exponential Distribution Function 84
   The Reliability 84
   Exponential Probability Plotting With Semilog Graph Paper 85
   Life Tests and Censored Data Sets 88
   Larger Sample Sizes: Wayne Nelson Exponential Plotting Paper 88
Exponential Goodness of Fit: The Lilliefors K–S Test 93
Multiple Failures and Redundancy 95
   The Gamma and Chi-Square Distributions 95
   The Poisson Distribution 97

Appendix to Chapter 5: Derivation of the Exponential Distribution: Relationships Between Exponential, Gamma, Chi-Square, and Poisson Distributions 99

An Example of a Poisson Process 99
   The Exponential Distribution 99
   Independence from Prior Data 100
   The Gamma Distribution: Multiple Failures 101
   Chi-Square Distribution 102
   Poisson Distribution 103
Problems 104
References 105

Part III: Reliability-Confidence Assessment 107

Chapter 6: Statistical Assessment: The Bernoulli and Poisson Processes 109

Introduction 109
   The Bernoulli Parameter Is p 109

The Poisson Parameter Is λ	110
Complete Data Sets	110
Censored Data Sets	110
The Maximum-Likelihood Method	111
The Likelihood Function	111
Properties of Maximum-Likelihood Estimates	111
Attributes Data	112
The m.l.e. of the Reliability	112
Confidence Level for the Reliability	112
Censored Attributes Data	113
Exponential Distributions	114
Maximum-Likelihood Assessment: The Epstein–Sobel Technique	114
Ordered Failure Times: The Maximum-Likelihood Estimate	114
Epstein–Sobel Confidence Limits	116
Minimum Variance Unbiased Estimation	117
Poisson Process	117
The Two-Parameter Exponential Distribution	118
General Considerations	118
Best Linear Unbiased Estimates	119
Problems	121
References	122

Chapter 7: Bayesian Reliability Analysis: Attributes and Life Testing — 125

Introduction	125
Lower Confidence Limits	125
Continuous Prior and Posterior Distributions: Sufficient Statistics	126
Current Data	126
Conjugate Distributions	126
Varieties of Poisson-Process Prior Distributions	127
Structure of a Bayesian Analysis	127
The Discrete Case	127
Bayesian Reliability	130
Attributes Data	130
The Beta Probability Distribution	130
Fixed Number of Tests: Binomial Sampling	131
Fixed Number of Occurrences: Negative Binomial Sampling	132

Bayesian Reliability Assessment for Attributes Data	132
Time-to-Failure Data	133
Gamma Prior Distribution	133
Fixed r	133
Fixed t	134
Bayesian Reliability Assessment: The Poisson Process	135
Negative-Log Gamma Prior Distribution on the Reliability	136
Similarities to Beta Distributions	137
The Diffuse n.l.g. Initial Uniform Prior	138
Reliability Assessment with the n.l.g. Prior Distribution	139
Appendix to Chapter 7: The Bayesian and Epstein–Sobel Models for Reliability Assessment of the Poisson Process: Interval Estimation	141
Similarities and Differences of the ES and Bayesian Models	141
Summary of Assessment Formulas	142
Interval Estimation	144
Problems	146
References	147

Chapter 8: Normal Theory Tolerance Analysis   149

Introduction	149
Tolerance Limits and Tolerance-Limit Factors	150
Appropriateness of Normal Theory	150
Historical Note	150
One-Sided Analysis	150
Upper Specification Limit	150
Relationship to Noncentral-t Tolerance-Limit Factors	151
Lower Specification Limit	153
Two-Sided Tolerance	154
Tolerance-Limit Factors	154
Problems in Developing Two-Sided Tolerance-Limit Factors	155
Hald's Approximate Two-Sided Tolerance-Limit Factors	156
Problems	156
References	158

Chapter 9: Hazard Analysis and the Weibull Distribution   161

Introduction	161

Life Testing, Hazard, Mortality Tables, and the Weibull Distribution	161
Applications of the Weibull Distribution	162
Historical Note	162
Hazard	163
Constant Failure Rate: The Exponential Distribution	163
Discrete Hazard Analysis: Halley's Method	163
Probability Functions and Probability Plotting	164
The Exponential Distribution	164
Elements of Weibull Probability Plotting	164
Weibull Probability Density Function	165
Two-Parameter Weibull Distribution	167
Weibull Probability Graph Paper	167
Hazard Functions and Hazard Plotting	169
Weibull Hazard and Cumulative Hazard	170
Plotting and Fitting Weibull Hazard Paper	171
Precise Weibull Estimation and Confidence Assessment	176
Purpose of this Section	176
Mann Best Linear Invariant Estimation	176
Linear Estimation of a Location and Scale Parameter	176
Confidence Bounds for the Reliable Lifetime and the Reliability	180
Problems	182
References	188

Part IV: Maintainability and Availability 191
Chapter 10: Maintainability and Availability 193

Introduction	193
Relationship to Queueing	193
Maintainability	194
Availability	194
Maintainability	196
Exponential TTR: Point Estimation and Confidence Assessment	196
Lognormal TTR	198
Inherent Availability A_I: Exponential TTR and TBF	201
Exponential TTR and Exponential TBF: The Prior Distributions	202

Approximate Confidence Bounds for A_I by Monte Carlo
Simulation — 202
Observed Availability A_O: Exponential Failure and Exponential
Repair — 205
    Differences between A_O and A_I — 205
    The Simulation Program for A_O — 206
Observed Availability A_O: Weibull TTF and Lognormal TTR — 208
    Weibull Time to Failure — 208
    Lognormal Time to Repair — 208
References — 210

Part V: System Reliability — 213

Introduction to System Reliability — 213

Chapter 11: Inclusion-Exclusion and Related Matters — 217

Introduction — 217
    Boolean Source-to-Terminal Networks — 218
    Network Graphs: Minimal Paths and Minimal Cuts — 218
Inclusion-Exclusion — 218
    Example: Bridge Network Reliability by IE — 218
Minimized Inversion of Paths or Cuts — 221
    The DeMorgan Theorems — 221
Series and Parallel Reductions — 223
Approximations — 225
    The Esary–Proschan Lower Bound — 225
    Approximation with a Subset of the Minimal Cuts — 226

Chapter 12: Sum of Disjoint Products — 229

Introduction — 229
Logical and Probabilistic Aspects of Disjointness — 230
Example: The Bridge Circuit — 231
The ALR Algorithm — 232
    The Outer Loop — 232
    The Inner Loop — 233
    Features of the Algorithm — 233
Reliability Importance — 239
    Importance: A Partial Derivative — 239

Chapter 13: Topological Reliability — 241

Introduction — 241
   Principal Differences Between TR and IE — 242
Concepts of Topological Reliability — 242
   Terminology — 243
The Rules of Tree Search — 246
   The Weight Restriction — 246
The Four Processing Rules — 247
   Backtracking and Stopping — 248
Deriving the System Formula — 248
   Node Labels and Processing Order — 248
   The System-Reliability Formula — 248
   Relationship to the IE Formula — 249
Extensions of TR: Unified Approach: k-Terminal Network — 252
   Introduction — 252
   Example: Source-to-k-Terminal Problem — 253

Problems and References for Part V — 257

Problems on System Reliability — 257
References on System Reliability — 262

Solutions for Problems: All Chapters — 267

Chapter 1 — 267
Chapter 2 — 267
Chapter 3 — 268
Chapter 4 — 269
Chapter 5 — 270
Chapter 6 — 270
Chapter 7 — 271
Chapter 8 — 271
Chapter 9 — 272
Part V — 276

Additional References — 278

xiv ■ *Table of Contents*

Appendix 281

Table A.1. Fixed Sample Size (n) and Confidence Level (γ): Reliability (p) Assessed as a Function of the Number of Successes (r) or the Number of Failures ($n-r$) — 281
Table A.2. Standard Normal Distribution Function $F_N(z) = P(Z \leq t)$ — 283
Table A.3. Plotting Points $\hat{F}_i$ for Probability Plotting — 285
Table A.4. Percentage Points of the χ^2 Distribution — 286
Table A.5. One-Sided Tolerance Limit Factors for a Normal Distribution — 287
Table A.6. Two-Sided Tolerance Factors for a Normal Distribution — 288
Table A.7. Table of Weights for Best-Linear Weibull Estimates — 290
Table A.8. Confidence Bounds for Two-Parameter Weibull Distributions for Censored Samples of Size 3(1)25 — 316
   a. Percentiles of the Distribution of $V_{.90}$ — 316
   b. Percentiles of the Distribution of $V_{.95}$ — 323
   c. Percentiles of the Distribution of $V_{.99}$ — 330

Index 337

PREFACE

This book covers quantitative reliability assessment at a technical depth within the capabilities of quality assurance and reliability engineers, for classroom instruction for both industrial short courses and college courses, as well as a reference for industrial applications. It follows the same general outline as the first edition, which appeared in 1973, but with some significant enhancements and additions. The following features should be noted.

1. The four standard widely used parametric reliability models are included (binomial-beta, normal-lognormal, exponential-gamma, and Weibull), as well as tolerancing, maintainability, and availability; hazard analysis including nonparametric methods; Bayesian techniques; goodness of fit; graphing reliability distributions; and system reliability.
2. The quantitative assessment viewpoint is maintained throughout. This means not only providing a numerical probability of success, but also an interval estimate for this probability. Thus, for most topics discussed in this book, a technique is described for obtaining the confidence level for any desired value of the reliability.
3. There are many examples and problems throughout. The problems at the ends of the chapters are based largely on industrial experience; solutions are also provided.
4. Several different types of blank reproducible probability graph paper are provided along with examples and instructions on how to plot and interpret the graphs, for normal and lognormal, exponential, Weibull, and Weibull hazard plotting.
5. There are simple procedures for both binomial-beta and exponential-gamma Bayesian assessment. Binomial assessment is by a one-page appendix table that covers a wide variety of possible cases; exponential-gamma assessment requires a chi-square table. The relationship between the Bayesian exponential time-to-failure model and the widely used Epstein–Sobel model is also developed. The negative-log-gamma prior distribution on the reliability is discussed, as well as its role in the Bayesian model.
6. Exact methods for linearized Weibull estimation and confidence-bound assessment are provided, along with extensive tables of weights for values obtained from life-test data.

7. Maintainability assessment in the exponential case is provided with a Bayesian approach similar to that used for nonrepairable systems and components. Availability assessment includes both Bayesian methods and computerized Monte Carlo simulation.
8. Part V, which was not included in the first edition, consists of three chapters on system reliability that incorporate the results of recent research, including the classical method of inclusion-exclusion, approximations, disjoint products, reliability importance, and topological reliability.
9. An attempt was made to keep chapters relatively free of notation. Technical appendices are provided for several chapters.
10. Normal-theory reliability-confidence assessment includes both one-sided and two-sided tolerancing.

ACKNOWLEDGMENTS

This book is the result of more than 30 years of work in the field of reliability and the opportunities that have been given to me to both teach the subject and perform research in it. The first edition, which was issued in 1973 by Hayden Books, was an expansion of a syllabus for a short course I presented on behalf of the Institute for Advanced Technology, which was at that time a part of the Control Data Corporation. Over time, it became clear that there was a need for a revision to add new topics, more tables, and more problems, as well as to clean up the rough spots. In 1993 the Quality Press of the American Society for Quality Control decided to publish this text, and the Institute for Advanced Technology transferred the copyright.

The technical influences on a person's work come from many different directions. To acknowledge the contributions of the sources that provided some of the main ideas of this book, I should like to concentrate on three aspects of reliability that are covered differently here than in most other books on the subject: Bayesian methods, exact Weibull methods, and system reliability. This description will also help to explain some of the differences between this book and its first edition.

My interest in Bayesian methods was initially piqued by a casual contact in 1957 with R. A. Fisher, and subsequent correspondence with him. While Sir Ronald was not identified as a member of the Bayesian school, his continuing polemics on the uses and interpretations of subjective probability, Bayes Theorem, and decision theory made it an interesting subject to pursue. The opportunity to learn more about and apply Bayesian methods in reliability came during the early 1960s when I was with the Mathematics and Statistics Group of the Rocketdyne Division of North American Rockwell Corporation in Canoga Park, California, working on a research contract with Wright–Patterson Air Force Base monitored by Dr. H. Leon Harter. The contents of chapter 7 on Bayesian reliability analysis reflect some of the knowledge gained as a result of these contacts.

During the period at Rocketdyne, my coworker Dr. Nancy Mann developed exact methods of estimating parameters of a Weibull distribution from life-test data based on a transformation of the extreme-value distribution. In order to use these techniques and to establish confidence bounds for the reliability with them, it is necessary to employ very large sets of tables which are available in monograph form but rarely found published in a book. In chapter 9 of this book, in

xviii ■ *Acknowledgnemts*

addition to better known graphical Weibull and Weibull hazard methods, we incorporate both the Mann BLI (best linear invariant) exact method for Weibull estimation and the Mann–Fertig–Scheuer method for obtaining confidence bounds; the complete sets of tables for both methods are also printed here as Tables A.7 and A.8.

System reliability became a particular interest of mine after participating in the Saturn space-propulsion vehicle program of the Space Division of Rockwell during the late 1960s. I began by documenting a computer program for system-reliability calculation developed at Rockwell that incorporated the classical method of inclusion-exclusion. While on the faculty of the College of Business Administration at Oklahoma State University from 1970 to 1986, in an atmosphere that was friendly to both faculty research and good teaching, under contractual sponsorship of both Wright–Patterson Air Force Base and the U.S. Air Force Office of Scientific Research, some of my graduate students developed a computer program for system reliability-confidence assessment that combined inclusion-exclusion, Bayesian methods, and Monte Carlo simulation.

My work in system reliability has continued, and throughout the 1970s and 1980s. I have a number of publications on the subject, mostly in the *IEEE Transactions on Reliability*. I became acquainted with disjoint products (based upon a paper by J. A. Abraham) a technique for system-reliability calculation which is more efficient than inclusion-exclusion, and developed alternatives to Abraham's algorithm in conjunction with Dr. John M. Wilson of Loughborough University, Loughborough, England. In 1981 during an academic sabbatical with Dr. R. Barlow at the University of California at Berkeley, I became acquainted with Dr. A. Satyanarayana, who originated topological methods of system reliability analysis. Part V of this book, on system reliability, has three principal parts: chapter 11 on inclusion-exclusion, chapter 12 on disjoint products, and chapter 13 on topological reliability.

I am very grateful to the students who helped develop and edit this book in many ways, and to their employers who cooperated to make this help possible. Early versions of this second edition were tested in classes of Reliability 512, a course in the Master of Science of Quality Assurance program at California State University, Dominguez Hills, a program directed by Steve Kozich. Many of the problems at the ends of the chapters were adapted from student term projects for that course, which were often based on reliability engineering problems at their own places of employment. However, the number of persons who made contributions directly or indirectly to this text is so large that it is impractical to list them all; my regrets and apologies go to all those who are not named here who have made such contributions.

Most of all, I thank my wife and helpmate, Rochelle, who provided help, inspiration, and moral support at every step of the way through changes of jobs, moving, shifts in the direction of work, and in preparation of both editions of this book.

INTRODUCTION

Components vs. systems

This book deals with elementary techniques for assessing or demonstrating the reliability, maintainability, or availability of a component or system. Particularly because of the availability of microminiaturized elements, there isn't always a sharp dichotomy between component and system; physical size doesn't matter. A component, for example, could be a complex integrated assembly of components and subsystems, such as a computer memory chip. For purposes of this book, what counts most in distinguishing between a component and a system is how the assessment is performed. It shall be assumed that a component is assessed based upon data obtained in testing it as a unit. For a system, on the other hand, a reliability evaluation is more likely to be a prediction, based upon two types of data: failure data obtained by testing the components, and a system formula derived by analyzing the system logic.

Reliability and confidence

Reliability (R) is a measure of the quality of a device or black box; it is the opposite of what is often called the *quality level* in the classic literature of quality assessment, the proportion of defective units in the population. The meaning of R is *probability of performing successfully under ordinary operating conditions*. R is usually assessed from observed success-or-failure data or field information relative to performance, under either actual or field conditions, test or simulation. Since random test results can vary, the estimated value of R can be different from one set of data to another, even without any substantial changes in physical characteristics of the devices, services, software, and so forth, being tested. Therefore, associated with the estimate of R, there is a secondary associated measure of the quality of the estimate, called the *confidence level*, which depends upon both the amount of data available (number of trials, and so on) and the results observed.

Parametric reliability assessment

Frequently, it is assumed that the component success-or-failure data are governed by a parametric probability distribution. This means that the probability law governing the value of some random variable that defines success or failure can be described completely by a mathematical formula. When that is the case, there is a group of related distributions called a *family,* for which each member of the family differs from all the others only in the value of constants called *parameters.* For example, every member of the family of normal distributions is identified by the values of the population mean (μ) and standard deviation (σ); every member of the family of binomial distributions is identified by the probability of success on a single trial (p) and the number of trials (n).

The principal steps in parametric assessment are first, identify the family, justifying the selection either by experience, a statistical test, a logic model, or an assumption. Then from the failure-history data, calculate the parameters of the distribution in that family which is in some sense a best estimate. For every value of the random variable x governed by that distribution, there is a cumulative probability p. If it is required, for example, that x be at least a certain specified value x_L, the associated probability p_L is the reliability, and the confidence level is the probability γ, based upon all of the available data, that the reliability is at least p_L; p_L is the *lower confidence limit* (LCL) on p at confidence level γ.

Families of parametric distributions

In this book, I emphasize certain well-known families of distributions with applications to a very large proportion of the parametric applications of reliability. Four groups are included.

1. *The binomial distributions* are used for attributes, zero-one or go–no go testing. For example, if a normally closed valve is supposed to open on demand, the reliability of the valve is the probability that it opens if there is forward flow, and then closes again when the flow ceases.
2. *The normal distributions* are used for continuous valued variables when the requirement upon which a reliability estimate is based is that the value of the variable shall be within predefined limits. An example is the pressure buildup in a boiler: the boiler does not operate properly if the pressure is too low, and it bursts if the pressure is too high. The reliability is the probability that the operating pressure is between two specified limits, the lower limit or the operating margin, and the upper limit or the safety margin.

3. *The exponential and Poisson distributions* are used when failures occur at a constant level of intensity that does not vary with the accumulated service life. In a seminal paper, D. J. Davis gives some applications in connection with payroll-checkwriting errors, linotype machine failures, radar set failures, and many other examples.
4. *The Weibull distributions* are used when the rate at which failures occur changes monotonically with the accumulation of service life. For example, the failure rates for semiconductors decrease with continued usage at proper operating temperatures, while the failure rates for ball bearings tend to increase.

The choice of a family of distributions is based upon a combination of experience and convenience. For example, the exponential distribution is appropriate if the rate or intensity at which failures occur is constant. If the failure rate were changing rather than constant, the Weibull family would be preferred to the exponential. Frequently, the choice of a family can be aided substantially by graphical techniques or a statistical test, called a *goodness-of-fit* test. Graphical techniques are particularly convenient for normal, exponential, and Weibull families because special types of graph paper are available. This book includes special graph paper for every one of these families and instructions on how to plot data points and estimate parameters.

Maximum likelihood and Bayesian methods

For both the exponential-Poisson and the binomial models, the book covers reliability-confidence assessment using maximum likelihood in chapter 6, and Bayesian techniques in chapter 7. Maximum likelihood obtains those estimates of the parameters that are more probable than any other possible values in the light of the observed data. Bayesian methods consider estimation for the viewpoint of a distribution on the reliability; the observed data modify the shape of the distribution from the form it had assumed in prior experience.

Although both the maximum likelihood and Bayesian techniques approach estimation from different points of view, they are closely related mathematically and give practically the same results for large data sets. In the binomial case, reliability-confidence assessment for either of these two methods is based upon tables of the binomial or beta distributions; as a substitute for these bulky tables, a special one-page table (Table A.1), for assessing reliable components for sample sizes up to

200 is provided. For the exponential-Poisson case, assessment is done either by the Epstein–Sobel technique or a Bayesian modification of Epstein–Sobel, using χ^2 (chi-square) tables.

Normal theory assessment

The normal and lognormal distributions are discussed in chapter 4. Chapter 8 covers assessment for both one-sided and two-sided limits on the value of a normally distributed random variable representing a performance characteristic. Both tabular and graphic methods are discussed.

The lognormal distribution is a form of normal distribution with the logarithm of the random variable representing performance normally distributed. Applications have been found for the lognormal with skewed data, such as times to failure, pressures, and so on. There is also evidence that the lognormal fits many types of time-to-repair data quite well. Chapter 10, on availability and maintainability, emphasizes uses of the lognormal, as well as of the Weibull distribution.

Weibull distributions

The Weibull distribution, named after the Swedish scientist W. Weibull, is especially useful for analysis of time-to-failure data. Weibull probability functions are closely related to the exponential. It is relatively simple to use Weibull graph paper to perform approximate goodness-of-fit analysis and parameter estimation. Because a Weibull distribution has one parameter more than the simple exponential functions, called the *shape parameter*, the analyst can ascertain from the value of this parameter whether the failure rate is decreasing over time, constant, or increasing.

Weibull estimation and goodness-of-fit determination have been greatly aided by the development of two different types of auxiliary techniques for estimating the parameters. The first of these is *hazard analysis*, which relates time to failure to cumulative failures with a relatively simple mathematical function that can be plotted on calibrated log-log graph paper. A major advantage of hazard calculation and plotting is that it can be used with multiple censoring, such as frequently occurs with historical data sets, where a significant number of items in operating condition are prematurely removed from service at different times, possibly for administrative reasons.

The second auxiliary family of techniques discussed in this book for estimating Weibull distributions is linearized weighting of the times to failure. The Mann best linear invariant (BLI) method incorporates an extensive set of tables of weights to calculate Weibull estimates from observed life-test data, including also censored life tests that were terminated before all items had failed. The estimates obtained are functions of linear combinations of the logarithms of the times to failure. This procedure is further supplemented by goodness-of-fit testing of the estimates, as well as confidence-bound assessment of the reliability, for the same censored or complete data sets as the Mann table of weights. The advantage of the linearized weighting method is precision that may not be otherwise obtainable; the parameter estimates do not differ from one analyst to another, as is the case with probability plotting that is subject to visual error.

Chapter 9 covers Weibull estimation, both by probability plotting and hazard plotting, as well as by the BLI method, and confidence-bound assessment of the reliability. Both Weibull and Weibull-hazard graph paper are included. A complete set of tables of BLI weights appears in Table A.7; and Table A.8 is a set of tables for confidence-bound assessment. Chapter 9 also includes a section on nonparametric hazard analysis, by modifications of a 300-year-old method that is still in use by insurance companies and actuaries for analysis of mortality data.

Maintainability and availability

Maintainability (M) and availability (A) are assessed for repairable systems or components. Both M and A, like R, have the dimensions of a probability, a number in the range zero to one. The principal input data are failure times and repair times. The difference between M and A is that M is a measure of the effectiveness of repair performance for the period of restoration to service, while A is a measure of total performance effectiveness, the ratio of actual successful operating time (including planned or preventive maintenance) to scheduled time (including emergency maintenance).

Maintainability is concerned with a single random variable, the repair time for a failed system or component. Thus, assessing M for a repairable element can be compared to assessing R for a nonrepairable component. If the time to an event is governed by a parametric normal, exponential, or Weibull process, confidence levels are assigned parametrically by a formula or by consulting a table, comparable to what is done with R.

Availability, however, is concerned with two random events, failure and repair. Therefore, assigning confidence levels to A cannot readily be done by plugging numbers into a formula. Monte Carlo simulation is employed, based upon estimated values of the parameters of both the time-to-failure (TTF) and time-to-repair (TTR) distributions. When both the TTF and TTR are exponential, the problem can be recast into a Bayesian framework, so that not only the times to events are simulated, but also the parameters of the TTF and TTR distributions. The relevant formulas are similar to those obtained for the analysis of queues and waiting lines.

Chapter 10 covers M and A assessment as follows: M: exponential or lognormal TTR; A: exponential TTR and TTF, lognormal TTR and Weibull TTF. Confidence assessment is incorporated into the discussion for each case, including the use of Bayesian techniques.

System reliability

Part V, which was not included in the first edition of this book, covers system reliability, which is often called *reliability prediction*. The type of problem considered in Part V differs from that of the first four parts of this book in the sense that the earlier chapters deal with the reliability of a single component, for the most part based upon life-test data. Part V takes the viewpoint of a complex system consisting of a set of interdependent elements that all function together in order to achieve a specified objective.

Typically, a system has both essential elements, without which it cannot function, and redundant elements such that a working component can take over the function performed by a failed component or one that is undergoing repair or preventive maintenance. For the most part, analysts treat the system as a network with every element being a binary component that has two values, success or failure. Since the question to be answered is the probability that the system proper succeeds, the analysis of this problem lends itself to Boolean algebra and logic, with the number 1 denoting success and 0 denoting failure. The reliability of the system is estimated based upon the logical organization of the elements, their respective probabilities of success or failure, and an algorithm for converting this information into the numerical value of the system probability of success or failure.

Part V has three chapters, each chapter covering a particular algorithmic technique. Chapter 11 deals with inclusion-exclusion (IE), frequently called the classical method of deriving a system-reliability formula, and also includes some of the basic concepts and definitions (such as minimal path, minimal cut, fault tree, and inversion) and elements of Boolean logic.

Chapter 12 is concerned with the sum of disjoint products (SDP) and shows how to use the minimal paths to derive a formula for the system reliability; SDP formulas are frequently an order of magnitude or more smaller than the formulas obtained by IE. It is also convenient to use disjoint system formulas to derive formulas for the relative importance values of the components.

Chapter 13 covers topological reliability (TR), which employs mathematical graph theory and computerized search methods to derive a system formula that is closely related to IE, but in a considerably shorter form and without any of the duplications that are a necessary part of the IE procedure. Although TR uses mathematical techniques that are not well known to most reliability analysts, it has a greater potential than either IE or SDP for handling general types of system reliability problems that are not restricted to a particular type of network.

Part I

Basic Probability Theory

Reliability is defined as the probability of success. Part I of this book covers elementary probability theory and consists of two chapters titled respectively "Basic Concepts of Probability" and "Probability Distributions." These chapters help establish the definitions of and relationships between concepts such as reliability, confidence, independence and/or dependence, Bayes theorem, discrete and/or continuous distributions, random variables, density functions and/or distribution functions, and standardized forms, as well as measures of both central tendency and variation. If you already have a statistics or probability background, this section is a review, and there is no need to spend much time on it if your objective is to use other parts of the book for a reliability assessment problem.

1

Basic Concepts of Probability

A reliability analyst works with data resulting from the outcomes or results of tests or experiments, and the probabilities associated with subsets of these outcomes. This chapter introduces some basic concepts of probability, using a finite set approach. The discussion includes events, partitions, joint probabilities, independence, dependence, and Bayes theorem. The material is elementary, however you can use it as a refresher if you already are familiar with the subject.

Attributes and outcomes

The simplest case to consider is attributes, also known as binary or zero-one testing (that is, successes and failures only). If a trial has a favorable or desired result, the value "one" is assigned, "zero" if the result is unfavorable. If n trials are performed, where n is any positive integer, an outcome is a string of n 1s and 0s, each 0 or 1 designating the result of the corresponding trial. The total number of outcomes possible is 2^n because there are two possible results at each of the n different trials.

EXAMPLE 1.1.

Representing an outcome: A fair coin is tossed three times. Let x_i denote the result where

$x_i = 1$, if the result of trial i is a head, *or*
$x_i = 0$, if a tail,
$i = 1, 2,$ or 3.

Then if the first two tosses both result in heads and the third is a tail, the outcome is represented as 110. This is *not* a number, such as decimal one hundred ten, or six in the binary system. It means that in the first two trials there was one type of result, and in the third trial the opposite result.

The universal set

The list of all possible outcomes of an event is the universal set U. Every outcome x is an element of U. This is shown by the element-inclusion symbol ϵ, $x \in U$. The opposite of inclusion is exclusion, with the symbol $\notin$. A lowercase letter, such as $a, b, c, x,$ and so on, represents an element of a set.

EXAMPLE 1.2.

Representing the universal set: Under the conditions of Example 1.1, U has $2^3 = 8$ elements.

111: three heads,
110: two heads followed by a tail,
101: a head, a tail, then another head,
011, 100, 010, 001, 000

Subsets and events

A subset of U, also called an *event*, contains either some, none, or all of the elements of U. It is customary to designate any set other than the universal set by means of a capital letter, such as A, or else by a bracketed group of elements, such as

$$A = \{x_1, \ldots, x_n\}.$$

Because we are dealing with sets, the equality sign ($=$) has a different meaning than the usual one. It implies that *every* element of the set on the right-hand side of the equality sign is also a member of the set A, and that every element in A is listed inside the brackets on the right-hand side.

EXAMPLE 1.3.

Representing an event: Under the conditions of Examples 1.1 and 1.2, let the event A denote that the first of the three tosses of the coin is a head. Then we have
$$A = \{111, 110, 101, 100\}.$$
The set-inclusion symbol $\supseteq$ designates subsets. The statement
$$A \supseteq B$$
implies that every element of the set B is also necessarily included in the set A. Of course, if every element of A were also included in B, it would be true that $A = B$. Likewise, since every element of the set A is also included in the universal set, it is also true that
$$U \supseteq A.$$

Unit events

It is convenient to be able to represent an event corresponding to a single outcome. A *unit event* (also known as *simple event* or *elementary event*) is a subset containing only one outcome.

EXAMPLE 1.4.

Representing a unit event: In Example 1.1, the unit event corresponding to the outcome 110 is the set $\{110\}$. Corresponding to the universal set U with eight such outcomes in Example 1.2, there are eight unit events.

The complement of any set A is the subset of the universal set consisting of elements not contained in A.
$$\overline{A} = \{x \in U: x \notin A\}$$

In particular, the complement of U is the *empty* or *null* set $\emptyset$, the set with no elements in it.
$$\emptyset = \{x \notin U\}$$

A Venn diagram is frequently used to represent a universal set and the events that are its subsets. A square or rectangle denotes U, and circles or other geometric shapes within the rectangle denote events.

EXAMPLE 1.5.

In the Venn diagram in Figure 1.1, B is a subset of A and $\overline{A}$ includes all of U except the elements of A.

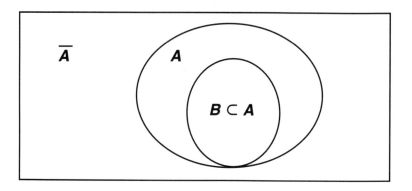

Figure 1.1. The set B is a proper subset of A.

Counting events

The number of different possible events that may be formed of the elements of the universal set is usually very large. In the simplest possible case, attributes testing, for n trials 2^n outcomes are possible; since a subset of these outcomes may have as few as none of these outcomes and as many as all of them, there are 2^{2^n} events. For example, for $n = 4$, $2^{2^4} = 65{,}536$; for $n = 5$, $2^{2^5} = (65{,}536)^2$.

EXAMPLE 1.6.

Counting events: With a single trial ($n = 1$) and attributes testing, two outcomes are possible: $x = 1$ and $x = 0$. There are 4 (that is, 2^2) events: U, $\{1\}$, $\{0\}$, $\emptyset$. With $n = 2$, the 16 events are U, $\{11,10,01\}$, $\{11,10,00\}$, $\{11,01,00\}$, $\{10,01,00\}$, $\{11,10\}$, $\{11,01\}$, $\{11,00\}$, $\{10,01\}$, $(10,00\}$, $\{01,00\}$, $\{11\}$, $\{10\}$, $\{01\}$, $\{00\}$, $\emptyset$.

Compound events

Unions and intersections

An event may be defined by its relationships to other events. The inclusive-or union of two or more events contains all the elements in *either* A or B.

$$A \cup B = \{x \in U: x \in A \text{ or } x \in B\}$$

The intersection of A and B contains the outcomes in *both* A and B.

$$A \cap B = \{x \in U: x \in A \text{ and } x \in B\}$$

Disjoint (mutually exclusive) events

If A and B have no outcomes in common, $A \cap B = \emptyset$, and A and B are said to be *disjoint*. The simplest examples of disjoint events are the 2^n unit events, since no unit event shares any outcomes in common with other unit events.

Partitions

A set which consists of disjoint subsets of U and which is exhaustive is called a *partition*. Symbolically, a partition is a set

$$C = \{A_1, \ldots, A_k\}$$

such that

$$A_i \cap A_j = \emptyset, \, i \neq j, \, i,j = 1, \ldots, k$$

and

$$\bigcup_{i=1}^{k} A_i = U.$$

EXAMPLE 1.7.

In Figure 1.2, the set A is the circle and the sets B_1, B_2, B_3, B_4, respectively, are the four rectangles. Thus

$$\{B_1, B_2, B_3, B_4\}$$

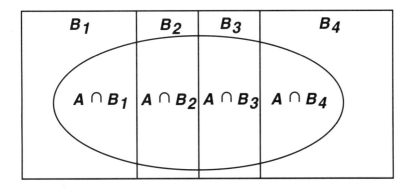

Figure 1.2. B is a partition of the universal set.

is a partition, because the B's are disjoint and exhaustive. Likewise

$$\{A \cap B_1, A \cap B_2, A \cap B_3, A \cap B_4\}$$

is a partition of A because all of A is covered, and the sets $A \cap B_i$ are disjoint.

EXAMPLE 1.8.

Mutually exclusive events: In the game of craps, played in Las Vegas, Atlantic City, and other gambling venues, two six-sided dice are thrown. Each of the six faces of a die has a different number of spots, from one to six. The result of a single throw is the sum of the number of spots on the two faces that turn up. For a single throw, let $x_i = 1, \ldots, 6$ represent the number of spots on the up face of die i, $i = 1,2$. The universal set for a single throw of two dice has 36 elements, each element having two numbers. So,

$$U = \{11, 12, 13, 14, 15, 16; 21, 22, 23, 24, 25, 26; 31, 32, 33, 34, 35, 36;$$
$$41, 42, 43, 44, 45, 46; 51, 52, 53, 54, 55, 56; 61, 62, 63, 64, 65, 66\},$$

where the first number for each element is the number of spots on the up face for the first die, and the second number is the number of spots on the up face for the second die.

Let A denote the event "the sum is 7"

$$A = \{16, 25, 34, 43, 52, 61\},$$

and let B be the event "the sum is 4"

$$B = \{13, 22, 31\}.$$

The event $A \cap B$ includes all outcomes in which both a 4 and a 7 occur. Because A and B cannot both occur on the same throw, $A \cap B = \emptyset$.

EXAMPLE 1.9.

Partitioning: In Example 1.8 the result of a throw of the dice can be any one of the eleven numbers from 2 through 12, inclusive. Since no two of these numbers can occur on the same throw,

$$\{[2], [3], [4], [5], [6], [7], [8], [9], [10], [11], [12]\}$$

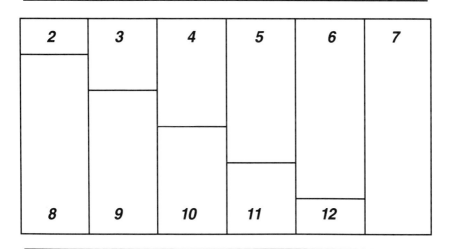

Figure 1.3. A Venn diagram with areas proportional to the probabilities for the sum of a roll of two dice.

is a partition. In Figure 1.3, the area associated with each of the different possible values obtainable with a throw of the dice corresponds to its relative frequency or probability, as defined in the next section.

Probability

Acceptable assignment of probabilities

Corresponding to each of the k unit events $\{x_j\}, j = 1, \ldots, k$, respectively, subsets of an event $A = \{x_j, j = 1, \ldots, k\}$, there is a nonnegative number $P(x_j)$, which is not less than zero, $P \geq 0$, and at the same time not greater than one, $P \leq 1$, called the *probability* of x_j, such that the sum of the probabilities of the k unit events is the probability of A.

$$P(A) = \sum_{j=1}^{k} P(x_j)$$

When $A = U$ is the universal set, the assignment of probabilities is acceptable if $P(A) = 1$. It is always advisable to check that there is an acceptable assignment, by verifying that the total of all probabilities is one; otherwise the inferences made from the data would be incorrect.

Joint events and dependence

The probability of the intersection of two or more events depends upon the probabilities of the intersecting events. For two events A and B, the conditional probability of B given A

$$P(B|A) = \frac{P(A \cap B)}{P(A)},$$

is defined only if the probability of A is nonnegative, $P(A) > 0$. Multiply both sides by $P(A)$ to obtain the joint probability

$$P(A \cap B) = P(A) \cdot P(B|A).$$

Independence

If A and B are independent, the probability that B occurs is not affected by A

$$P(B|A) = P(B);$$

therefore, with independence,

$$P(A \cap B) = P(A) \cdot P(B).$$

Additive probabilities of disjoint events

When A and B have no outcomes in common, the probability of the union $A \cup B$ is the sum of $P(A)$ and $P(B)$. The reason for this is that by adding up the probabilities of all of the outcomes in $A \cup B$, since both A and B account for all outcomes and none are counted twice, the probability of the union is the sum of the probabilities of A and B.

EXAMPLE 1.10.

Probability of an event consisting of two disjoint events: In the game of craps, let A and B be defined as in example 1.8. Then

$$P(A) = \frac{6}{36}; \ P(B) = \frac{3}{36}; \ P(A \cup B) = \frac{9}{36}; \ P(A \cap B) = 0.$$

Conditional probability

When an outcome y consists of a specified set of results for n prior trials, the probability $P(y)$ depends upon the sequence of the results of the earlier trials. In an experiment with n trials, let x_j, $j = 1, \ldots, n$, denote the result of the jth trial. For example, in binary (attributes) testing $x_j = 1$ denotes a success and $x_j = 0$ a failure. Then the conditional probability

$$P(x_j|x_1, \ldots, x_{j-1})$$

is the probability of result x_j, given that x_1 occurred on the first trial, x_2 on the second trial, and so on. The probability of the outcome

$$y = x_1, \ldots, x_n$$

is the product of a string of conditional probabilities

$$P(y) = P(x_1) P(x_2|x_1) \ldots P(x_n|x_1, x_2, \ldots, x_{n-1}).$$

Trials $1, \ldots, n$ are said to be independent if

$$P(x_j|x_1, \ldots, x_{j-1}) = P(x_j), j = 1, \ldots, n. \tag{1.1}$$

If the trials are totally independent of one another, $P(y)$ is the product of the probabilities of all n trials:

$$P(y) = P(x_1) \cdot \ldots \cdot P(x_n).$$

Formulas for calculating probabilities of outcomes or events are less complicated when successive trials are independent than if the trials are dependent.

EXAMPLE 1.11.

Joint probability with dependence: Two cards are drawn from a deck of 52 playing cards that includes four aces. The first one is not replaced after it is drawn. What is the probability of drawing two aces? Let

$x_i = 1$, if card i is an ace, $i = 1, 2$
$x_i = 0$, otherwise.

The probability of the outcome 11 (two aces) is

$$P(11) = P(x_1 = 1) \cdot P(x_2 = 1|x_1 = 1)$$
$$= \frac{4}{52} \cdot \frac{3}{51} = \frac{1}{221}.$$

EXAMPLE 1.12.

Joint probability with independence: The conditions are the same as in Example 1.11 except that the first card is replaced in the deck. The probability that both cards are aces is

$$P(x_1 = 1) \cdot P(x_2 = 1) = \frac{4}{52} \cdot \frac{4}{52} = \frac{1}{169}.$$

Bayes theorem

Bayes theorem is a form of conditional probability for making inferences about a partition of unknown events, given that a known event happens. It is convenient to explain Bayes theorem by starting with the probability of a partition.

Let A be any event and let $\{B_1, \ldots, B_n\}$ be a partition of U. Observe that

$$C = \{A \cap B_1, \ldots, A \cap B_n\}$$

is a partition of A, because $A \cap B_i$ and $A \cap B_j$ have no elements in common if $i \neq j$, and because every element in A is included in exactly one of the sets in the partition C. Therefore, the probability of A is the sum of the probabilities of the sets in C:

$$P(A) = \sum_{i=1}^{n} P(A \cap B_i)$$

$$= \sum_{i=1}^{n} P(B_i) \cdot P(A|B_i).$$

EXAMPLE 1.13.

Probability of a partition: In the game of craps (see Example 1.8), if a 4, 5, 6, 8, 9, or 10 is thrown (the point) on the first toss, the dice are thrown repeatedly until either the point is repeated or else a seven is obtained. If the point is repeated, the player wins; if a seven comes first, the house wins. Find the probability that the player will win by making the point.

Let A_4 denote "4 on first throw," and define A_5, A_6, A_8, A_9, and A_{10} similarly. Also let B_4 be the event "4 on final throw," with similar definitions for B_5, B_6, B_8, B_9, and B_{10}. The set of winning combinations is the partition

$$C = \{A_4 \cap B_4, A_5 \cap B_5, A_6 \cap B_6, A_8 \cap B_8, A_9 \cap B_9, A_{10} \cap B_{10}\}.$$

The probabilities of the 6 elements of the partition C can be summarized as follows:

$$P(A_4 \cap B_4) = P(A_{10} \cap B_{10}) = \frac{3}{36} \cdot \frac{3}{9} = \frac{1}{36};$$

$$P(A_5 \cap B_5) = P(A_9 \cap B_9) = \frac{4}{36} \cdot \frac{4}{10} = \frac{2}{45};$$

$$P(A_6 \cap B_6) = P(A_8 \cap B_8) = \frac{5}{36} \cdot \frac{5}{11} = \frac{25}{396}.$$

The probability that the player wins by making the point is the sum of the probabilities of the events in C.

$$P(C) = 2 \cdot \left(\frac{1}{36} + \frac{2}{45} + \frac{25}{396}\right) = 0.2707$$

If it is assumed or known that event A occurs, the probability that B_i also occurs is given by

$$P(B_i|A) = \frac{P(A \cap B_i)}{P(A)}, i = 1, \ldots, n \qquad (1.2)$$
$$= \frac{P(B_i) \cdot P(A|B_i)}{\sum_{i=1}^{n} P(B_i) \cdot P(A|B_i)}.$$

Equation 1.2 is known as Bayes theorem, so named in honor of Rev. Thomas Bayes, who applied it over two centuries ago to certain elementary problems of statistical inference. The probabilities

$$P(B_i), i = 1, \ldots, n,$$

in Equation 1.2 are known as the prior probabilities of the events B_i. These are the unconditional probabilities of the elements of the partition of U. The conditional probabilities

$$P(B_i|A), i = 1, \ldots, n,$$

are called the *posterior probabilities*. The key idea that makes this concept work is the knowledge that the occurrence of A improves our ability to evaluate which one of the B_i occurs.

EXAMPLE 1.14.

Bayes theorem: Under the conditions of Example 1.13, a player wins by making the point. We do not know which of the six numbers 4, 5, 6, 8, 9, or 10 was the point. What is the probability that the point is a 6?

The prior probabilities are the normalized probabilities of the six numbers (*normalized* means that the sum is 1).

14 Chapter 1 ■ *Basic Concepts of Probability*

Event	P (Event)	Prior P
A_4	$\dfrac{3}{36}$	$\dfrac{1}{8}$
A_5	$\dfrac{4}{36}$	$\dfrac{1}{6}$
A_6	$\dfrac{5}{36}$	$\dfrac{5}{24}$
A_8	$\dfrac{5}{36}$	$\dfrac{5}{24}$
A_9	$\dfrac{4}{36}$	$\dfrac{1}{6}$
A_{10}	$\dfrac{3}{36}$	$\dfrac{1}{8}$

The corresponding joint probabilities (probability of the given number on the first roll of the dice, followed by the same number on the final roll) are as follows:

Point	Joint probability
4	$\dfrac{1}{8} \cdot \dfrac{3}{9} = \dfrac{330}{7920}$
5	$\dfrac{1}{6} \cdot \dfrac{4}{10} = \dfrac{528}{7920}$
6	$\dfrac{5}{24} \cdot \dfrac{5}{11} = \dfrac{750}{7920}$
8	$\dfrac{5}{24} \cdot \dfrac{5}{11} = \dfrac{750}{7920}$
9	$\dfrac{1}{6} \cdot \dfrac{4}{10} = \dfrac{528}{7920}$
10	$\dfrac{1}{8} \cdot \dfrac{3}{9} = \dfrac{330}{7920}$
Total	$\dfrac{3216}{7920}$

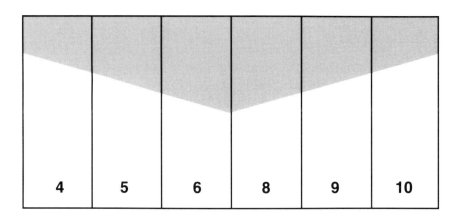

Figure 1.4. A representation of Example 1.14.

The required probability is $\dfrac{750}{7920} \div \dfrac{3216}{7920} = 0.2332$.

Figure 1.4 represents this analysis in a kind of Venn diagram. Since the thrower of the dice wins by making the point, the universe is a partition with six numbers. The shaded portion of Figure 1.4 for each number is the probability of winning with that number. According to Bayes theorem, the probability that both the first and final rolls of the dice result in a 6 is the ratio of the size of the shaded area for a 6 to the size of the entire shaded area.

This chapter includes some basic concepts of probability. Probabilities were calculated for the elements of a finite universe. In reliability analysis, the events are numerical valued results of tests of a performance characteristic and are called *random variables*. In chapter 2 the formation of probability distributions for random variables will be considered.

Problems

1.1. Draw a Venn diagram for the event $A \supseteq B \supseteq C$.

1.2. Show that for the sequence of numbers $2^{2^1}, 2^{2^2}, 2^{2^3}, \ldots, 2^{2^n}$, each number in the sequence represents the number of events possible if there are n elements in the universal set U, and every term is the square of its immediate predecessor: 2, $2^2 = 4$, $2^4 = 4^2 = 16$, $2^8 = 16^2 = 256$, $2^{16} = 256^2 = 65{,}536$, and so on.

1.3. A baseball player approximates his chances for every time at bat as follows: probability 0.3 of getting a hit, 0.1 of a base on balls, and 0.6 of a putout. Consider the four times a player is at bat in a game as four independent trials, and compute the probability of
   a. One walk and three hits
   b. One walk, one hit, and two putouts
   c. Four putouts

   Answer the following questions.
   d. How many different arrangements of hits, putouts, and bases on balls are possible for a player's four times at bat?
   e. A superstar hitter has a probability of 0.5 of getting on base every time he comes to bat, because opposing pitchers will often intentionally give him a base on balls. What is the probability that a superstar player will get on base twice in a game if he comes up to bat four times?

1.4. A spacecraft failure is attributed to either one of two valves, A or B. Valve A is actuated six times in a mission and valve B is actuated three times. The estimated failure rate of valve A is four times per million actuations, and the failure rate of valve B is two times per million actuations. What is the probability that the failure was due to valve A?

1.5. The game of Mini-keno is played with five cards numbered 1 to 5, respectively. The house draws three cards without replacing them. A player bets $1 and selects two numbers. The player gets back $3 ($2 plus the original $1) if both selections are included in the three drawn by the house. What is the probability that
   a. The player wins by selecting two cards correctly?
   b. The player selects exactly one card correctly?
   c. Neither one of the player's selections is correct?

1.6. Alter the first sentence of Example 1.14 to read "Under the conditions of Example 1.13, a player has won"; this means that there is a possibility of having thrown a 7 or an 11 on the first throw in addition to the possibility of having repeated a point of 4, 5, 6, 8, 9, or 10. What is the probability that the thrower of the dice won with a 6?

References

Bayes, Thomas. 1763. An essay toward solving a problem in the doctrine of chances, with Richard Price's foreword and discussion. *Philosophical Transactions of the Royal Society*: 370–418. Reprinted by The Graduate School, U.S. Department of Agriculture, *Facsimiles of Two Papers by Bayes*. Edited by W. Edwards Deming.

Boole, George. [1854] 1958. *An investigation of the laws of thought.* Reprint, New York: Dover Publications.

Bradley, James V. 1976. *Probability; decision; statistics.* Englewood Cliffs, N.J.: Prentice Hall.

Cramer, Harald. 1946. *Mathematical methods of statistics.* Princeton, N.J.: Princeton University Press.

Feller, William. 1957–68. *An introduction to probability theory and its applications.* 2 vols. New York: Wiley.

Fry, Thornton C. 1928. *Probability and its engineering uses.* New York: D. Van Nostrand.

Goldberg, Samuel. 1960. *Probability, an introduction.* Englewood Cliffs, N.J.: Prentice Hall.

Hausner, Melvin. 1971. *Elementary probability theory.* New York: Harper & Row.

Keynes, John Maynard. 1921. *A treatise on probability.* Reprint. London: Macmillan & Company Ltd.

Kolmogorov, A. N. [1933] 1956. *Foundations of probability.* Reprint of English edition, New York: Chelsea Publishing Company.

Kyburg, Henry E., Jr., and H. E. Smokler. 1964. *Studies in subjective probability.* New York: Wiley.

Laplace, Pierre Simon. 1951. *A philosophical essay on probabilities.* Translated from the 6th French edition by F. W. Truscott and F. L. Emory. New York: Dover Publications.

Mises, Richard von. [1928] 1957. *Probability, statistics and truth.* 2d rev. English ed. Translated by H. Geiringer based on 1951 definitive German edition. New York: Macmillan.

Ramsey, F. P. 1931. *The foundations of mathematics and other logical essays.* London: Routledge & Kegan Paul.

Ross, Sheldon M. 1980. *Introduction to probability models.* 2d ed. New York: Academic Press.

Shook, Robert C., and H. J. Highland. 1969. *Probability models with business applications.* Homewood, Ill.: R. D. Irwin.

Trivedi, K. S. 1982. *Probability and statistics with reliability, queuing, and computer science applications.* Englewood Cliffs, N.J.: Prentice Hall.

Uspensky, J. V. 1937. *Introduction to mathematical probability.* New York: McGraw-Hill.

Venn, John. [1866] 1962. *The logic of chance.* 4th ed. New York: Chelsea Publishing Company.

Walpole, Ronald E., and R. H. Myers. 1989. *Probability and statistics for engineers and scientists.* 4th ed. New York: Macmillan.

Note: Almost every technical library has a section of books on probability and statistics containing additional background readings on probability. The books listed here, some of which have significant historical importance because of their influence on the development of the science of probability, are representative of the large number of excellent books available on the subject.

2

Probability Distributions

Introduction

In a probability distribution, probabilities are assigned to the elements of a partition formed by classifying events according to a numerical ordering. The function that assigns numerical values to the unit events is a random variable (r.v.). This chapter covers how to form probability distributions for both discrete-valued and measurable or continuous-valued random variables, with common applications for reliability assessment.

Probability functions

Customarily an r.v. is represented by capital letters, such as X and Y, and specific numerical values of the r.v. by the corresponding lowercase letters, x, y, and so on. X represents a range of values; x is a number in that range. If the r.v. is discrete-valued, the probability $P(x)$ of the event $X = x$ is the sum of the probabilities of the outcomes that are assigned the value x. The probability function assigns probabilities to every value in the range. The sum of these probabilities is 1; this follows from the fact that the unit events are disjoint with nonnegative probabilities and the entire range of values is covered.

Distribution functions

A convenient way to represent a probability distribution is the distribution function (d.f.). For every different value of x, the d.f. is the probability that $X \leq x$. The symbol F is used for the d.f.

$$F(x) = P(X \leq x)$$

Because probabilities are nonnegative and the r.v. is ordered, if $x \geq y$, it follows that $F(x) \geq F(y)$.

Continuous-valued random variables

Continuous distributions, for example the normal distribution, represent measured or averaged sets of data. The principal difference between a continuous r.v. and a discrete r.v. is that the continuous range includes an infinitely large set of values. Therefore it is necessary to employ calculus. The d.f. is obtained by integration. Let Δx be an infinitesimally small interval between any two successive values of a continuous r.v. The probability density function (p.d.f.) is

$$f(x) = \lim_{\Delta x \to 0} \frac{F(x) - F(x - \Delta x)}{\Delta x};$$

that is, f is the derivative of F. It follows that

$$F(x) = \lim_{\Delta x \to 0} [F(x - \Delta x) + f(x) \Delta x].$$

Thus, F is the accumulated sum of the areas for an infinite number of very small intervals of height f and uniform width Δx, for all $X \leq x$. If a is the smallest possible value x can take on,

$$F(x) = \lim_{\Delta x \to 0} \sum_{a}^{x} f(x) \Delta x.$$

If the number b is the highest value in the range,

$$F(b) = \int_{a}^{b} f(x)\,dx = 1.$$

Table 2.1. Probabilities for a pair of dice.

x	$P(x)$	$F(x)$
2	$\frac{1}{36}$	$\frac{1}{36}$
3	$\frac{2}{36}$	$\frac{3}{36}$
4	$\frac{3}{36}$	$\frac{6}{36}$
5	$\frac{4}{36}$	$\frac{10}{36}$
6	$\frac{5}{36}$	$\frac{15}{36}$
7	$\frac{6}{36}$	$\frac{21}{36}$
8	$\frac{5}{36}$	$\frac{26}{36}$
9	$\frac{4}{36}$	$\frac{30}{36}$
10	$\frac{3}{36}$	$\frac{33}{36}$
11	$\frac{2}{36}$	$\frac{35}{36}$
12	$\frac{1}{36}$	$\frac{36}{36}$

EXAMPLE 2.1.

Discrete distribution: the game of craps. In Example 1.8 there are 36 elements in the universal set U. The r.v. X is the sum of the dots for the up faces of the two dice that are rolled. Table 2.1 gives both the probability function and the distribution function for the r.v.

Families of related distributions

A parametric probability distribution for an r.v. is defined by means of a formula. A standard formula identifies a family of related distributions.

Uniform distributions

Every value of the r.v. of a uniform distribution has the same probability. The parameters are the lower limit of the range, a, and the upper limit, b. In most applications of the uniform distribution, such as in computerized simulation, $a = 0$ and $b = 1$. The p.d.f. is

$$f(x) = \frac{1}{b-a}, \quad a \leq x \leq b,$$
$$= 0, \text{ otherwise.}$$

This means that the p.d.f. is a rectangle with a height of $1/(b-a)$ in the range from a to b, and that the width is $b-a$. The total area of the rectangle, the product of width times height, is 1.

EXAMPLE 2.2.

The uniform distribution of F. A variable that depends upon an r.v. is also an r.v. The distribution function F is an r.v. because it is a function of an r.v. and it is uniformly distributed with p.d.f. 1 over the range zero to one. A graph of the uniform distribution from $\{0,1\}$ is shown in Figure 2.1, giving both the p.d.f. and the continuous distribution function (c.d.f).

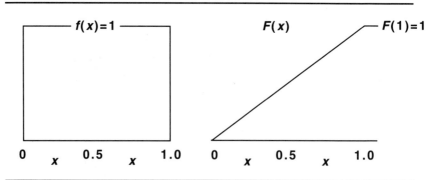

Figure 2.1. p.d.f. and d.f. for a uniform distribution in (0,1).

Binary random variables: The binomial distribution

A binary r.v. takes one of two values, zero (0) or one (1). The probability of a 0 is $1-p$ and the probability of a 1 is p. Since probability is assigned to only two numbers, there are just two places to accumulate probability. The c.d.f. is: $F(0) = 1-p$; $F(1) = 1$. The graph of the p.d.f. is a bar chart, and the graph of the distribution function F is a two-step ladder, shown in Figure 2.2.

Binary distributions are fundamental for the analysis of attributes data, such as zero-one, success-or-failure, or go–no go processes. The most frequently used distribution in this family is the binomial distribution, for the probability of the number of successes, r, in n trials.

EXAMPLE 2.3.

Coin tossing: An example of a binary r.v. is the result of a toss of a single coin. If the coin is fair, the probability of a head, p, is one-half, as is the probability of a tail, $1-p$.

Even though they are very simple, binary distributions are fundamental for analyzing attributes data, such as those arising in zero-one or go–no go processes. For example, the binomial distribution, which specifies the probability for the number of successes in n trials,

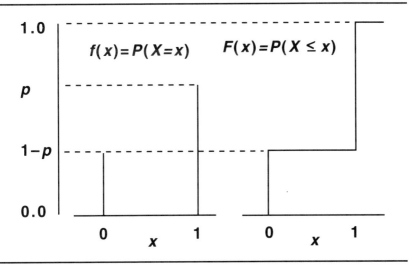

Figure 2.2. p.d.f. and d.f. for a binary distribution.

arises from n independent binary r.v.'s, all having the same parameter p. Likewise, in analyzing systems, it is frequently necessary to treat the success or failure of each component as a binary r.v.

Exponential distributions

An exponential distribution has a single parameter, the nonnegative number $\lambda > 0$. The c.d.f. is

$$F(x) = 1 - e^{-\lambda x}, x > 0,$$

where e is the base of natural logarithms

$$e = \lim_{y \to 0} (1+y)^{1/y} = 2.71828.$$

The p.d.f. $f(x)$ is obtained by differentiating $F(x)$

$$f(x) = \lambda\, e^{-\lambda x}, x > 0.$$

Exponential distributions are used frequently for the analysis of time-dependent data, such as queuing and maintenance processes, as well as in reliability.

EXAMPLE 2.4.

Exponential distribution for $\lambda = 1$: Graphs of the exponential p.d.f. and c.d.f. for $\lambda = 1$ are shown in Figure 2.3. Variations of Figure 2.3 are obtained by changing λ, rescaling the graph, without changing the shape. For example, in Figure 2.3, the vertical-axis intercept of $f(x)$ is at $\lambda = 1$; in general, the intercept is λ. The scale parameter of an exponential

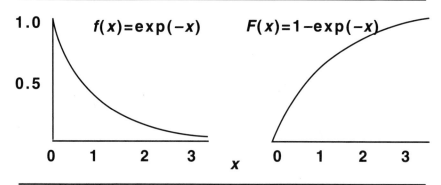

Figure 2.3. p.d.f. and d.f. for exponential distribution ($\lambda = 1$).

distribution is called λ, and all exponential distributions are characterized as having the same shape.

Measures related to parameters of distributions

If the family is known, a distribution is completely specified by its parameters. Alternatively, the distribution is identified by characteristics such as the mean, median, standard deviation, and so on.

Mean: The expected value

The most well-known characteristic of a probability distribution is the mean; also known as the *expected value, EX*, it is the weighted sum of all the possible values the r.v. can assume, with the probabilities as weights. If the r.v. is discrete, with a set of n values $\{x_i\}$ having probabilities $\{f(x_i)\}$

$$EX = \sum_{i=1}^{n} x_i f(x_i). \tag{2.1}$$

The mean of a continuous r.v. is

$$EX = \int_{-\infty}^{+\infty} x f(x)\, dx.$$

Variance and standard deviation

The variance measures dispersion. The *variance* is the average (mean) of the squared differences between the values and the mean

$$\sigma^2 = E(X - EX)^2,$$

and the *standard deviation* σ is the square root of σ^2.

There is a convenient shortcut for obtaining the variance of any distribution. It is derived algebraically from the definition by first expanding the square of the term inside the parentheses and then rearranging terms.

$$\begin{aligned} E(X - EX)^2 &= E(X^2 - 2X\,EX + (EX)^2) \\ &= E(X^2) - 2(EX)^2 + E(EX)^2 \\ &= E(X^2) - (EX)^2 \end{aligned} \tag{2.2}$$

In this derivation, the expected value of a constant is the constant.

$$E\,(EX)^2 = (EX)^2 \tag{2.3}$$

EXAMPLE 2.5.

Mean and variance of a binary distribution: The mean of a binary r.v. is

$$EX = 0 \cdot (1-p) + 1 \cdot p = p.$$

The average squared value, $E(X^2)$, is identical to EX.

$$E(X^2) = 0 \cdot (1-p) + 1 \cdot p = p.$$

By Equation 2.2 the variance is

$$\sigma^2 = p - p^2 = p(1-p).$$

EXAMPLE 2.6.

Mean and variance of a uniform distribution: The mean of a uniform distribution is

$$EX = \frac{1}{b-a} \int_a^b x \, dx = \frac{b+a}{2}.$$

The mean squared value of a uniform r.v. is

$$E(X^2) = \frac{1}{b-a} \int_a^b x^2 \, dx = \frac{b^2+ab+a^2}{3}.$$

The variance is

$$\sigma^2 = \frac{b^2+ab+a^2}{3} - \frac{(b+a)^2}{4} = \frac{(b-a)^2}{12}.$$

EXAMPLE 2.7.

Mean, variance, and standard deviation of an exponential distribution: For an exponential distribution with scale parameter λ the mean is

$$EX = \lambda \int_0^\infty x e^{-\lambda x} dx = \frac{1}{\lambda}.$$

The mean squared value is

$$E(X^2) = \lambda \int_0^\infty x^2 e^{-\lambda x} dx = \frac{2}{\lambda^2}.$$

The variance is

$$\frac{2}{\lambda^2} - \frac{1}{\lambda^2} = \frac{1}{\lambda^2},$$

and the standard deviation is identical to the mean, $\frac{1}{\lambda}$.

Percentage points

Since an r.v., X, represents numbers in the scale of the real number system, it is convenient to associate every value x with its cumulative probability on the c.d.f. This is like having two related well-ordered sets: X, a set of numbers, and the c.d.f., F, a number between 0 and 1. For any distribution and a specified value α of F, the $100 \cdot (1-\alpha)$ percentage point is the value x_α such that

$$P(X \le x_\alpha) = \alpha.$$

The best known percentage point is the median, $x_{.50}$, with 50 percent of the values of X less than or equal to the median, and 50 percent greater than the median.

Reliability and percentage points

The concepts of reliability and percentage points are closely related. If a device or system is operating successfully, the observed or measured value of a performance characteristic X, an r.v., is within acceptable limits. If it is required that X exceed some specified lower limit x_L, the reliability is $R = P(X > x_L)$, and x_L is the $100 \cdot (1-R)$th percentage point of the distribution.

EXAMPLE 2.9.

Reliability for a uniform distribution: Suppose that the distribution is uniform in the range a to b, and that the required operating limits are the range c to d, where $c \ge a$ and $d \le b$. The reliability R is the probability that X is between c and d.

$$R = F(d) - F(c) = \frac{d-c}{b-a}$$

EXAMPLE 2.10.

Reliability for an exponential distribution: When the distribution is exponential, it is usually required that X (for example, lifetime) exceed some specified lower limit x_L. The reliability is

$$R = P(X \ge x_L) = 1 - F(x_L).$$

The mode

The *mode* is the value with the highest probability or probability density. Although the rest of this book does not discuss the mode explicitly, a related concept called *maximum likelihood* is covered.

EXAMPLE 2.11.

The mode: The uniform distribution does not have a mode because every value has exactly the same probability. The mode of a binary distribution is either zero or one. The mode of an exponential distribution is zero.

Random sample summary statistics

Data collection frequently involves obtaining some related observations, called a *sample*, in order to find information about the family of distributions from which the data came. In this section, we describe two measures related to the population mean and variance, the sample mean and the sample variance.

Assume that the data are obtained under desirable conditions not always attained in practice: first, that the sampling procedure is *unbiased*, meaning that the chance of selection is proportional to the probability; and second, that successive observations are *independent*, meaning that the probability of selection of one observation or unit event at one stage does not influence the probability of selection of other observations. A sample that satisfies these conditions is a random sample.

Under random sampling, for any value of the r.v. X, the expected value is equal to the population mean μ

$$EX = \mu,$$

and the expected squared difference between X and μ is equal to the population variance

$$E(X-\mu)^2 = \sigma^2.$$

The sample mean

The *sample mean* $\overline{X}$ is the sum of the values in the sample divided by the number of observations. If the random sample consists of n observations $X_1, X_2, \ldots, X_n$,

$$\overline{X} = \frac{X_1 + X_2 + \ldots + X_n}{n} = \sum_{i=1}^{n} \frac{X_i}{n}.$$

$\overline{X}$, which is a function of the sample data, is an unbiased estimator of the parameter μ if the expected value of $\overline{X}$ is equal to μ. Suppose that the n observations $X_1, \ldots, X_n$, are obtained by random sampling. Then the sample mean is unbiased because

$$E\bar{X} = E\left\{\frac{X_1 + X_2 + \ldots + X_n}{n}\right\} = \frac{n\mu}{n} = \mu.$$

The variance of the sample mean

Because it is a function of random variables, $\bar{X}$ is an r.v. with a probability distribution. The previous derivation shows that the mean of this distribution for an unbiased sample is the population mean μ. The variance of the sample mean $\sigma^2_{\bar{X}}$ however, is smaller than the population variance σ^2 and is inversely proportional to the size of the sample. By the definition of a variance,

$$E(\bar{X} - \mu)^2 = E\left\{\frac{X_1 + X_2 + \ldots + X_n}{n} - \mu\right\}^2$$
$$= \frac{1}{n^2} E\{(X_1-\mu) + (X_2-\mu) + \ldots + (X_n-\mu)\}^2. \quad (2.4)$$

Expanding this expression and applying the expectation operator E to each term separately, there are two types of terms:

$$E(X_i-\mu)^2, \ i = 1, \ldots, n.$$

These n terms are all identically equal to σ^2 because the sampling is unbiased and

$$E(X_i-\mu)(X_j-\mu) = 0, \ i \neq j.$$

These terms are all zero because of independence; Equation 2.4 becomes

$$\sigma^2_{\bar{X}} = \frac{n\sigma^2}{n^2} = \frac{\sigma^2}{n}. \quad (2.5)$$

EXAMPLE 2.12.

The variance of the mean as a function of n: A random variable X has a variance σ^2 of one (1). A graph of $\sigma^2_{\bar{X}}$ is Figure 2.4.

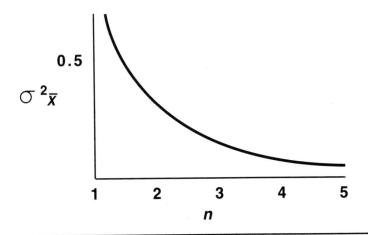

Figure 2.4. The variance of the sample mean.

The sample variance: Correction for bias

The *sample variance* s^2 is a summary statistic derived from the sample data, and is similar to the population variance σ^2. The definition is

$$s^2 = \frac{\sum_{i=1}^{n}(X_i-\bar{X})^2}{n}. \tag{2.6}$$

The sample variance is not an unbiased estimate of the population variance. It is usually advisable to correct for bias by multiplying Equation 2.6 by the factor

$$\frac{n}{n-1}.$$

To prove this, consider all of the terms of the form $(X_i-\mu)$

$$X_i-\mu = (X_i-\bar{X}) + (\bar{X}-\mu), \quad i = 1, \ldots, n.$$

Therefore,

$$\sigma^2 = E\frac{\sum_{i=1}^{n}(X_i-\mu)^2}{n} = E\frac{\sum_{i=1}^{n}\{(X_i-\bar{X}) + (\bar{X}-\mu)\}^2}{n}$$

$$= E\frac{\sum_{i=1}^{n}(X_i-\bar{X})^2}{n} + 2E(\bar{X}-\mu)\frac{\sum_{i=1}^{n}(X_i-\bar{X})}{n} + E(\bar{X}-\mu)^2. \tag{2.7}$$

The first term of Equation 2.7

$$E \frac{\sum_{i=1}^{n}(X_i - \bar{X})^2}{n}$$

is Es^2. The second term is zero. Why? Because the numerator of the fraction is zero. The third term is $\sigma^2_{\bar{X}} = \sigma^2/n$. Equation 2.7 simplifies to

$$Es^2 = \frac{n-1}{n} \sigma^2.$$

The unbiased estimate of σ^2 is

$$\frac{n}{n-1} s^2. \tag{2.8}$$

Degrees of freedom

The factor $(n-1)$ in the foregoing discussion is called the number of *degrees of freedom*. The reason for this is that in random sampling, the sample mean is used instead of the universe mean when μ is unknown in order to obtain an estimate of the variance. Thus, the number of independent observations, or degrees of freedom, is $n-1$.

Standardized random variables

Sometimes it is convenient to standardize random variables, particularly in connection with the use of tables. A standardized r.v. Z, obtained by subtracting the mean from X and dividing the difference by the standard deviation, is defined by

$$Z = \frac{X - \mu}{\sigma}. \tag{2.9}$$

The mean of a standardized r.v. is zero and its variance is one.

$$EZ = \frac{1}{\sigma} E(X - \mu) = 0$$

$$E(Z - \mu)^2 = E(Z^2) - (EZ)^2 = E\left(\frac{X - \mu}{\sigma}\right)^2 - 0$$

$$= \frac{1}{\sigma^2} E(X - \mu)^2 = 1$$

32 Chapter 2 ▪ Probability Distributions

Tables of the standardized normal distribution that are found in every statistics book are known as Z-tables.

EXAMPLE 2.14.

Standardized binary random variable: The mean of a binary r.v. is p and the standard deviation is

$$\sqrt{p(1-p)}.$$

The standardized value is

$$Z = \frac{X-p}{\sqrt{p(1-p)}}.$$

For $X = 0$ the standardized value is

$$-\sqrt{\frac{p}{1-p}}. \tag{2.10}$$

For $X = 1$ the standardized value becomes

$$\sqrt{\frac{1-p}{p}}. \tag{2.11}$$

Problems

2.1. Problem of the Chevalier de Méré: What is the probability that, in 24 independent throws of two fair dice, a 12 (boxcars, or double-six) will not be obtained?

Answer: In a single throw, the probability of no double-six is 35/36. Because the throws are independent, the probability for 24 throws is

$$\left(\frac{35}{36}\right)^{24} = 0.509.$$

(This is the famous problem posed by the Chevalier de Méré to the French mathematician Pascal in the latter part of the 17th century.)

2.2. Draw the graph of the p.d.f. for a triangular distribution, an isoceles triangle between $X = -1$ and $X = 1$, with $f(-1) = f(1) = 0$ (an isoceles triangle has two opposite sides equal). What are $f(0)$, $F(0)$, $F(0.5)$ and $X_{0.75}$?

Answers: 1, 0.5, 0.875, and 0.293. $F(0.5) = 0.875$ can be obtained either geometrically or by adding the integral of $(1-x)dx$ from 0 to 0.5

Chapter 2 ■ Probability Distributions   33

to $F(0)$. $X_{0.75}$ is obtained by integrating $(1-x)dx$ from 0, which is $X_{0.50}$, to $X_{0.75}$ and solving the resulting quadratic formula

$$\frac{X^2}{2} - X + 0.25 = 0.$$

2.3. Generalize Problem 2.2 to a triangular distribution which is an isoceles triangle from $X = a$ to $X = b$.
  a. What are the parameters of this distribution?
  b. What is the mode?
  c. What is the p.d.f. at the mode?

2.4. Scaling the exponential distribution: From an offset or photoreproduced copy of Figure 2.3, reevaluate the vertical scales of both the p.d.f. and the d.f. in terms of a general parameter λ, without redrawing the figure.

2.5. Scaling the uniform distribution: From a photoreproduced copy of Figure 2.1, reevaluate the vertical scales of both the p.d.f. and the d.f. in terms of the lower limit a and the upper limit b without redrawing the figure.

2.6. Standardizing the binary distribution: For a binary distribution, the probability of success is 0.6 (note also that this is the way Figure 2.2 is drawn). Using a photoreproduced copy of Figure 2.2, rescale the X-axis with the standardized expressions in Equations 2.10 and 2.11.

2.7. In deriving the general formula for the variance of the mean for a sample of size n, Equation 2.5, a key step is

$$E(X_i-\mu)(X_j-\mu) = 0, \; i \neq j,$$

from the assumption that X_i and X_j are independent. Note that

$$E(X_i-\mu)(X_j-\mu) = E\,X_iX_j - \mu\,EX_i - \mu\,EX_j + \mu^2$$
$$= E\,X_iX_j - \mu^2.$$

2.8. Mean and variance of a standardized binary distribution: Using the standardized values, Equations 2.10 and 2.11, and the probabilities p and $1-p$, show that the mean of a standardized binary distribution is zero. Use Equation 2.2 to show that the variance is one.

2.9. Standardized uniform distribution: Using the results of Example 2.6, give the form of a standardized uniform r.v. with lower limit a and upper limit b.

2.10. Show that e, the base of natural logarithms, is the limit of a sequence $(3/2)^2$, $(4/3)^3$, $(5/4)^4$, $(6/5)^5$, ..., $\left(\dfrac{n}{n-1}\right)^{n-1}$.

2.11.a. What is the variance for the sum of the roll of two dice?

Hint: $\sigma^2 = E(X^2) - (EX)^2$

2.11.b. What is the standard deviation for the sum of the roll of two dice?

2.12. In Problem 1.5, what is the variance of the number of wins in Mini-keno

a. In a single play of the game?
b. In five plays of the game?
c. In ten plays of the game?

2.13. In Problems 1.5 and 2.12,

a. What is the expected value of the payoff for a $1 bet?
b. What is the variance of the payoff for a $1 bet?

Hint: $= \sigma^2 = E(X^2) - (EX)^2$

c. What is the standard deviation of the payoff for a $1 bet?

2.14. In the 1990 California lottery 6/49 game, the holder of a $1 ticket selects six numbers out of 49 numbers available. The lottery also obtains six of these 49 numbers by a randomizing device. What is the probability of selecting the same six numbers drawn by the lottery?

References

Bradley, James V. 1976. *Probability; decision; statistics.* Englewood Cliffs, N.J.: Prentice Hall.

Dixon, W. J., and F. J. Massey, Jr. 1957. *Introduction to statistical analysis.* 2d ed. New York: McGraw-Hill.

Feller, William. 1957–68. *An introduction to probability theory and its applications.* 2 vols. New York: Wiley.

Gnedenko, B. V., and A. Khinchin. 1961. *An elementary introduction to the theory of probability.* Trans. by W. R. Stahl. Edited by J. R. Roberts. San Francisco: Freeman.

Goldberg, Samuel. 1960. *Probability, an introduction.* Englewood Cliffs, N.J.: Prentice Hall.

Hahn, G. J., and S. S. Shapiro. 1960. *Statistical models in engineering.* New York: Wiley.

Hald, A. 1952. *Statistical theory with engineering applications.* New York: Wiley.

Hausner, Melvin. 1971. *Elementary probability theory.* New York: Harper & Row.

Lehmann, E. H. 1959. *Testing statistical hypotheses.* New York: Wiley.

Neyman, J. 1950. *First course in probability and statistics.* New York: Holt, Rinehart, and Winston.

Parzen, E. 1960. *Modern probability theory and its applications.* New York: Wiley.

Ross, Sheldon M. 1980. *Introduction to probability models.* 2d ed. New York: Academic Press.

Shook, Robert C., and H. J. Highland. 1969. *Probability models with business applications.* Homewood, Ill.: R. D. Irwin.

Trivedi, K. S. 1982. *Probability and statistics with reliability, queuing, and computer science applications.* Englewood Cliffs, N.J.: Prentice Hall.

Walpole, Ronald E., and R. H. Myers. 1989. *Probability and statistics for engineers and scientists.* 4th ed. New York: Macmillan.

Note: Almost every technical library has a section of books on probability and statistics containing additional background readings on the subject of probability distributions. The books listed herein are representative of the large number of excellent books available on the subject.

Part II

Families of Probability Distributions

Part II of this book adapts reliability analysis to the case of a nonrepairable component whose failure law may be governed by a relatively simple and well-known distribution such as the binomial, normal, or exponential. It consists of chapters 3, 4, and 5, respectively titled "Attributes, Go–No Go, or Zero-One Processes," "Normal Distributions," and "Exponential Times to Failure." In each of these chapters, fundamental concepts of the failure process that give rise to the distribution or related distributions are introduced and some representative problems related to that process are solved.

Because the binomial distribution is discrete, whereas the normal and exponential distributions are continuous and may require calculus, reliability-confidence assessment is simpler for attributes data using the binomial distribution than it is for time-dependent data with the exponential distribution or variable-dependent data with the normal distribution. Accordingly, a reader having attribute failure data may find that chapter 3, combined with Table A.1, yields a solution to the problem of establishing a confidence interval on the reliability, possibly at the cost of using a larger sample size than would be necessary with an alternative method.

Since they both deal with continuous distributions, the format of both chapters 4 and 5 differs from that of chapter 3. For example, we do not do reliability-confidence assessment for the normal and exponential processes, but leave that to Part III of this book. In chapters 4 and 5, in addition to describing the failure processes that give rise to the normal and exponential distributions, we supply both graphical and algebraic techniques for goodness of fit. Each chapter also has a technical appendix which describes distributions emanating from the same process that are related to topics covered later in this book.

3

Attributes, Go–No Go, or Zero-One Processes

Introduction

In this chapter, distributions are described that govern probabilities for random variables derived from binary trials. This includes the binomial family, with the sample size (number of trials) fixed and the number of events of a specified type (successes or else failures) as the r.v.; and the negative binomial family, the r.v. being the number of trials to achieve a specified number of events of a certain type. The relationships of these two families to the beta distributions are also discussed, as well as the use of tables of the percentage points of the beta distribution for reliability-confidence assessment.

Binary trials

Sampling binary random variables when the probability of success or alternatively of failure does not change from trial to trial, and successive trials are independent is known as a *Bernoulli process*. The trials are known as Bernoulli or binomial trials. The number of trials is a finite number, but because it can be very large, combinatorial mathematics is used to count events.

Orderings or permutations

If the results of n binary trials are scrambled and then rearranged into n fixed positions, n choices are available for the first position, then $n-1$ for the second, $n-2$ for the third, and so on, leaving one choice for the last position. The total number of different orderings or permutations possible is

$$n(n-1)(n-2)\ldots 1 = n! \tag{3.1}$$

The number $n!$ is known as *factorial n*. The reader is reminded that $0! = 1$.

Indistinguishable orderings of n binary trials

Suppose that r trials result in successes (1's) and the remaining $(n-r)$ are failures (0's). Instead of rearranging all n components, first permute only the 1's and then only the 0's. There are

$$r!\,(n-r)!$$

such orderings possible, because for every one of the $r!$ different permutations of all the 1's there are $(n-r)!$ permutations of all the 0's.

Combinations: Sets of outcomes with the same result

If the n trials result in exactly r successes, there are $n!$ permutations and $r!(n-r)!$ of these are indistinguishable from one another. Therefore, there are exactly

$$\frac{n!}{r!\,(n-r)!} = \binom{n}{r}$$

different outcomes with r successes. The number $\binom{n}{r}$, called the *binomial coefficient*, is also known as the number of combinations of n things taken r at a time. Because of symmetry,

$$\binom{n}{r} = \binom{n}{n-r}.$$

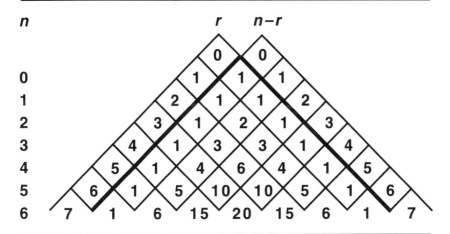

Figure 3.1. The remarkable Pascal's triangle.

EXAMPLE 3.1.

Pascal's triangle: The numerical values of binomial coefficients can be obtained from a remarkable triangle named after Blaise Pascal. Figure 3.1 is a representation of this triangle for $n \leq 7$. The rows represent n and the intersecting diagonal columns are respectively for r and $n-r$. The sum of two adjacent entries in any one row is the entry between those two in the next row. An enlarged version of the triangle is in many books on probability and statistics.

EXAMPLE 3.2.

Iterative formula for Pascal's triangle: Successive terms of the triangle are generated by the formula

$$\binom{n-1}{r} + \binom{n-1}{r-1} = \binom{n}{r}. \tag{3.2}$$

Equation 3.2 is proved as follows:

$$\begin{aligned}
\binom{n-1}{r} + \binom{n-1}{r-1} &= \frac{(n-1)!}{r!\,(n-1-r)!} + \frac{(n-1)!}{(r-1)!\,(n-r)!} \\
&= \frac{(n-1)!\,(n-r)}{r!\,(n-r)!} + \frac{r\,(n-1)!}{r!\,(n-r)!} \\
&= \frac{n!}{r!\,(n-r)!} = \binom{n}{r}.
\end{aligned}$$

Binomial probability

Since the probability p of success on each of n successive Bernoulli trials does not change from trial to trial, the probability of a failure is $1-p$. Because of independence, the probability of an outcome, r successes and $n-r$ failures, is the product of the probabilities of the results of the trials. However the successes and failures are ordered, the probability of an outcome is

$$p^r (1-p)^{n-r}.$$

Because there are $\binom{n}{r}$ outcomes with the same result, the probability of exactly r successes and $n-r$ failures is the well-known binomial formula

$$f_b(r|p, n) = P(X = r|p, n)$$
$$= \binom{n}{r} p^r (1-p)^{n-r}, \; r = 0, 1, \ldots, n.$$

Binomial distribution function

The distribution function d.f. is the accumulation of binomial probabilities

$$F_b(r|p, n) = P(X \le r|p, n)$$
$$= \sum_{i=0}^{r} f_b(i|p, n).$$

EXAMPLE 3.3.

Distribution function for a binomial r.v.: A fair coin is tossed five times. What is the probability of *at most* three heads? Since the coin is fair, the probability of a head on a single trial is $p = 1-p = 0.5$. Pascal's triangle shows that there are 26 outcomes for which $r \le 3$.

$$\binom{5}{0} + \binom{5}{1} + \binom{5}{2} + \binom{5}{3} = 1 + 5 + 10 + 10 = 26$$

Therefore the answer is $26/32 = 0.8125$. The probability that *at least* four heads is obtained in five trials is

$$1 - \frac{26}{32} = \frac{6}{32}.$$

Chapter 3 ■ *Attributes, Go–No Go, or Zero-One Processes*

Tables of the binomial distribution

Tables of the binomial distribution tend to be bulky because there are three different arguments, n, r, and p, and there is no simple way to standardize, as is true of the normal distribution. Reliability analysts usually operate with data about reliable elements; the reliability p, or probability of success, is close to 1 but not 1. Under those circumstances r, the number of successes in n trials, would be close to n.

Most published binomial tables cannot effectively be used for reliability assessment when n is large, say 50 or more, p is close to one, and r is close to n. Table A.1, however, is a special two-page table designed for this particular purpose; it is explained later in this chapter and also in chapter 7.

EXAMPLE 3.4.

Symmetry of the binomial distribution: The d.f. on the number of successes r is the complementary d.f., $1-$d.f., of the number of failures $n-r$. Let $q = 1-p$ be the probability of a failure; then the probability of r or fewer successes,

$$\binom{n}{0}(1-p)^n + \binom{n}{1}p(1-p)^{n-1} + \ldots + \binom{n}{r}p^r(1-p)^{n-r},$$

is identical term by term to the probability of $n-r$ or more failures,

$$\binom{n}{n-r}q^{n-r}(1-q)^r + \binom{n}{n-r+1}q^{n-r+1}(1-q)^{r-1} + \ldots + \binom{n}{n}q^n.$$

This symmetry makes it possible to cut the size of binomial tables to approximately one-half of what they would be with a full range of values. The identity formula is

$$F_b(r|p, n) = 1 - P(X \leq n-r-1|1-p, n)$$
$$= 1 - F_b(n-r-1|1-p, n).$$

Because of this identity, a binomial table can have a half-range of values for p, $0 \leq p \leq 0.5$, together with a full range for n and r. Other methods of shortening binomial tables are possible. Even shortened binomial tables with large n and p nearly 1 are too bulky for most needs in reliability assessment.

Mean and variance of the binomial distribution

The binomial is a distribution on the sum of n independent binary random variables. Let $X_1, \ldots, X_n$ be the n binary values, 1 or 0 respectively. Example 2.5 shows that for each binary r.v. X_i the mean is

$$E(X_i) = p, \; i = 1, \ldots, n,$$

and the variance is

$$E(X_i - p)^2 = p(1 - p).$$

Introduce the binomial r.v. Y, the number of successes or the sum of the X_i's,

$$Y = X_1 + X_2 + \ldots + X_n. \tag{3.3}$$

The mean of Y is

$$EY = E(X_1 + X_2 + \ldots + X_n) = np.$$

The variance of Y is

$$\begin{aligned} E(Y-np)^2 &= E(X_1 + X_2 + \ldots + X_n - np)^2 \\ &= E\{(X_1-p) + (X_2-p) + \ldots + (X_n-p)\}^2 \\ &= E(X_1-p)^2 + E(X_2-p)^2 + \ldots + E(X_n-p)^2 = np(1-p). \end{aligned} \tag{3.4}$$

In this derivation the cross-product terms

$$E(X_i-p)(X_j-p) = 0, \; i \neq j$$

all vanish because of independence.

We can now obtain the mean and variance of the fractional form of a binomial r.v., which is directly related to reliability. Let Y/n be the proportion of successes in a sample of size n. The mean of Y/n is

$$E\left(\frac{Y}{n}\right) = \frac{1}{n} E(Y) = p.$$

This is like saying that the proportion of successes in a sample of any size is an unbiased estimate of the reliability. The variance is

$$\begin{aligned} E\left(\frac{Y}{n} - p\right)^2 &= \frac{1}{n^2} E(Y-np)^2 \\ &= \frac{np(1-p)}{n^2} \\ &= \frac{p(1-p)}{n}. \end{aligned}$$

EXAMPLE 3.5.

A fair coin is tossed 10 times. What are the mean and variance of the number of heads and the proportion of heads?

Since the coin is fair, the probability p of a head is 0.5, the same as the probability of a tail, $1-p$. The mean number of heads is $10(0.5) = 5$; the variance is $10(0.5)(0.5) = 2.5$. The mean proportion of heads is 0.5; the variance of the proportion of heads is

$$\frac{(0.5)(0.5)}{10} = 0.025.$$

Negative binomial (Pascal) distribution

The Bernoulli process gives rise to a variety of forms of useful probability distributions. With the negative binomial (n.b.), also called Pascal distribution, the sample size n is the r.v, and r, the number of events of a specified type, is fixed. The negative binomial is useful in statistical applications such as reliability because variable sample sizes are permitted. As before, let p be the probability of success ($X_i = 1$) on the ith trial, and $1-p$ the probability of a failure ($X_i = 0$). The n.b. probability function is

$$f_{nb}(n|p, r) = \binom{n-1}{r-1} p^r (1-p)^{n-r}, r \geq 1; n = r, r+1, r+2, \ldots \quad (3.5)$$

Equation 3.5 is interpreted as follows: n is the sample size for exactly r successes. To have exactly r successes in n trials, the last, or nth, trial must be a success; up to and including the $(n-1)$st trial, there would have had to be $r-1$ successes, each with probability p, and $n-r$ failures, each with probability $1-p$.

Dual binomial and negative binomial relationship

The binomial and n.b. families have a dual relationship; this means that one may be derived from the other. Let N denote a negative binomial r.v. The distribution function

$$F_{nb}(n|p, r) = P(N \leq n|p, r)$$

is the probability that n or fewer trials are required for exactly r successes; but this is the same as the probability that with n fixed, r or more successes occur. Let R denote a binomial r.v.; then

$$F_{nb}(n|p, r) = P(R \geq r|p, n)$$
$$= 1 - P(R \leq r-1|p, n)$$
$$= 1 - F_b(r-1|p, n).$$

46   Chapter 3 ■ *Attributes, Go–No Go, or Zero-One Processes*

EXAMPLE 3.6.

Negative binomial distribution function: What is the probability that five or fewer tosses of a fair coin are needed to obtain exactly four heads? This is

$$\binom{3}{3}\left(\frac{1}{2}\right)^4 + \binom{4}{3}\left(\frac{1}{2}\right)^5 = \frac{6}{32};$$

this is identical to the probability that at least four heads are obtained in five binomial trials.

Because of this dual relationship, it is possible to look up n.b. probabilities in binomial tables. It is more convenient, however, to use n.b. tables such as Williamson and Bretherton.

Symmetry of the negative binomial distribution function

Negative binomial symmetry differs from the binomial. The n.b. is symmetrical with respect to p and $1-p$, but not with respect to r and $n-r$. For the binomial, the d.f. on r is identical to the complementary d.f. on $n-r$. For the n.b., both n and $n-r$ have the same distribution, since there is always a constant difference of r.

This symmetry makes it convenient to use the n.b. as a distribution on the number of trials necessary to achieve a specified number of failures. If $q = 1-p$ is the probability of a failure, and r is the number of failures in n trials, by analogy to the foregoing,

$$f_{nb}(n|q, r) = \binom{n-1}{r-1} q^r(1-q)^{n-r}, \text{ and}$$
$$F_{nb}(n|q, r) = 1 - F_b(r-1|q, n).$$

EXAMPLE 3.7.

Figure 3.2 portrays the negative binomial probabilities for the number of trials n to result in one failure when $p = 0.8$.

Average sample size: The negative binomial mean

One use for the negative binomial is to determine sample sizes; for example, the n.b. mean is a kind of average sample size. The mean is evaluated by a discrete form of the change-of-variables technique.

Because the assignment of probabilities is acceptable

$$\sum_{n=r}^{\infty} f_{nb}(n|p, r) = \sum_{n=r}^{\infty} \binom{n-1}{r-1} p^r (1-p)^{n-r} = 1.$$

Chapter 3 ■ Attributes, Go–No Go, or Zero-One Processes

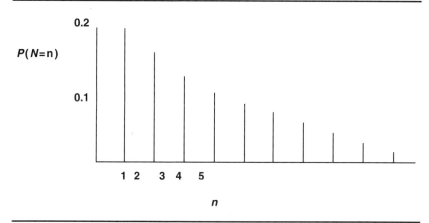

Figure 3.2. Negative binomial probabilities with $q = 0.2$ and $r = 1$.

Following Equation 2.1 the mean, or expected value, of the negative binomial is the sum of the products of the values of the r.v. by their respective probabilities. Let N be the r.v. for the sample size. The mean is

$$E(N) = \sum_{n=r}^{\infty} n \frac{(n-1)!}{(r-1)!(n-r)!} p^r (1-p)^{n-r}$$
$$= \frac{r}{p} \sum_{n=r}^{\infty} \frac{n!}{r!(n-r)!} p^r (1-p)^{n-r} = \frac{r}{p}. \qquad (3.6)$$

The summation in the second line of Equation 3.6 has a value of 1, by the change of variables: $n' = n+1$ and $r' = r+1$. The mean r/p is the average sample size for specified r and p.

EXAMPLE 3.8.

What is the average sample size if $p = 0.95$ and sampling is terminated at $r = 1$ failure?

For this purpose, we employ the alternative form of the negative binomial, namely, failures instead of successes, and obtain

$$E(N) = \frac{r}{1-p} = 20.$$

Lower confidence limit on the reliability: Discrete distributions

The reliability is the probability of a success on a single trial; therefore assessing the reliability for attributes data is the same as assessing p.

Since the value of p is unknown due to the inherent variability of the data, it can be specified only up to a confidence factor. Note that there are (or already have been) r successes in n trials. The lower confidence limit (LCL) for p at confidence level γ is the smallest value p_L such that the probability is approximately γ that if the next or $(n+1)$st trial would be a failure, the result, r successes in $n+1$ trials, is the worst that could occur; symbolically,

$$\gamma \approx F_b(r|p_L, n+1)$$
$$= 1 - F_{nb}(n+1|p_L, r+1). \quad (3.7)$$

The approximation symbol $\approx$ is included in this definition because the probability is a sum of terms from a discrete distribution. A relatively small change in the value of p, with r fixed, can cause a large deviation in γ. Later in this chapter, it is shown that with the beta distribution as a continuous distribution on p, a lower confidence limit on p for attributes data is obtained with γ defined by an equality rather than an approximation.

EXAMPLE 3.9.

LCL in coin tossing: A coin is tossed once and a head is obtained. It is not known whether the coin is fair. At a confidence level γ of 0.75, what is the LCL p_L on the probability of a head? Answer: p_L is 0.5, since if $p = 0.5$, the probability of either one head or none in two trials is 0.75. In other words, there is a 0.75 probability that the coin is fair.

EXAMPLE 3.10.

LCL in coin tossing (continued): In Example 3.9, what is the LCL at $\gamma = 0.5$? The answer is obtained from Equation 3.7.

$$\gamma = 1 - p_L^2, \text{ and}$$
$$p_L = \sqrt{1 - \gamma} = \sqrt{0.5} = 0.7071.$$

Beta distributions

For the beta distribution, the mathematical form is similar to that of the binomial and negative binomial distributions. Because the r.v. is continuous in the zero-to-one range, the beta serves the purpose of a distribution on the reliability p as a random variable instead of as a parameter. Using the beta percentage points, confidence intervals on the reliability can be set.

Consider the function

$$p^{a-1}(1-p)^{b-1}, \ 0 < p < 1, \tag{3.8}$$

known as the *kernel*, common to both the binomial and n.b. distributions, usually with $a-1 = r$ and $b-1 = n-r$. The kernel is normalized to a probability by dividing it by a constant such that the total probability is one. By the calculus technique of successive integration by parts,

$$\int_0^1 p^{a-1}(1-p)^{b-1} \, dp = \frac{(a-1)!\,(b-1)!}{(a+b-1)!}. \tag{3.9}$$

Equation 3.9 is known as the beta function $B(a,b)$. Dividing Equation 3.8 by Equation 3.9 the beta p.d.f. is obtained.

$$f_\beta(p|a,b) = \frac{(a+b-1)!}{(a-1)!\,(b-1)!} p^{a-1}(1-p)^{b-1}, \ 0 < p < 1 \tag{3.10}$$

The incomplete integral of Equation 3.10,

$$F_\beta(p|a,b) = \int_0^p \frac{(a+b-1)!}{(a-1)!\,(b-1)!} x^{a-1}(1-x)^{b-1} \, dx \tag{3.11}$$

is the beta d.f. also known as the incomplete beta function $I_p(a,b)$, and it has been tabled by Karl Pearson, H. L. Harter, and others.

The dual relationship between the beta and binomial families can be shown by integrating Equation 3.11 by parts. This yields

$$F_\beta(p|a,b) = \sum_{i=a}^{a+b-1} \frac{(a+b-1)!}{i!\,(a+b-1-i)!} p^i (1-p)^{a+b-1-i}$$
$$= 1 - F_b(a-1|p,\,a+b-1).$$

Because of the dual relationship between the binomial and n.b. families

$$F_b(a-1|p,\,a+b-1) = 1 - F_{nb}(a+b-1|p,\,a);$$

there is also a dual relationship between the n.b. and the beta

$$F_\beta(p|a,b) = F_{nb}(a+b-1|p,\,a).$$

Confidence limits for the reliability

If the observed result is r successes in n trials, the lower confidence limit p_L at confidence level γ is defined by Equation 3.7.

$$\gamma \approx F_b(r|p_L, n+1)$$
$$= 1 - F_{nb}(n+1|p_L, r+1). \qquad (3.12)$$

Because of this dual relationship, the LCL p_L is a percentage point of a beta distribution

$$F_b(r|p_L, n+1) = 1 - F_\beta(p_L|r, n-r+1);$$

therefore,

$$\gamma = 1 - F_\beta(p_L|r, n-r+1).$$

Beta distribution percentage points for reliability assessment

Because of the pairwise dual relationships between the beta, binomial, and n.b. distributions, tables of any one of the three distributions could be substituted for either one of the other two in finding the value of the LCL at a specified confidence level. In practice, to find the LCL as a percentage point of the beta distribution from binomial tables often is not feasible. For example, the Harvard binomial tables were developed for general use and do not include much data for either large or small values of p. On the other hand, the Weintraub binomial tables can be used effectively for $p < 0.1$ or $p > 0.9$.

Table A.1 is an abbreviated table of beta-distribution percentage points for assessing reliable elements when the sample sizes are between 2 and 200. Six selected confidence levels are used: 0.5, 0.75, 0.9, 0.95, 0.99, and 0.995. The LCL tabulations are provided only for those cases where r, the number of successes, is no more than five fewer than n, the sample size. By interpolating on the column labeled "$n-r$," the assessment can be done with an error less than 0.001 for values of n not included in the table. A much larger set of tables was published by the U.S. Naval Ordnance Test Station (NOTS) at China Lake, California; however, there is a discrepancy between the NOTS tables and Table A.1, in that the parameters r and n are offset by one, since they use

$$\gamma \approx F_b(r-1|p_L, n)$$

instead of

$$\gamma \approx F_b(r|p_l, n+1).$$

Chapter 3 ■ *Attributes, Go–No Go, or Zero-One Processes* 51

Nevertheless, the NOTS tables can be used for the type of assessment described in this text if these offsets are compensated for by increasing both parameters by one.

EXAMPLE 3.11.

Using the assessment tables: For 147 successes in 150 trials, what is the LCL on the reliability for the confidence level γ of 0.75?
Answer: 0.966.

EXAMPLE 3.12.

Approximating the LCL for a value of n not included in the table: For 175 trials and 4 failures, what is the LCL at $\gamma = 0.99$?
For $n = 150$, $\gamma = 0.99$, and $n - r = 4$ the LCL is 0.925; for $n = 200$, $\gamma = 0.99$, and $n - r = 4$, the LCL is 0.943. For $n = 175$, $\gamma = 0.99$, and $n - r = 4$ the LCL is approximately halfway between these two, or 0.934.

Problems

3.1. For n trials, it can be shown that the sum of the binomial coefficients, which corresponds to the total number of outcomes for n binary events, is 2^n, by expanding $(1+1)^n$; do so.

3.2. Use Table A.1 to find the confidence level for a reliability of 0.95 if there are $r = 97$ successes in $n = 100$ trials. It is instructive to compare this result with those obtained from other widely used books of tables. The Harvard binomial tables do not tabulate for $r = 97$ with $n = 100$, because the distributions for large values of n are concentrated in the middle probability ranges, in the vicinity of 0.5. The Pearson *Tables of the Incomplete Beta Function* do not tabulate for any usable data beyond $n = 50$.

3.3. What is the minimum sample size for 0.99 reliability with 90 percent confidence?

3.4. What is the minimum sample size with *no failures* to guarantee 90 percent reliability with 90 percent confidence?
The general formula for this calculation is $n \approx \dfrac{\log(1-\gamma)}{\log R} - 1$.
Derive this formula from first principles, using Equation 3.7.

3.5. Approximately how many binomial trials are necessary with one failure to demonstrate 90 percent reliability with 90 percent confidence?

3.6. The probability that a transformer will survive 100 hours of continuous operation in an accelerated temperature test environment is 0.9. If five transformers are tested in the accelerated environment, what is the probability that at least four will survive 100 hours of continuous operation?

3.7. In the aborted attempt to rescue the American hostages from the American embassy in Tehran, Iran, in 1979, there were eight helicopters. Hypothetically, the attempt would have been successful if six or more of these helicopters had been able to complete the mission, and the probability of success for any single helicopter was 0.7. Under these circumstances, what is the calculated probability that the mission would have succeeded?

3.8. In MIL-STD 105E, the average quality level (AQL) is the maximum percentage of defective items permitted in a lot. An inspection plan requires sampling and inspecting 10 parts from a production line every hour. If one or more defective parts are found, the lot is rejected; this means that the entire output for that hour is inspected for defects, and the defective items are replaced by good ones. Otherwise the lot is accepted; this means that there is no further inspection of the hour's output. Complete the following chart by giving the probability of accepting a lot, for each AQL listed.

AQL (%)	P(Accept)
0.5	
1.0	
2.0	
4.0	
6.0	

3.9. If the AQL (defined in Problem 3.8) is 6 percent and there were no limitations on the size of a sample, and if the sampling plan called for rejection of a lot with two failures, what is the average sample size before the lot is rejected?

References

Bradley, James V. 1976. *Probability; decision; statistics.* Englewood Cliffs, N.J.: Prentice Hall.

Burington, R. S., and D. C. May. 1970. *Handbook of probability and statistics with tables.* 2d ed. New York: McGraw-Hill.

Cook, J. R., M. T. Lee, and J. P. Vanderbeck. 1960. *Binomial reliability tables.* China Lake, Calif.: U.S. Naval Ordnance Test Station.

Doty, L. A. 1989. *Reliability for the technologies.* 2d ed. New York: Industrial Press.

Goldberg, Samuel. 1960. *Probability, an introduction.* Englewood Cliffs, N.J.: Prentice Hall.

Harter, H. L. 1964. *Tables of the incomplete gamma-function ratio and of percentage points of the chi-square and beta distributions.* U.S. Government Printing Office.

Hausner, Melvin. 1971. *Elementary probability theory.* New York: Harper & Row.

Pearson, K. 1934. *Tables of the incomplete beta-function.* New York: Cambridge University Press.

Ross, Sheldon M. 1980. *Introduction to probability models.* 2d ed. New York: Academic Press.

Shook, Robert C., and H. J. Highland. 1969. *Probability models with business applications,* Homewood, Ill.: R. D. Irwin.

Tables of the cumulative binomial probability distribution. 1955. Cambridge, Mass.: Harvard University Press.

Trivedi, K. S. 1982. *Probability and statistics with reliability, queuing, and computer science applications.* Englewood Cliffs, N.J.: Prentice Hall.

Walpole, Ronald E., and R. H. Myers. 1989. *Probability and statistics for engineers and scientists.* 4th ed. New York: Macmillan.

Weintraub, S. 1963. *Tables of the cumulative binomial probability distribution for small values of* p. New York: Free Press.

Williamson, E., and M. H. Bretherton. 1963. *Tables of the negative binomial probability distribution.* New York: Wiley.

4

Normal Distributions

Introduction

This chapter covers the normal *Gaussian* distribution, possibly the most widely used of all probability families, and selected families that are related to or derived from the normal. The normal distribution is easy to work with because there are only two parameters, the mean and the variance; this makes it convenient to standardize in the way described in Chapter 2. A table of cumulative standard normal probabilities is widely available; that table is also included in this text as Table A.3.

Many phenomena are approximately normal. LaPlace (1749–1827) and Gauss (1777–1855) found that the normal distribution approximates the errors of astronomical measurements. DeMoivre (1667–1754) found that the normal is the limit of the binomial for large sample sizes. Many psychologists believe that the distribution of intelligence quotients (IQs) is normal. The 3σ specification or tolerance limits that engineers commonly set for operational characteristics of equipment and hardware is based on the assumption of normality. It is also widely known by the so-named central limit theorem that for every family of distributions, including those that are not normal, the distribution of random sample means tends toward normality with increasing sample sizes. A binomial distribution for small sample sizes is closely approximated by the normal if appropriate correction factors are used.

Normal-theory reliability-confidence assessment is covered in this text in two stages; first in this chapter and then in chapter 8. Chapter 4 is an extended introduction, including the normal, standard normal, and lognormal distributions, tables, plotting normal graphs, small sample approximations, and goodness of fit. There is also a technical appendix on advanced topics, including the change-of-variables technique, and the χ^2 (chi-square) and t-distributions. Confidence assessment in this chapter overlaps with chapter 3; since a normal distribution approximates the binomial, lower confidence limits for attributes data can be found with normal tables. In chapter 8, however, normal-theory assessment is performed with both one- and two-sided tolerance limits.

Standardized normal distribution

The p.d.f. and the d.f.

Let X be a normally distributed random variable with mean μ and variance σ^2; this is abbreviated $X{:}N(\mu, \sigma^2)$. A standardized r.v. Z is defined in chapter 2 as

$$Z = \frac{X-\mu}{\sigma}. \tag{4.1}$$

Since X is a normal r.v., Z is also normal. Because the mean of Z is zero and the variance is one, as was shown in chapter 2, Z is represented in the abbreviated form $Z{:}N(0, 1)$, with p.d.f.

$$f(z) = \frac{1}{\sqrt{2\pi}} \exp\left(-\frac{z^2}{2}\right), \quad -\infty < z < \infty, \tag{4.2}$$

over an infinite range of values for the standardized normal r.v. $Z{:}N(0, 1)$; and the p.d.f. for any normal r.v. $X{:}N(\mu, \sigma^2)$ is

$$f(x) = \frac{1}{\sigma\sqrt{2\pi}} \exp\left\{-\frac{(x-\mu)^2}{2\sigma^2}\right\}, \quad -\infty < z < \infty, \tag{4.3}$$

as is shown in the technical appendix to this chapter.

Since the normal distribution is continuous, probabilities for ranges of an r.v. are accumulated by mathematical integration. The distribution function for $Z{:}N(0,1)$ is

$$F(z) = P(Z \le z) = \int_{-\infty}^{z} \frac{1}{\sqrt{2\pi}} \exp\left(-\frac{z^2}{2}\right) dz;$$

and the d.f. for $X:N(\mu, \sigma^2)$ is

$$F(x) = P(X \le x) = \int_{-\infty}^{x} \frac{1}{\sigma\sqrt{2\pi}} \exp\left\{-\frac{(x-\mu)^2}{2\sigma^2}\right\} dx.$$

Shape, shift, and scale

In substituting $Z:N(0, 1)$ for $X:N(\mu, \sigma^2)$, or vice versa, there is no change of shape of the distribution because Z differs from X only by a shift, subtracting μ, and a rescaling, dividing by σ. No matter what μ and σ are, the shape of the distribution of probabilities for values of X is represented by a picture or graph of the distribution of Z, and the cumulative probabilities of X can be obtained from a table of the cumulative probabilities of Z such as Table A.2.

The well-known bell shape of a normal distribution is a graph of the probability density, Figure 4.1. Another way to represent the normal shape is by a graph of the distribution function in the nondecreasing ogive form, Figure 4.2.

Standard normal probability table: Symmetry

A normal distribution is symmetric about the mean. This makes it possible to derive the probabilities for the lower one-half from those of the upper one-half, or vice versa, by a simple transformation. For example,

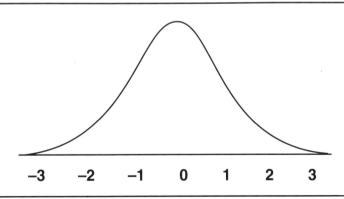

Figure 4.1. Probability density of $Z:N(0,1)$.

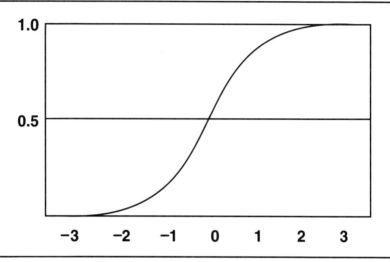

Figure 4.2. Distribution function of $Z{:}N(0,1)$.

Table A.2 contains the cumulative probabilities for $-3.09 \leq Z \leq +3.09$. To obtain the d.f. for complementary values of Z, use the formula

$$F(z) = 1 - F(-z). \qquad (4.4)$$

EXAMPLE 4.1.

z for a fixed F: For the standard normal $Z{:}N(0,1)$ what are the values, z, such that: (1) $F(z) = 0.95$ and (2) $F(z) = 0.05$? For (1) $z = 1.645$, by interpolation, and for (2) $z = -1.645$.

EXAMPLE 4.2.

F for a fixed z: What are

$F(1)$? 0.8413
$F(-1)$? 0.1587
$F(2)$? 0.9772
$F(-2)$? 0.0228

EXAMPLE 4.3.

The closing time for valves of a certain type that are usually open is normally distributed with a mean μ of 100 milliseconds (ms) and a variance σ^2 of 100 ms². (The mean has the same units as the original data, milliseconds, and the variance is in units of milliseconds-squared. Why?) A closing time of <80 ms or >120 ms is defective. What is the reliability, the probability that a valve is not defective?

The standard deviation σ is 10. The reliability is the probability that the closing time is between 80 and 120 ms. To solve this problem, use the transformation to a standard form of an r.v., Equation 2.9,

$$z = \frac{x-\mu}{\sigma}.$$

For $X = 80$, $z = -2$; for $X = 120$, $z = 2$. From the answers to Example 4.2, it can be seen that the reliability is

$$0.9772 - 0.0228 = 0.9544.$$

Probability plotting and graphing

Normal probability graph paper

Normal probability graph paper makes it possible to analyze data in an insightful way. The striking features of this graph paper are: the d.f. of any normal distribution is represented as a straight line; the parameters μ and σ can be approximated; and a visual display of the p.d.f. is obtained. If the distribution is perfectly normal, the line is exact; if the data are nearly normal, the straight line approximating the normal d.f. is the center of a scatter of data points. If the data are not normal, the line is curved, indicating either that the data fit a distribution other than the normal or that there are outlying observations.

Figure 4.3 is a blank sheet of normal graph paper especially designed for use with this text; reproduction is permitted. The bottom section taking up most of the page is for plotting the data points and the approximating straight line; the upper section provides the visual p.d.f., which is bell shaped if the d.f. line is straight. The graph is scaled both in percentages and standard deviations. The left-hand edge is the scale for percentage of cases under; the right-hand edge is the scale for percentage of cases over; and the equally spaced dots along both edges indicate fractions of the standard deviation: 0, $\pm 0.5\sigma$, $\pm 1.0\sigma$, and so on, for plotting the p.d.f. The vertical lines are scaled by the user in units of the data.

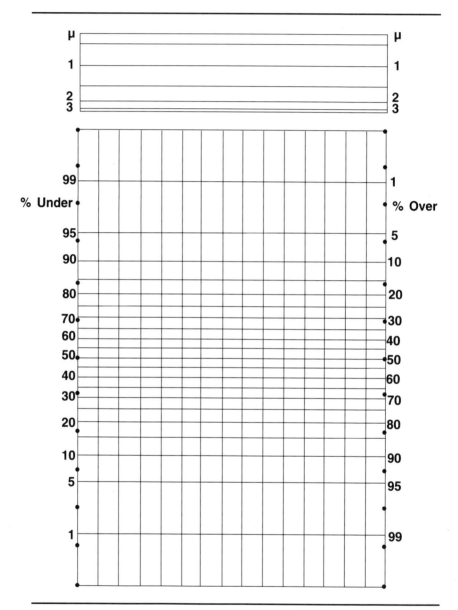

Figure 4.3. Normal probability graph paper.

EXAMPLE 4.4.

The probability plot of the normal distribution of valve closing times in Example 4.3 is in Figure 4.4. The most visible features are the straight line representing the distribution function and the bell-shaped curve for the p.d.f. The p.d.f. curve is derived from the d.f. A point is obtained by squaring off at a right angle the intersection of a horizontal line drawn from one of the dots along an edge to the d.f. line with a vertical line to the horizontal line on the upper chart corresponding to the dot.

The parameters of the normal distribution of valve closing times are obtained from the plot. The value of μ is clearly 100; there are two ways this can be seen. First, it is the 50th percentile of the straight d.f. line; and second, the intersecting vertical line goes to the highest point of the p.d.f. curve. The value of σ, 10, can also be estimated several different ways. The simplest way is as the horizontal difference between μ and a point of intersection with the 1.0σ vertical line; σ is also the horizontal difference between 1.0 and 2.0, 2.0 and 3.0, and so on, because of equal spacing.

Plotting points

The principal use for probability graph paper is to determine whether a set of observed data fits a particular parametric form such as the normal. Since the parameters of the distribution are unknown and the data are variable, there wouldn't be an idealized plot like Figure 4.4, with a straight-line d.f. and no scatter.

Plotting points are approximate substitute values for the distribution function. Given an ordered set of data, called the *order statistics,* the plotting point for an observation depends both on its rank in the ordered set and the total number of cases. There are a number of plotting point procedures; some of the better ones have complicated formulas. The following way to calculate a plotting point is simple and satisfies requirements. Suppose there are n ordered observations and let X_i be the ith one; the plotting point for X_i is

$$\hat{F}_i = \frac{i}{n+1} . \tag{4.5}$$

The reasons Equation 4.5 is an acceptable plotting point formula are:

1. It is symmetrical; this means that

$$\hat{F}_i = 1 - \hat{F}_{n+1-i} . \tag{4.6}$$

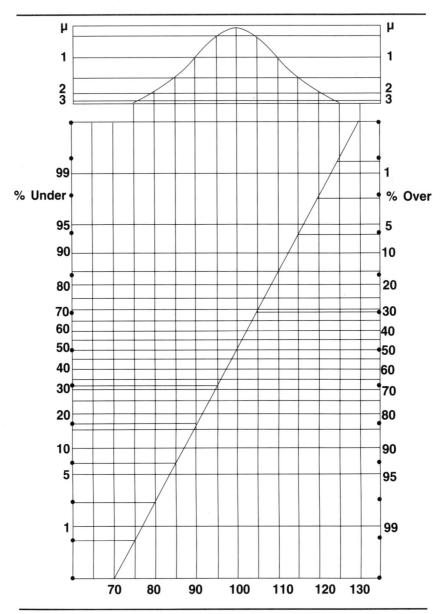

Figure 4.4. Probability plot of Example 4.3.

2. There is greater weight for the observations in the middle of the ranking than for those at the extremes.
3. The calculation is simple.

Table A.3 gives the plotting points for $n \leq 30$. Other simple and well-known plotting point procedures are symmetrical, but introduce more distortion than Equation 4.5 by giving too much weight to the values at the extremes, causing the d.f. line to be too flat.

EXAMPLE 4.5.

Graphing sample data: The following weights, in grams, of 10 apples are given by L. H. C. Tippett.

106, 107, 223, 125, 139, 119, 81, 131, 136, 123

By rearranging the data, the order statistics are

81, 106, 107, 119, 123, 125, 131, 136, 139, 223.

Figure 4.5 shows curvature. The distribution is close to normal for the first nine observations with μ about 121 and σ about 24. The last order statistic, 223, is an outlier, or unusual observation, and bends the line.

Probability plotting with grouped data

The probability plots above all deal with ungrouped data. Normal probability plots are used with grouped data for larger samples in connection with quality control and reliability applications. In those cases, the endpoints of the class intervals become the plotting points.

Goodness of fit: The Lilliefors K–S statistic

Goodness of fit (g.o.f.) deals with the relationship between observed and hypothetical probability distributions. The connection between g.o.f. and reliability assessment is that in order to be able to make confidence statements, it is necessary to know whether the probability distribution derived from the data is an appropriate one for making inferences. Among the better-known techniques for g.o.f. analysis are the chi-square, Kolmogorov–Smirnov (K–S), and Cramer–von Mises tests.

General procedure of a goodness-of-fit test

A g.o.f. test is a statistical hypothesis that a set of observed values are a random sample from a specified distribution. A g.o.f. test is based on the differences between the empirical d.f. derived from the sample data and the hypothetical d.f. it is supposed to fit. If the test statistic, which

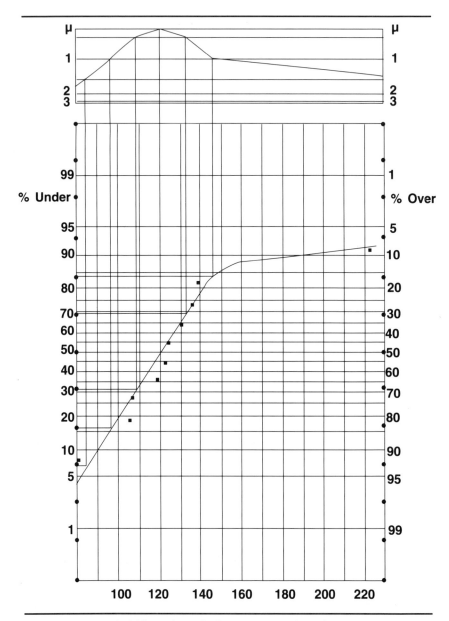

Figure 4.5. Probability plot of Tippett's "apple" data.

is a function of these differences, is small enough, the null hypothesis that there is no significant difference is not rejected; if the statistic is large, beyond some critical value, the null hypothesis is rejected in favor of an alternative that the difference is significant and therefore the fit is poor.

The Lilliefors K–S test for an unspecified normal population

The well-known K–S test of g.o.f. tests the differences between the empirical d.f. and the d.f. of a known population and does not require that the population have a particular parametric form. Lilliefors modified the K–S test for an unspecified normal population; this means that the population is normal, but the parameters are not known. The method provides critical values for a test of the greatest absolute difference between the empirical d.f. of an order statistic and the d.f. of a corresponding *rank-order statistic*. Lilliefors also provided a similar alternative version of the K–S test for an unspecified exponential population, which is discussed in chapter 5 of this text.

The raw data for a Lilliefors K–S test of an unspecified normal population are n ordered observations X_i, $i = 1, \ldots, n$. For each i the rank-order statistic is

$$\frac{i}{n}. \tag{4.7}$$

From Chapter 2 the mean of the data is

$$\overline{X} = \frac{\sum_{i=1}^{n} X_i}{n};$$

and the unbiased estimate of the variance is

$$\hat{\sigma}^2 = \frac{\sum_{i=1}^{n} (X_i - \overline{X})^2}{n-1}.$$

For the ith order statistic X_i, the empirical d.f. $F(X_i)$ is the same as the d.f. of the corresponding standardized normal Z_i;

$$F(X_i) = F(Z_i) = F\left(\frac{X_i - \overline{X}}{\hat{\sigma}}\right). \tag{4.8}$$

The Lilliefors g.o.f. test statistic D is the largest absolute difference between Equation 4.8 and an adjacent rank-order statistic. The formula is

Table 4.1. Critical values of D for an unspecified normal distribution.

Sample size n	Critical level		
	0.10	0.05	0.01
4	0.352	0.381	0.417
5	0.315	0.337	0.405
6	0.294	0.319	0.364
7	0.276	0.300	0.348
8	0.261	0.285	0.331
9	0.249	0.271	0.311
10	0.239	0.258	0.294
11	0.230	0.249	0.284
12	0.223	0.242	0.275
13	0.214	0.234	0.268
14	0.207	0.227	0.261
15	0.201	0.220	0.257
16	0.195	0.213	0.250
17	0.189	0.206	0.245
18	0.184	0.200	0.239
19	0.179	0.195	0.235
20	0.174	0.190	0.231
25	0.165	0.180	0.203
30	0.144	0.161	0.187
>30	$(0.805n)^{-1/2}$	$(0.886n)^{-1/2}$	$(1.031n)^{-1/2}$

Reprinted with permission of the American Statistical Association. Lilliefors, H. W., "On K-S for Normality with Mean and Variance Unknown," *Journal of the American Statistical Association*, vol. 62, pp. 399–402, 1967.

$$D = \max\left\{\left|F(X_i) - \frac{i-1}{n}\right|, \left|\frac{i}{n} - F(X_i)\right|\right\}.$$

Table 4.1 is a table of the critical values of D for $n \geq 4$ at the 0.10, 0.05, and 0.01 levels of significance.

EXAMPLE 4.6.

g.o.f. test of apple weight data: Perform a Lilliefors K–S test of the data from Example 4.5 on the 10 apple weights. By calculation, the mean weight is 129.0 grams and the standard deviation is 37.205 grams. The worksheet for calculating D is Table 4.2.

The largest absolute difference between either the ith or the $(i-1)$st rank-order statistic and $F(X_i)$ is 0.2940, identical to the critical value at the 0.01 level of significance for $n=10$ in Table 4.1. Since the probability is only 0.01 that a difference this large or larger could occur if the distribution were normal, there is a definite indication of a lack of normality.

Table 4.2. Worksheet for calculating D.

i	$\dfrac{i}{n}$	X_i	$\dfrac{X_i - 129.0}{37.205}$	$F(X_i)$	D
1	0.10	81	−1.290	0.0985	
2	0.20	106	−0.618	0.2682	
3	0.30	107	−0.591	0.2773	
4	0.40	119	−0.269	0.3940	
5	0.50	123	−0.161	0.4360	
6	0.60	125	−0.108	0.4570	
7	0.70	131	0.054	0.5215	
8	0.80	136	0.188	0.5745	
9	0.90	139	0.269	0.6060	0.2940
10	1.00	223	2.527	0.9942	

Approximate confidence limits for attributes data

A confidence statement is an inference about a percentage point of a probability distribution. For attributes data, confidence limits on the reliability can be obtained with any one of the three distributions discussed in chapter 3—binomial, negative binomial, or beta—because all three are dual to one another in pairs, and any one of them can be derived from either one of the other two. This section of chapter 4 covers approximating a binomial distribution with the normal using the Yates correction factor and calculating approximate confidence limits for the reliability by means of normal tables.

The standardized form

In chapter 3 it was shown that with parameter p, the average number of successes in n binomial trials is np; and that the variance of the number of successes is $np(1-p)$. Let the r.v. X, $0 \le X \le n$, denote the number of successes in n trials; the standardized form is

$$Z = \frac{X - np}{\sqrt{np(1-p)}}. \tag{4.9}$$

Approximate reliability-confidence limits: Yates correction factor

The distribution of the number of successes X can be approximated with normal tables. Even if the binomial distribution is nearly normal, there is an error in this approximation because the binomial is discrete and the normal is continuous. *Yates correction factor,* also known as the finite population correction factor, adds the number 0.5 to the numerator of Equation 4.9 and introduces just enough additional accuracy to give satisfactory results, even for small samples.

Let X denote the number of successes and let

$$F_b(x|n, p) = P(X \leq x|n, p)$$

be the binomial distribution function and let

$$F_N(z) = P(Z \leq z)$$

be the standard normal d.f. The approximation, including Yates correction, is

$$F_b(x|n, p) = F_N(z) \approx F_N\left(\frac{X - np + 0.5}{\sqrt{np(1-p)}}\right).$$

EXAMPLE 4.7.

(*See also* Example 3.3): A fair coin is tossed five times. What is the probability that at most three heads are obtained?

The approximating normal deviate is

$$z = \frac{3 - 5(0.5) + 0.5}{\sqrt{5(0.5)(0.5)}} = 0.894.$$

From Table A.2, the approximate probability is $F(0.894) = 0.8144$. The exact probability calculated in Example 3.3 is 0.8125.

EXAMPLE 4.8.

Large sample reliability assessment (*See also* Example 3.11): Given that there were 147 successes in 150 trials, what is the approximate confidence level for a reliability of 0.966? That is, 0.966 is the LCL on the reliability at what approximate confidence level?

By Eq. 3.12,

$$\gamma = F_b(r|p_L, n+1) = F_b(147|0.966, 151)$$
$$\approx F_N\left(\frac{147 - (0.966)(151) + 0.5}{\sqrt{151(0.966)(0.034)}}\right) = F_N(0.733) = 0.768.$$

The exact confidence level is 0.75 from Table A.1.

Example 4.8 shows that it is possible to calculate a reasonably accurate confidence level on the reliability for binomial success-or-failure data by the normal approximation. The accuracy of the approximation depends upon the skewness of the binomial. If the LCL is in the middle range, say between 0.2 and 0.8, there is little skewness and the approximation is very accurate. If the LCL is outside this range, the approximation may not be as good. Since the binomial variance increases with larger sample sizes, the approximation is better for large samples than for small ones, even if the LCL is not in the middle range.

EXAMPLE 4.9.

Small sample reliability assessment: For $p = 0.5$ and 12 successes in 15 trials, what is the confidence for a reliability of 0.5?
From Equation 3.12,

$$\gamma = F_b(12|0.5, 16)$$

$$\approx F_N\left(\frac{12-(0.5)(16)+0.5}{\sqrt{16\,(0.5)(0.5)}}\right) = F_N(2.25) = 0.9878.$$

From Table A.1 for 12 successes in 15 trials and the LCL = 0.510, the confidence level is 0.99.

The lognormal distribution

Fitting reliability data to the normal distribution often does not produce useful results. The twin requirements of symmetry and a doubly infinite range are usually not satisfied. For example, the most frequently used r.v. in reliability is time to an event, such as time to failure, service time, or repair time. If time is the r.v., there can be no negative-valued observations and the dispersion is right skewed, with a high concentration of small values, and a tendency for a spread out toward the large ones.

The lognormal distribution is a member of the normal family of distributions, for a right-skewed positive-valued r.v. The name *lognormal* is explained by the fact that the logarithms are normally distributed; this makes it possible to perform the same kinds of analyses as with the normal, such as plotting, goodness of fit, estimating parameters, assessment, and so on.

70   Chapter 4 ■ *Normal Distributions*

Lognormal probability plotting

In this text lognormal probability plotting is done with standard normal probability plotting paper, Figure 4.3. This is feasible because the logarithm of the r.v. is normally distributed, even though the values are not normally distributed. The only adjustment necessary is to make the horizontal scale of abscissas logarithmic.

An alternative to Figure 4.3, or any other standard normal plotting paper, is commercially available lognormal plotting paper. The problem with lognormal paper is that because the horizontal scale is in decade cycles, you often cannot show the detail necessary to make the plot useful without either stretching or compressing the scale; this involves much computation and interpolation, with many possibilities for error. Using standard normal plotting paper is simpler.

Interpreting a lognormal plot and g.o.f. test

EXAMPLE 4.10.

Graph the logarithms of the weights of the 10 apples in Example 4.6, test for goodness of fit and interpret the results.

The weights range from 81–223 grams. The straight d.f. line on standard normal probability paper suggests a good lognormal fit for the eight middle observations; however, the heaviest apple is too large to fit that distribution and the lightest apple may be too small. Figure 4.6 was drawn in such a way as to show the skewness caused by the largest observation.

The p.d.f. chart at the top of the graph is at its highest point for log $X = 2.08$, implying that 2.08 is the estimated mean logarithm. Since the mean of a normal distribution is also the median, the median logarithm is 2.08. The antilogarithm of 2.08, which is 120, is the median of the lognormal distribution, so that our estimated median weight for an apple is 120 grams.

Note this relationship between the measures of central tendency: the median of the logarithms and the median of the lognormal distribution coincide; the median logarithm is also the logarithm of the median of the lognormal distribution. That is not true for the means; the mean logarithm is *smaller* than the logarithm of the mean of the lognormal distribution.

An eyeball estimate of the standard deviation of the logarithms is 0.105; this is obtained by subtracting the logarithm 1σ below the *mean* for the straight portion of the line, 1.97, from the mean logarithm, which

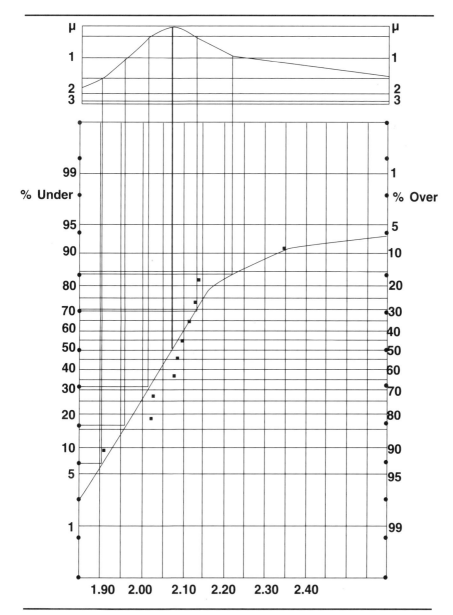

Figure 4.6. Lognormal plot of Tippett's apple data.

is approximately 2.075. Since the antilog of 0.105 is 1.27, any value on the straight line portion of the graph is 27 percent greater than one 1σ below it and 27 percent lower than a value 1σ above. For example, the

Table 4.3. Worksheet for lognormal g.o.f. analysis.

i	X_i	$\log X_i$	$\dfrac{i}{n}$	$\dfrac{\log(X_i) - 2.097}{0.112}$	$F(X_i)$	D
1	81	1.9085	0.10	−1.683	0.046	
2	106	2.0253	0.20	−0.640	0.261	
3	107	2.0294	0.30	−0.604	0.273	
4	119	2.0755	0.40	−0.192	0.424	
5	123	2.0899	0.50	−0.063	0.475	
6	125	2.0969	0.60	−0.001	0.500	
7	131	2.1173	0.70	0.181	0.572	
8	136	2.1335	0.80	0.326	0.628	
9	139	2.1430	0.90	0.411	0.659	0.241
10	223	2.3483	1.00	2.244	0.988	

±1σ range is from the antilog of 1.97, or 93, to the antilog of 2.18, or 151. This ±1σ has a different meaning for the lognormal distribution than for the normal. In the case of the normal distribution, σ defines a range for probabilities; for the lognormal, σ defines the ratio of the limits of the range.

Goodness of fit

A goodness-of-fit test for a lognormal distribution tests for the normality of the logarithms. Table 4.3 is a worksheet for a Lilliefors K–S g.o.f. test, similar to Table 4.2. The calculated mean of the logarithms for all 10 observations is 2.097; and the calculated standard deviation of the logarithms is 0.112.

D is smaller than in Example 4.7; this means that there is a different critical level at which the data do not pass the Lilliefors K–S test. D is larger than the critical number at the 0.10 level, 0.239, and smaller than the 0.05 critical number, which is 0.258. There is a slightly better fit to the lognormal distribution than to the normal, but not a very good one.

Appendix to Chapter 4
Advanced Topics and Related Distributions

Some Properties of Related Distributions
Normalizing the probability

It can be shown that the integral of the nonnegative function $\exp(-z^2/2)$, which is symmetrical about zero, is

$$\int_{-\infty}^{\infty} \exp\left(-\frac{z^2}{2}\right) dz = \sqrt{2\pi}; \tag{4.10}$$

$\exp(-z^2/2)$ is *normalized*, converted to a p.d.f., by dividing both sides of Equation 4.10 by the right-hand side factor of Equation 4.10;

$$(\sqrt{2\pi})^{-1} \int_{-\infty}^{\infty} \exp\left(-\frac{z^2}{2}\right) dz = 1. \tag{4.11}$$

Thus, the standard normal distribution $Z{:}N(0, 1)$ has an acceptable assignment of probabilities, because the p.d.f., Equation 4.2, is nonnegative, and the integral of the p.d.f., Equation 4.11, is unity.

Change-of-variables technique

In order to derive the p.d.f. for any normal distribution, the change-of-variables technique is employed. Let $F(Z)$ be the d.f. for the standard normal $Z{:}N(0, 1)$ and $F(X)$ the d.f. for the general normal $X{:}N(\mu, \sigma^2)$. Since $X = \sigma Z + \mu$, there is a one-for-one relationship between Z and X; this means there is a unique z for every x, and vice versa, and that

$$F(X(Z)) = F(Z).$$

Differentiating both sides with respect to Z,

$$F'(X) \frac{dX}{dZ} = F'(Z). \tag{4.12}$$

Since $F(X)$ and $F(Z)$ are both nondecreasing, the derivative dX/dZ in Equation 4.12, called the Jacobian in honor of the mathematician Carl Jacobi, is nonnegative, that is, either positive or zero-valued. $F'(X)$ and $F'(Z)$ are both probability densities and are usually written in the p.d.f form with lowercase letters $f(x)$ and $f(z)$, respectively. At a particular value x, the p.d.f. is

$$f(x) = f(z(x)) \Big/ \frac{dx}{dz}$$
$$= \frac{1}{\sigma\sqrt{2\pi}} \exp\left\{-\frac{(x-\mu)^2}{2\sigma^2}\right\},$$

since $\dfrac{dx}{dz} = \sigma$.

Mean and variance of a normal distribution

The mean EZ of the standard normal $Z{:}N(0, 1)$ is zero and the variance $E(Z^2)$ is 1, as was shown in the discussion on standardized probability distributions in chapter 2. It can therefore be proved that the mean of $X{:}N(\mu, \sigma^2)$ is μ, and the variance is σ^2, as follows:

$$EX = E(\sigma Z + \mu) = \sigma\, EZ + \mu = \mu;$$
$$E(X-\mu)^2 = E(\sigma Z + \mu - \mu)^2 = \sigma^2 E(Z^2) = \sigma^2.$$

Chi-square distributions

The chi-square (χ^2) distribution is a type of gamma distribution with many uses in both statistics and reliability. In the normal-theory context, chi-square distributions are closely related to both the Student-t and F distributions, and chi-square tables are the principal reference for analysis of variance. For time-dependent failure processes with constant failure rates, as discussed in chapter 5, chi-square tables are used instead of the cumbersome tables of the gamma distribution for reliability-confidence assessment. The term chi-square is awkward, but the name is well established.

Distribution of the normal sum of squares

Assume that there are ν independent observations or values,

$$X_i,\ i = 1, \ldots, \nu,$$

all from the normal distribution $N(\mu, \sigma^2)$, with p.d.f. Equation 4.2. The ν standardized random variables

$$Z_i = \frac{(X_i - \mu)}{\sigma},\ i = 1, \ldots, \nu,$$

are all normally distributed from the standard normal $Z{:}N(0,1)$. Let Y denote the sum of squares of the Z_i's.

$$Y = Z_1^2 + \ldots + Z_\nu^2 \qquad (4.13)$$

The r.v. Y has the χ^2 p.d.f. for ν degrees of freedom.

$$f_{\chi^2(\nu)}(y) = \frac{y^{\nu/2-1} \exp(-\nu/2)}{2^{\nu/2}(\nu/2-1)!}, \; 0 < y. \qquad (4.14)$$

A derivation of the p.d.f. Equation 4.14 is performed by mathematical induction, with $\nu = 1$, then for $\nu = 2$, then for any ν, with a change of variables and an integration at each step. The full derivation would require extensive text and is deleted.

It has already been shown that

$$E(Z_i^2) = 1, \; i = 1, \ldots, \nu.$$

It follows that the mean of the χ^2 distribution

$$EY = \sum_{i=1}^{\nu} E(Z_i^2) = \nu$$

is equal to the number of independent observations, or *degrees of freedom*. Usually the sample mean is substituted for the population mean μ. For a sample of size n the estimated sum of squares corresponding to Equation 4.13 is

$$Y = \sum_{i=1}^{n} \left(\frac{X_i - \bar{X}}{\sigma}\right)^2 = \frac{ns^2}{\sigma^2},$$

where the estimated sample variance is

$$s^2 = \sum_{i=1}^{n} \frac{(X_i - \bar{X})^2}{n}.$$

From the derivation of Equation 2.8, the expected value of Y is

$$E\left(\frac{ns^2}{\sigma^2}\right) = \frac{n}{\sigma^2} Es^2 = n-1.$$

To summarize, when the sample mean is used instead of μ, there are $n-1$ independent observations or degrees of freedom rather than n, Y has the χ^2 distribution with $n-1$ degrees of freedom, and the mean of the χ^2 distribution is also equal to $n-1$.

Percentage points of chi-square distributions

A table of percentage points of the χ^2-distribution for up to 80 degrees of freedom appears as Table A.4.

EXAMPLE 4.11.

Percentage-points lookup: A sample of 10 observations is obtained from a normal population with an unknown mean and a known variance of 4. The sample variance s^2 is 3.3372. What is the probability of obtaining a value of the sample variance as high as this but no higher?
By Equation 4.13, the value of $Y (=\chi^2)$ is

$$\frac{10(3.3372)}{4} = 8.343.$$

This is the median, or 50th percentage point, of the χ^2 distribution for 9 degrees of freedom; the probability desired is 0.5.

EXAMPLE 4.12.

Confidence limits for a sample variance: In Example 4.11, what are the upper and lower 80 percent confidence limits of the sample variance s^2, with 10 percent excluded in each tail?
The 10 percent lower limit s_L^2 is obtained by solving

$$10\left(\frac{s^2}{\sigma^2}\right) = \chi^2_{.10}(9),$$

which is the 10th percentage point of the χ^2 distribution for 9 degrees of freedom. The value desired is

$$s_L^2 = \frac{4(4.168)}{10} = 1.6672.$$

By the same procedure, the 90 percent upper confidence limit s_U^2 is

$$10\left(\frac{s^2}{\sigma^2}\right) = \chi^2_{.90}(9),$$

$$s_U^2 = \frac{4(14.684)}{10} = 5.8736.$$

The range 1.6672 to 5.8736 is the central 80 percent confidence interval.

Student-t distributions

Like the chi-square distributions, the Student-t distributions have a variety of applications for data that are normal or approximately normally distributed. Student-t is used when both the mean and the variance are unknown. A table of values is provided in practically all statistics texts for the t-statistic, which is similar in form to the standard normal Z: $N(0, 1)$ but differs in that the sample standard deviation s is used as a scale factor in place of the population standard deviation σ. Because s is an r.v. and not a parameter, the extent of this variability is inversely related to the sample size and the distribution of the t-statistic depends on the amount of data.

In chapter 2 it is shown that that an unbiased estimate of the variance is

$$\hat{\sigma}^2 = \frac{n}{n-1} s^2,$$

and that the variance of the r.v. $\overline{X}$ is

$$\sigma^2_{\overline{X}} = \frac{\sigma^2}{n}.$$

An unbiased estimate of the variance of $\overline{X}$ is therefore $s^2/(n-1)$. The t-statistic is defined as

$$t = \sqrt{n-1}\left(\frac{\overline{X}-\mu}{s}\right);$$

it does not depend upon the value of σ, and there are $\nu = n-1$ degrees of freedom. The t-statistic is not tabled in this book, but we consider an alternate form called the noncentral-t, which is used in chapter 8 to set one-sided tolerance limit factors for normally distributed data.

Problems

4.1. *Trimming* a data set is providing a substitute value for one that appears to be abnormal; generally, trimming is done only after the cause of the abnormality is ascertained. Suppose that Tippett's apple data is trimmed by substituting a weight of 156 grams for the outlying observation of 223 grams. Graph the revised data set, with the revised set of 10 data points, on normal probability paper and perform a Lilliefors K–S goodness-of-fit test.

4.2. Ten pressure switches of the same type were tested to determine the pickup point, the pressure at which a switch would pick up. These were 504, 505, 495, 507, 488, 500, 498, 528, 485, and 504 psia (pounds per square inch in atmosphere). Graph this data both on normal probability paper and lognormal probability paper (use normal paper with a logarithmic scale) and perform Lilliefors goodness-of-fit tests both for a normal distribution and a lognormal distribution. Which distribution, normal or lognormal, gives a better fit to these data and why?

4.3. The following times to burnout, in seconds, were obtained in testing a new rocket motor: 125, 134, 148, 152, 158, 163, 176, 190, 194, 221, 230, and 250.

  a. Graph these data both on normal probability paper and lognormal paper.
  b. Perform Lilliefors goodness-of-fit tests both for a normal distribution and a lognormal distribution. Which distribution is better?
  c. What are the calculated parameters of the normal distribution?
  d. What are the calculated parameters of the lognormal distribution?
  e. Suppose that it is a requirement that a rocket motor should be able to sustain a burn for at least 150 seconds. What is the calculated reliability
     • based on the calculated normal distribution?
     • based on the calculated lognormal distribution?

4.4. A numerically controlled milling machine has had the following repair times, in days per repair, during a six-month period: 0.25, 0.5, 0.5, 1.0, 1.5, 2.0, 3.5, 5.6, 10.5, and 20 days.

  a. Find the estimates of the parameters of the lognormal distribution of repair times (base-10 logarithms): est. $\mu=$ ___, est. $\sigma=$ ___.
  b. Graph the lognormal distribution of repair times. Is the fit good?
  c. Calculate the Lilliefors D-statistic. Is the fit good?
  d. What is the estimated median repair time based on the calculated mean logarithm?
  e. What is the repair time that is one logarithmic standard deviation above the estimated median repair time obtained in 4.4d?

4.5. A sensor with amplified digital readout records the distance of movement of a particular location of a part in a specialized gauge in digital counts, where each count is 25×10^{-6} inches (25 millionths of an inch). After the gauge is calibrated, 17 consecutive readings (after reordering according to distance of movement) are, in counts,

1579, 1598, 1613, 1613, 1617, 1618, 1618, 1619,
1620, 1620, 1621, 1621, 1626, 1627, 1627, 1629, and 1638.

a. What are the mean and standard deviation of these readings?
b. Graph the distribution of these readings on normal probability paper.
c. What is the value of the Lilliefors K–S statistic D?
d. Is the deviation from normality significant at the 5 percent critical level?

4.6. With a sample of size 300 and 10 observed failures, what is the approximate confidence level for a reliability of 0.95? For this purpose use the Yates finite sample approximate correction factor of 0.5; also keep in mind the assumption that there is one additional test, 301 total, and the last test is a failure.

4.7. The same milling machine considered in Problem 4.4 had the following times to failure, in hours, for nine randomly selected repair times.

3.4, 11.2, 20.2, 26.0, 38.0, 48.9, 66.9, 89.9, and 249.1

a. Graph this data as a lognormal plot on a photoreproduced copy of Figure 4.3, after marking the horizontal scale logarithmically.
b. What is the calculated mean of the logarithms?
c. What is the calculated standard deviation of the logarithms?
d. What is the mean logarithm obtained from your plot?
e. What is the standard deviation of the logarithms from your plot?
f. Perform a Lilliefors K–S goodness-of-fit test. Is the fit good at the 10 percent level? What is the calculated value of D?
g. Based upon your answer to 4.7b, what is the median TTF?
h. Based upon your answers to 4.7b and 4.7c, what is the antilog (that is, time to failure) of the number that is one logarithmic standard deviation below the mean of the logarithms?

4.8. A robotic calibration system reads the deviations of readings, in microinches, of 0.5-inch gage blocks that are installed and used in the field and periodically rechecked against a master gage block. For a six-month period, 20 deviations recorded are respectively:

1.4, 1.5, 1.7, 1.7, 1.7, 1.8, 1.8, 1.8, 1.8, 1.8,
1.9, 1.9, 1.9, 1.9, 2.0, 2.0, 2.1, 2.1, 2.3, and 2.6.

a. What are the mean and standard deviation of the deviations?

b. Graph the data on normal probability graph paper. Is there a good fit?
c. What is the calculated D-statistic?
d. What does the D-statistic say about goodness-of-fit?

References

Abramowitz, M., and I. A. Stegun, eds. 1964. *Handbook of mathematical functions.* Washington, D.C.: National Bureau of Standards. U.S. Government Printing Office.

Aitchison, J., and J. A. C. Brown. 1957. *The lognormal distribution.* New York: Cambridge University Press.

Billinton, R., and R. N. Allan. 1983. *Reliability evaluation of engineering systems.* New York: Plenum Press.

Bradley, James V. 1976. *Probability; decision; statistics.* Englewood Cliffs, N.J.: Prentice Hall.

Cochran, W. G. 1962. The chi-square test of goodness of fit. *Annals of Mathematical Statistics* 28:315–45.

Darling, D. A. 1957. The Kolmogorov–Smirnov, Cramer–von Mises tests. *Annals of Mathematical Statistics* 28:823–38.

Doty, L. A. 1989. *Reliability for the technologies.* 2d ed. New York: Industrial Press.

Fisher, R. A., and F. Yates. 1957. *Statistical tables for biological, agricultural, and medical research.* 5th ed. New York: Hafner.

Goldberg, Samuel. 1960. *Probability, an introduction.* Englewood Cliffs, N.J.: Prentice Hall.

Hahn, Gerald J., and S. S. Shapiro. 1967. *Statistical models in engineering.* New York: Wiley.

Hausner, Melvin. 1971. *Elementary probability theory.* New York: Harper & Row.

Lilliefors, H. W. 1967. On the Kolmogorov–Smirnov test for normality with mean and variance unknown. *Journal of the American Statistical Association* 62:399–402.

Lindgren, B. W. 1962. *Statistical theory.* New York: Macmillan.

Pearson, E. S., and H. O. Hartley, eds. 1962. *Biometrika tables for statisticians.* Vol 1. 2d ed. London and New York: Cambridge University Press.

Ross, Sheldon M. 1980. *Introduction to probability models.* 2d ed. New York: Academic Press.

Shook, Robert C., and H. J. Highland. 1969. *Probability models with business applications.* Homewood, Ill.: R. D. Irwin.

Simon, Leslie E. 1941. *An engineers' manual of statistical methods.* New York: Wiley.

Smirnov, N. V. 1965. *Tables of the normal probability integral, the normal density and its normalized derivative.* New York: Macmillan.

Student. 1908. The probable error of a mean. *Biometrika* 5:1–25.

Technical and Engineering Aids for Management (TEAM). Box 25, Tamworth, NH 03886.

Tippett, L. H. C. 1961. *Statistics.* New York: Oxford University Press.

Trivedi, K. S. 1982. *Probability and statistics with reliability, queuing, and computer science applications.* Englewood Cliffs, N.J.: Prentice Hall.

Walpole, Ronald E., and R. H. Myers. 1989. *Probability and statistics for engineers and scientists.* 4th ed. New York: Macmillan.

5

Exponential Times to Failure

Introduction

This chapter specializes in analysis of time-to-failure or usage-to-failure data under the assumption of a constant failure rate that is independent of the cumulative usage or operating time. The collection of families of distributions generated by this assumption is known as the Poisson process. This chapter is an extended introduction to reliability for the Poisson process, including distributions: the exponential, gamma, Poisson, and chi-square distributions and their interrelationships; and probability plotting, goodness of fit, and incomplete data sets and life testing. The technical appendix to the chapter contains mathematical details. Confidence assessment for a Poisson process with maximum likelihood is covered in chapter 6, and with Bayesian techniques in chapter 7.

The chi-square distribution, discussed in the technical appendix of chapter 4, is a form of the gamma distribution with familiar tables that are widely used. Because the tables are concise and the distributions are all interrelated, the chi-square tables are a convenient reference source of information about percentage points and probabilities of the Poisson process.

Failure of a single component

The exponential distribution function

For any given component, the cumulative operating time T before a failure occurs is a random variable. T is measured in its usual units of hours, days, minutes, seconds, months, or years. With usage-to-failure data, T also can be the number of operations or events (openings, closings, or bendings) to failure. $T = 0$ denotes a starting point of time from which measurement begins, such as just after the component is acquired or just after a repair. The probability that the failure occurs before $T = t$ is the distribution function

$$F(t) = P(T \le t).$$

Let λ be the constant failure rate, failures per time unit of operation of a single component; λ is independent of t because the failure rate doesn't change. The technical appendix to this chapter contains a derivation of the formulas for the one-parameter exponential d.f.

$$F(t) = 1 - \exp(-\lambda t)$$

and the derivative of $F(t)$, the probability density function

$$f(t) = \lambda \exp(-\lambda t).$$

The reliability

A time-to-failure device is often evaluated by how long it can operate before a failure occurs. Time is not the only measure of usage. For example, new automobiles are sold with warranties for so many miles of usage (say 50,000 miles) or else so many years (say 5 years) whichever comes first, without failure of the transmission, engine, or some other essential component. The reliability $R(t)$, the probability of survival to t, is the probability that the failure does not occur before t. By definition, $R(t)$ is the complement of the d.f.,

$$R(t) = P(T > t) = 1 - F(t) = \exp(-\lambda t). \tag{5.1}$$

If it is required that the equipment or device operate for a specified period of time t_o without failure, as is typically the case with consumer warranties on appliances and military and aerospace procurement, the reliability formula is

$$R(t_o) = \exp(-\lambda t_o). \tag{5.2}$$

Exponential probability plotting with semilog graph paper

Because Equation 5.1, is quite simple, a probability plot for an exponential distribution can be made with ordinary commercially available semilog graph paper. Time t is on the horizontal axis and probability is on the vertical axis. By taking the natural logs of both sides of Equation 5.1, the natural log of $R(t)$ is proportional to t, with λ as the constant of proportionality.

$$\ln R(t) = -\lambda t$$

This means that $-\ln R(t)$ is an increasing linear function of t, with a slope of λ; $-\ln R(t)$ is always positive because $R(t)$, a positive number which is less than one, has a negative logarithm.

Figure 5.1 is an example of one-cycle semilog paper similar to the commercial product but with fewer divisions. It was prepared especially for this text and reproduction is permitted. The graph is scaled three different ways vertically: along the right-hand edge, F represents the cumulative d.f., and it also corresponds to the natural scale of one-cycle semilog graph paper; at the left-hand edge, there are the scales of both $R = 1 - F$ and $-\ln R$. Both F and $-\ln R$ increase downward, and R increases upward.

EXAMPLE 5.1.

Fitting an exponential plot: Five small strips (known as coupons) of a certain type of aluminum are subjected to a highly corrosive life test environment. The first coupon cracked at 35 hours, the second at 100 hours, the third at 240, the fourth at 440, and the fifth at 720 hours. From Table A.3, the corresponding five plotting points are, respectively, .167, .333, .500, .667, and .833. The data are plotted in Figure 5.2, including a straight line of visual best fit. This line must include the origin ($t = 0$, $F = 0$) in the upper left-hand corner, because at $t = 0$ there have been no failures and the one-parameter exponential distribution does not allow for shifts of the distribution function away from that point.

1. What is the failure rate?
2. What is the mean time between failures (MTBF)?
3. What is the reliability if it is a requirement that failure not occur before 400 hours of exposure?

These three questions are answered here in order.

86 Chapter 5 ■ *Exponential Times to Failure*

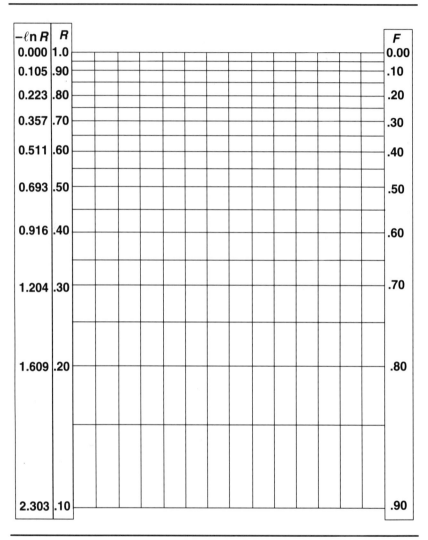

Figure 5.1. Semilog graph for exponential probability plotting.

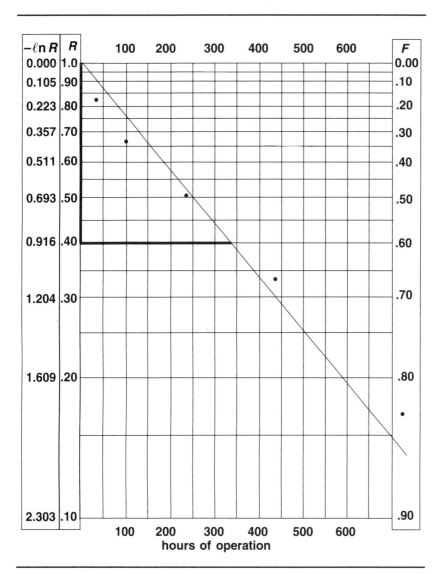

Figure 5.2. Probability plot of Example 5.1.

1. An estimate of λ is obtained by triangulation off of the downward-sloping straight line of Figure 5.2. There are two thick lines at a right angle, a vertical line from the origin (0,0) to ($t = 0, F = 0.60$) and a horizontal line to ($t = 340, F = 0.60$); λ is the slope of the hypotenuse of the triangle formed by these lines:

$$\lambda = \frac{0.916}{340} = 0.00269, \text{ or } 0.269 \text{ failures per } 100 \text{ hours.}$$

2. It is shown in the technical appendix to this chapter that the mean of the exponential distribution is the inverse of the failure rate, or $1/\lambda$. The estimated MTBF is the inverse of 0.00269, or 371 hours.

3. The reliability R for $\lambda = 0.00269$, $t_o = 400$, is 0.34, by Equation 5.2.

Life tests and censored data sets

Life tests are frequently terminated before all n items have failed, because a test could run on indefinitely at considerable cost with no corresponding benefit of gaining additional information. This premature termination is called censoring. If a life test is censored at the rth failure, the first r-out-of-n data points are plotted, and the remaining unfailed $n-r$ points are projected as a part of the straight line formed by the first r failures.

EXAMPLE 5.2.

Censored life test: Suppose that the life test in Example 5.1 had been terminated at 440 hours, after the fourth failure. Plot the line of best fit and provide the estimates of λ, the MTBF, and the reliability.

Using the same technique as in Example 5.1, by triangulation on Figure 5.3, the failure rate is

$$\lambda = \frac{0.5978}{200} = 0.00299, \text{ or } 0.299 \text{ failures per } 100 \text{ hours.}$$

The MTBF is 335 hours and the estimate of the reliability is 0.3025 or 30 percent.

Larger sample sizes: Wayne Nelson exponential plotting paper

The procedure described for analyzing exponential life-test data by plotting on one-cycle semilog graph paper can be used only for samples

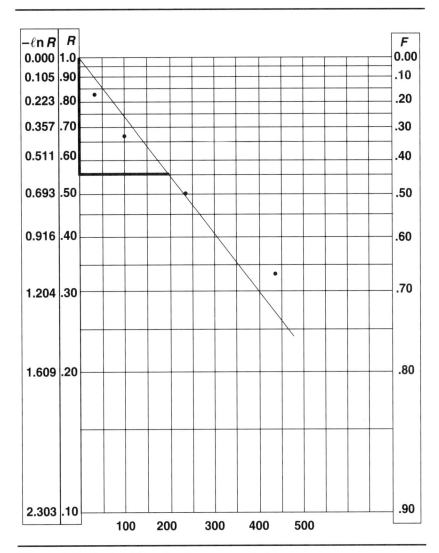

Figure 5.3. Probability plot for censored data.

smaller than 10, because one cycle does not conveniently provide plotting points for larger samples. Samples of between 10 and 100 observations require two cycles.

Wayne Nelson designed exponential graph paper that goes slightly beyond the second cycle, with the vertical scale reversed so that the origin (0,0) is at the lower left-hand corner instead of the upper left-hand corner and the d.f. line slopes upward instead of downward. Figure 5.4 is a modification of Nelson's exponential graph paper; reproduction is permitted. The vertical scale is F percent on the left-hand side or $-\ell n\ R$ on the right, and the distances between equally spaced divisions are equal differences in $-\ell n\ R$. The procedure for visually analyzing the exponential fit and calculating λ, the reliability, and the MTBF, for either complete or censored samples, is identical to that of Examples 5.1 and 5.2.

EXAMPLE 5.3.

Fan failure data: Nelson supplied data on hours of service for 70 diesel fans that had been removed from service. Although only 12 of the 70 actually failed, we shall treat every removal as an "end of life" for purposes of this illustration of an exponential plot with a larger sample size. Also, instead of plotting all 70 data points, we shall consider one-half of them, or 35 units altogether, by a systematic sample that selects every other observation. These 35 end-of-life points are, respectively, in hours,

460,	1150,	1600,	1850,	1850,	1850,	2030,	2070,	2080,
3000,	3000,	3100,	3450,	3750,	4150,	4150,	4300,	4300,
4850,	4850,	5000,	5000,	6100,	6100,	6450,	6700,	7800,
8100,	8200,	8500,	8750,	8750,	9900,	10,100,	and	11,500.

Source: W. Nelson, *Applied Life Data Analysis,* © John Wiley & Sons, Inc. Reprinted with permission of John Wiley & Sons, Inc.

In this list there are a number of ties; for example, the fourth, fifth, and sixth fans were all removed at 1850 hours. In those cases an average plotting point of 14 was used instead of 11, 14, and 17. Also the plotting point for the 10th and 11th observations (3000) were averaged at 30.

The plot and a visual line of apparent best exponential fit is in Figure 5.5. This example does not show a good fit to the exponential distribution because the data points curve. Up to about 9000 hours the data

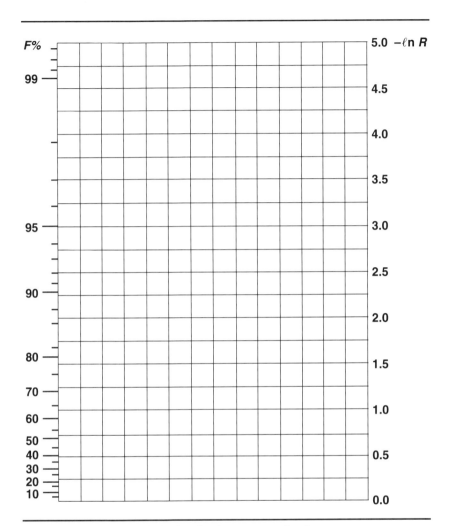

Figure 5.4. Exponential plot for larger sample sizes.

92 Chapter 5 ■ *Exponential Times to Failure*

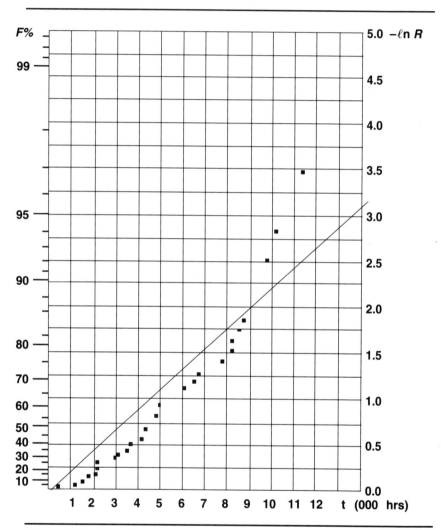

Figure 5.5. Plot of times of removal of fans from service.

are all below the line; after 9000 hours, the points are above the line. The curvature at the beginning, below the line, combined with that at the end, above the line, suggests decreasing removal rate up to 3000 hours, constant rate from 3000 to 7000 hours, and increasing rate thereafter. Since the line is not a good fit, no parameters can be calculated for it. For further discussion on variable rates, refer to chapter 6 on the two-parameter exponential distribution and chapter 9 on the Weibull distribution.

Exponential goodness of fit: The Lilliefors K–S test

A Kolmogorov-Smirnov (K–S) type of goodness-of-fit test for an unspecified exponential population was devised by Lilliefors, similar to his K–S test for an unspecified normal population discussed in chapter 4. This procedure has larger critical values of the test statistic D than those in chapter 4 because the exponential distribution has greater relative dispersion than the normal.

The raw data consist of n ordered time-to-an-event or time-to-failure observations t_i, $i = 1, \ldots, n$. For each i the corresponding rank-order statistic is the ratio of i to n. The mean time to an event is

$$\overline{T} = \frac{\sum_{i=1}^{n} t_i}{n}. \tag{5.3}$$

One noticeable difference between the g.o.f. procedure for the exponential distribution and that for the normal is that since the exponential is a one-parameter distribution, Equation 5.3, the mean of the data, is the only calculated statistic needed to provide empirical d.f. values to compare to the rank orders. For the normal-population test there are two calculated statistics, the mean and the standard deviation.

In the technical appendix to this chapter, it is shown that the failure rate λ is the inverse of the mean time between failures. The estimated failure rate is

$$\hat{\lambda} = \frac{1}{\overline{T}}.$$

For each i, the calculated or empirical distribution function is

$$F(t_i) = 1 - \exp(-\hat{\lambda} t_i) = 1 - \exp\left(\frac{-t_i}{\overline{T}}\right). \tag{5.4}$$

Table 5.1. Critical values of Lilliefors K–S statistic D.

Sample size n	Critical level		
	0.10	0.05	0.01
3	0.511	0.551	0.600
4	0.449	0.487	0.548
5	0.406	0.442	0.504
6	0.375	0.408	0.470
7	0.350	0.382	0.442
8	0.329	0.360	0.419
9	0.311	0.341	0.399
10	0.295	0.325	0.380
11	0.283	0.311	0.365
12	0.271	0.298	0.351
13	0.261	0.287	0.338
14	0.252	0.277	0.326
15	0.244	0.269	0.315
16	0.236	0.261	0.306
17	0.229	0.253	0.297
18	0.223	0.246	0.289
19	0.218	0.239	0.283
20	0.212	0.234	0.278
25	0.191	0.210	0.247
30	0.174	0.192	0.226
>30	$(0.96n)^{-1/2}$	$(1.06n)^{-1/2}$	$(1.25n)^{-1/2}$

Reprinted with permission of the American Statistical Association. Lilliefors, H. W., 1969, "On the Kolmogorov–Smirnov Test for Goodness-of-Fit with Mean Unknown." *Journal of the American Statistical Association*, vol. 64: 387–89.

The test statistic D is the largest absolute difference between a d.f. calculated by Equation 5.4 and one of the two adjacent rank order statistics. The formula is

$$D = \max_i \left\{ \left| F(t_i) - \frac{i-1}{n} \right|, \left| \frac{i}{n} - F(t_i) \right| \right\}.$$

A table of the critical values of D for $n \geq 3$, at the 0.10, 0.05, and 0.01 levels of significance, appears in Table 5.1.

EXAMPLE 5.4.

Perform a Lilliefors K–S test of an exponential population with the data on aluminum coupons in Example 5.1.

The visual plot of Figure 5.2 has already shown that the exponential fit is good; the g.o.f. test should help to confirm this. The calculated MTBF is 307 hours.

Table 5.2. Worksheet for Example 5.4.

i	i/n	t_i	$F(t_i)$	D
1	0.20	35	0.11	
2	0.40	100	0.28	
3	0.60	240	0.54	
4	0.80	440	0.76	0.16
5	1.00	720	0.90	

Table 5.2 is a worksheet for D. The calculated value of D, 0.16, is far below 0.406, the critical value at the 10 percent level for $n = 5$ in Table 5.1; this indicates a good fit and confirms what we already know.

Multiple failures and redundancy

So far this text has considered reliability analysis from a narrow point of view, success or failure of just one component. We have not yet discussed reliability in the context of a system with multiple components, all of which can perform the same function. This section extends the reliability applications of the Poisson process, to include redundancy, such as standby operation or m-out-of-n voting networks, where a system with several identical components all having the same failure rate can continue to operate even after one or more failures because there is backup for the failed element or elements.

The discussion in this section focuses on three probability distributions: gamma, chi-square, and Poisson. The random variable for the gamma distribution is the time to the nth failure; for the Poisson, it is the number of failures in a fixed period of time t. Although chi-square has many applications outside this context, it is just a special case of the gamma with a convenient table for probability lookups, such as Table A.4.

The gamma and chi-square distributions

In the technical appendix to this chapter, it is shown that for a Poisson process, for any integer n, $n \geq 1$, the probability that it requires at most t time for the nth failure to occur is the gamma-n distribution function $F_n(t)$. The relationship between the gamma-n and chi-square distribution functions is

$$F_n(t) = F_{\chi^2(2n)}(2\lambda t). \tag{5.5}$$

96   Chapter 5 ■ *Exponential Times to Failure*

This means that for the probability that no more than t time is required for the nth failure to appear, find in Table A.4 what percentage point $2\lambda t$ is of the chi-square distribution with $2n$ degrees of freedom. When $n = 1$, Equation 5.5 is just the d.f. for an exponential distribution and the corresponding chi-square has 2 degrees of freedom.

EXAMPLE 5.5.

Percentage point of a gamma distribution: With a failure rate λ of 0.0027 per hour, what is the probability that the first failure occurs at or before 39 hours?

In this case $2\lambda t$ is 0.211. For the $\chi^2(2)$ distribution the 10th percentage point is 0.211, and the probability desired is 0.10. The failure rate in this example, 0.0027, is taken from Example 5.1, which is probability plotted in Figure 5.2; the plot shows that the 10th percentage point is approximately 40 hours. This close relationship between these results and the plot is no accident: it can be shown by a change of variables that the chi-square distribution for two degrees of freedom is the exponential distribution with

$$t = \frac{\chi^2}{2\lambda}.$$

EXAMPLE 5.6.

Percentage point of a gamma distribution (continued): With the same failure rate as in Example 5.5, 0.0027 per hour, what is the probability that it requires at most 440 hours for the fourth failure to occur?

The chi-square value $2\lambda t$ is 2.376. With 8 degrees of freedom, this is about midway between 2.032, with a probability of 0.02, and 2.733, with a probability of 0.05; 440 hours, therefore, is the 3.5th percentage point and the required probability is approximately 0.035.

An intuitive argument showing the relationship between the 4 failure times and chi-square (8), based on the observed data is presented here. Note that the fourth failure in Example 5.1 was at 440 hours. At that time there had already been

$$35 + 100 + 240 + 440 + 440 = 1255$$

hours of operation on the five units under life test, because the first one failed at 35 hours, the second at 100, the third at 240, the fourth at 440, and the fifth was still OK at the time. Taking t as the total of 1255 hours,

$2\lambda t$ is 6.78, in the center of the chi-square(8) distribution, a little below 7.344, the 50th percentage point.

The Poisson distribution

The discrete Poisson distributions define probabilities for the number of failures n in a specified time period t. There is a dual relationship between the gamma and Poisson distributions that can be demonstrated intuitively. The gamma d.f. $F_n(t)$ is the probability that it takes t or less time for the nth failure to occur; and the Poisson d.f. $F_t(n)$ is the probability that n or fewer failures occur in t time. Then the reverse d.f., $1-F_n(t)$, of the gamma-n is the probability that more than t time is required for the nth failure, which is the same as the Poisson d.f. $F_t(n-1)$ that $n-1$ or fewer failures occur in t.

In the technical appendix to this chapter, it is shown that the Poisson d.f.

$$F_t(n) = P(N \le n) = 1 - F_{\chi^2_{(2n+2)}}(2\lambda t),$$

and the p.d.f.

$$f_t(n) = P(N = n) = F_{\chi^2_{(2n)}}(2\lambda t) - F_{\chi^2_{(2n+2)}}(2\lambda t).$$

Therefore the χ^2 tables can be used to obtain Poisson probabilities. The results, however, are approximate, because it is necessary to interpolate.

EXAMPLE 5.7.

Poisson d.f.: With $\lambda = 0.0027$, what is the probability that no more than four failures would occur in 1730 hours? First calculate

$$2\lambda t = 2(0.0027)(1730) = 9.342.$$

The 50 percent point on the $\chi^2(10)$ distribution is 9.342. The desired probability is $1-0.50 = 0.50$.

EXAMPLE 5.8.

Poisson p.d.f.: For the same data as in Example 5.7, what is the probability that exactly four failures occur in 1730 hours? The answer is

$$F_{\chi^2_{(8)}}(9.342) - F_{\chi^2_{(10)}}(9.342) = 0.69 - 0.50 = 0.19, \text{ approximately.}$$

The exact answer obtained by the Poisson p.d.f. formula is 0.186.

EXAMPLE 5.9.

Standby reliability: A computer center has two mainframe computers A and B, either of which can perform all of the work the center normally handles. On most days, A is designated primary and carries most of the load of computation, and B is secondary or standby, ready to carry the load if A fails, or the overload if A is overloaded; however these roles can be reversed and B could be primary with A standby. Both mainframes have the same failure rate, 0.356 failures per 24-hour day. If management specifies that the center shall be operational continuously for a 24-hour day, what is the reliability of the center (probability of continuous operation for one day)?

The reliability is the probability that at most one failure occurs per day. Why? Because if both computers fail, the center is down. By the relationship between the Poisson and chi-square distribution functions

$$F_t(1) = 1 - F_{\chi^2(4)}(0.712) = 1 - 0.05 = 0.95.$$

Appendix to Chapter 5
Derivation of the Exponential Distribution: Relationships between Exponential, Gamma, Chi-square, and Poisson Distributions

An example of a Poisson process

There is a convenient physical example for a Poisson process in the measurement of radioactivity. At any position in a radioactive field, the intensity of radiation is directly proportional to the frequency or rate of occurrence of the ticks of a Geiger counter. These ticks don't come at equally spaced intervals, but the average number of ticks per second or per minute is fairly constant as long as the Geiger counter remains in one place. Moving the counter around can cause the average frequency of ticking to change. The ticking that occurs as a result of letting the counter stay in one place is an example of a Poisson process.

The exponential distribution

Suppose that the Geiger counter is preset so that it makes just one tick, and then stops operating until it is recycled. Starting from the beginning of operation, the time required for the first tick to occur is a random variable, subject to the exponential distribution. The reason for the term *exponential* is that the probability function is a scaled form of the exponential function $\exp(-x)$. The scale factor is the constant intensity of the Poisson process that governs the occurrence of events.

The Geiger counter ticks in a radioactive field are like the failures of a component which is either repaired or replaced when it fails, if the rate at which failures occur is governed by a Poisson process. The random variable for the exponential distribution is the time T at which a failure occurs. For any particular point of time t, the probability that the failure occurs before t is the distribution function $F(t) = P(T \le t)$. The probability that the failure occurs after t is the survival probability, the reliability $R(t) = P(T > t) = 1 - F(t)$.

The probability density function, the d.f., and R for an exponential population are all derived based upon the assumption that the failure rate λ, the intensity of the process, is a constant and is independent of t. λ is a kind of conditional probability. Given that the failure does not occur before time t, with this probability of nonoccurrence being $1 - F(t)$, what is the density of concentration of the probability,

$$\frac{\Delta F}{\Delta t},$$

that the failure occurs in a small time interval, t to Δt? This is λ. By the conditional probability formula of chapter 1,

$$\lambda = \lim_{\Delta t \to 0} \frac{\frac{\Delta F}{\Delta t}}{1 - F(t)} = \frac{\frac{dF}{dt}}{1 - F(t)} = \frac{f(t)}{1 - F(t)}. \quad (5.6)$$

Because of the relationship between the numerator and the denominator of Equation 5.6, it becomes a differential equation:

$$\lambda = -\frac{\frac{d}{dt}(1 - F(t))}{1 - F(t)} = -\frac{d}{dt} \ell n(1 - F(t)).$$

To solve this differential equation, multiply both sides by $-dt$ and integrate over the range zero to t to obtain

$$-\int_0^t \lambda \, dx = \int_0^t d \, \ell n(1 - F(x)), \text{ so that}$$
$$-\lambda t = \ell n(1 - F(t)). \quad (5.7)$$

By taking the exponent of Equation 5.7 the probability functions are

$$\exp(-\lambda t) = 1 - F(t)$$
$$F(t) = 1 - \exp(-\lambda t) \quad (5.8)$$
$$f(t) = \lambda \exp(-\lambda t) \quad (5.9)$$
$$R(t) = \exp(-\lambda t). \quad (5.10)$$

Independence from prior data

A consequence of the fact that λ is constant is that no matter where the origin is placed for purposes of measuring time, the probabilities from Equations 5.8, 5.9, and 5.10 at any time t depend only upon the elapsed time from the origin to t. Let x be any time greater than zero. From Equation 5.9, the probability of survival to x is

$$\exp(-\lambda x)$$

and the conditional probability of survival to $t > x$ is

$$\exp(-\lambda t).$$

Therefore, the conditional probability of survival to t, given survival to x, is

$$R(t - x) = \frac{\exp(-\lambda t)}{\exp(-\lambda x)} = \exp\{-\lambda(t-x)\}.$$

From this the d.f. and the p.d.f. are obtained.

$$F(t - x) = 1 - \exp\{-\lambda(t-x)\}$$
$$f(t - x) = \lambda \exp\{-\lambda(t-x)\}$$

The gamma distribution: Multiple failures

The model is extended to include the possibility of multiple failures. Examples are a repairable system that is as good as new after it fails and is then repaired; a voting circuit that fails only if a specified number n of failures of identical components occur; or a standby system with n identical components, any one of which can perform the desired function, and the system fails only if all n fail. If all failures occur with the same failure rate λ and there is flawless switching, the p.d.f. on the time t to the nth failure is

$$f_n(t) = \frac{\lambda^n t^{n-1} \exp(-\lambda t)}{(n-1)!}, \tag{5.11}$$

sometimes called the p.d.f. for the gamma-n distribution. The reason for the term *gamma* is that the denominator is a gamma function

$$\Gamma(n) = \int_0^\infty x^{n-1} \exp(-\lambda x)\, dx = (n-1)!. \tag{5.12}$$

Equation 5.12, which is discussed in many calculus books, is derived using integration by parts. The gamma-n p.d.f., Equation 5.11, is a normalized gamma function derived by mathematical induction. For $n = 2$, let x be the time of a first failure and $t > x$ the time of a second failure. The joint probability of a failure at any specified x and also at t is

$$f(x)\,f(t-x) = \lambda \exp(-\lambda x)\,\lambda \exp\{-\lambda(t-x)\} = \lambda^2 \exp(-\lambda t),$$

which does not depend upon x; but this accounts for only one specified point of time before t. In order to account for all of the time prior to

t, integrate over the interval $(0, t)$ to obtain

$$f_2(t) = \int_0^t f(x)f(t-x)\,dx = \lambda^2 \exp(-\lambda t) \int_0^t dx = \lambda^2 t \exp(-\lambda t).$$

The induction step is: Assume that for $n - 1$ failures

$$f_{n-1}(x) = \frac{\lambda^{n-1} x^{n-2} \exp(-\lambda x)}{(n-2)!};$$

then the joint probability of the $(n - 1)$st failure at x and the nth at t is

$$f_{n-1}(x)f(t-x) = \frac{\lambda^{n-1} x^{n-2} \exp(-\lambda x)}{(n-2)!} \lambda \exp\{-\lambda(t-x)\}$$

$$= \frac{\lambda^n \exp(-\lambda t)}{(n-2)!} x^{n-2}.$$

The p.d.f. for the nth failure is the limiting sum of these joint probabilities for all $0 < x < t$, or

$$\frac{\lambda^n \exp(-\lambda t)}{(n-2)!} \int_0^t x^{n-2}\,dx = \frac{\lambda^n t^{n-1} \exp(-\lambda t)}{(n-1)!}.$$

The gamma distribution function has been tabled in a variety of ways. The best-known set of tables is Karl Pearson's *Tables of the Incomplete Γ-Function*, which has a very involved notation. Instead, it is more convenient to use H. L. Harter's tables of the chi-square distribution function.

Chi-square distribution

To obtain the chi-square distribution from the gamma, or vice versa, both a change of variables and of parameters is needed. The chi-square r.v. is

$$y = 2\lambda t$$

and the number of chi-square degrees of freedom is

$$\nu = 2n$$

Since there is a one-for-one relationship between t and y, the relationship between the gamma and chi-square d.f.'s is

$$F_{\chi^2(\nu)}(y) = F_n(t(y)). \tag{5.13}$$

By differentiating Equation 5.13, the two p.d.f.'s are related by

$$f_{\chi^2(v)}(y) = f_n(t(y)) \frac{dt}{dy}, \qquad (5.14)$$

with the Jacobian

$$\frac{dt}{dy} = \frac{1}{2\lambda}.$$

Substitutions in Equation 5.14 give the chi-square p.d.f. $f_{\chi^2(v)}(y)$. Gamma probabilities are looked up in chi-square tables by using the relationship

$$F_v(2\lambda t) = F_n(t). \qquad (5.15)$$

Poisson Distribution

Whereas the gamma is a continuous distribution on the time t to the nth failure, the Poisson is a discrete distribution on n failures in t time. There is a dual relationship between the gamma and the Poisson that is similar to the dual relationship between the binomial and negative binomial distributions. Let T_n be the time to the nth failure; the d.f.

$$F_n(t) = P(T_n \le t)$$

is the gamma probability that at most t time is required for the nth failure to occur. Therefore, $1 - F_n(t)$ is the probability that more than t time is required; reversing the roles of n and t, it is also the Poisson probability that at most $n - 1$ events occur in t time

$$1 - F_n(t) = P(T_n > t|n) = P(N \le n-1|t) = F_t(n-1).$$

Thus, Poisson probabilities can be looked up in chi-square tables by the relationships

$$F_t(n) = P(N \le n|t) = 1 - F_{n+1}(t) = 1 - F_{\chi^2(2n+2)}(2\lambda t).$$

The Poisson probability function is obtained by subtracting $F_t(n-1)$ from $F_t(n)$

$$f_t(n) = P(N = n|t) = P(N \le n|t) - P(N \le n-1|t)$$
$$= F_t(n) - F_t(n-1).$$

This becomes

$$f_t(n) = F_{\chi^2(2n)}(2\lambda t) - F_{\chi^2(2n+2)}(2\lambda t).$$

Problems

5.1. Show that the $\chi^2(2)$ distribution is the exponential distribution by performing an appropriate substitution and a change of variables in Equation 5.13.

5.2. Let $R(t) = 1 - F(t)$ be the probability of survival to t, and let $R'(t)$ be the derivative of $R(t)$. Show that the exponential failure rate λ is

$$\frac{-R'(t)}{R(t)}.$$

5.3. The failure rate λ for a Poisson process is the ratio of the average number of failures in a specified time period to the length of time in that period. The reciprocal of λ is known as the mean time between failures (MTBF). Show that the MTBF is the mean of an exponential distribution with parameter λ.

Hint: Perform the integration $\int_0^\infty \lambda \exp(-\lambda x)\, dx$.

5.4. In the famous set of data compiled by von Bortkewicz, it was found that over a 200-year period the average number of deaths per year in the Prussian army due to the kick of a horse was 0.61. These data fit a Poisson distribution quite well. What is the probability that 2 years would pass without a single death due to the kick of a horse?

5.5. Five transistors were put onto an accelerated life test. The failures were at 96 hours, 240 hours, 500 hours, 750 hours, and 1000 hours, respectively.

    a. Graph these data on exponential-type graph paper.
    b. How many total good hours of operating time are on this test?
    c. What is the average good operating time between failures?
    d. What is the value of the Lilliefors D-statistic?
    e. What is the critical value of the Lilliefors D-statistic at the 10 percent level?
    f. Do the answers to 5.5a and 5.5e confirm a good fit?

5.6. The following data represent the endurance times of 13 specimens (called coupons), all of the same type of aluminum, to stress corrosion. The data were obtained by immersing the coupons in a corrosive bath, and observing for each coupon the time to failure, that is, the number of hours to appearance of stress corrosion.

Chapter 5 ■ *Exponential Times to Failure* 105

The data were 66, 76, 82, 92, 94, 97, 100, 107, 111, 117, 122, 130, and 136 hours, respectively.

a. Graph these data on exponential plotting paper, and test for goodness of fit by the Lilliefors K–S test for an exponential distribution.
b. Does the result improve if 60 hours is subtracted from each observation?

5.7. A device is required to have a lifetime of 1000 hours. Six units are placed on a life test; failures occurred at 425, 1000, 1650, 2400, 3600, and 6500 hours, respectively.

a. What is the calculated mean time to failure (MTTF)?
b. Based on the calculated MTTF, what is the estimated reliability?
c. Perform a Lilliefors K–S goodness-of-fit test. What is the value of the D-statistic, and does it indicate good fit?

5.8. Twelve identical hydraulic pumps were inspected on a daily basis to determine the time, in hours, required for the pump seals to fail. The failure times were

1173, 1737, 830, 4018, 11,278, 1304, 2304, 3047, 5189, 18,102, 6912, and 9029.

a. What is the calculated MTTF?
b. Based on the MTTF calculated in 5.8a, what is the calculated value of the failure rate λ?
c. Assuming that the value of λ calculated in 5.8b is the parameter of the exponential distribution, what is the probability of a pump *not* exceeding a lifetime of 5000 hours?
d. After reordering the observations, graph the data on exponential graph paper. What value of λ do you obtain from the graph?
e. Perform a Lilliefors K–S test of goodness of fit to an exponential distribution. What is the calculated value of the D-statistic? Does this indicate a good fit at the 10 percent level of the significance?

5.9. Graph the service–time data of Problem 4.4 on exponential plotting paper and perform a Lilliefors g.o.f. analysis. What is the calculated value of the D-statistic? Do the exponential plot and the g.o.f. test show a good fit? Compare to the lognormal fit in Problem 4.4.

References

Ascher, H., and H. Feingold. 1984. *Repairable systems reliability.* New York: Marcel Dekker.

Bain, Lee J. 1978. *Statistical analysis of reliability and life-testing models.* New York: Marcel Dekker.
Barlow, Richard E., and F. Proschan. 1965. *Mathematical theory of reliability.* New York: Wiley.
Bazovsky, Igor. 1961. Chapter 4 in *Reliability theory and practice.* Englewood Cliffs, N.J.: Prentice Hall.
Billinton, R., and R. N. Allan. 1983. *Reliability evaluation for engineering systems.* New York: Plenum Press.
Doty, Leonard A. 1989. *Reliability for the technologies.* 2d ed. New York: Industrial Press.
Dreyfogle, F. W., III. 1992. *Statistical methods for testing, development, and manufacture.* New York: Wiley.
Epstein, B. 1960. Tests for the validity of the assumption that the underlying distribution of life is exponential. *Technometrics* 1:83–101.
———. 1960. Tests for the validity of the assumption that the underlying distribution of life is exponential. *Technometrics* 2:167–84.
Feller, William. 1968. *An introduction to probability theory and its applications.* Vol I. 3d ed.
Harter, H. L. 1964. *New tables of the incomplete gamma-function ratio and of percentage points of the chi-square and beta distributions.* U.S. Government Printing Office: Aerospace Research Laboratories. U.S. Air Force.
Lilliefors, H. W. 1969. On the Kolmogorov–Smirnov test for the exponential distribution with mean unknown. *Journal of the American Statistical Association* 64:387–89.
Massey, F., Jr. 1951. The Kolmogorov–Smirnov test for goodness-of-fit. *Journal of the American Statistical Association* 46:68–78.
Nelson, W. 1969. Hazard plotting for incomplete failure data. *Journal of Quality Technology* 1:27–52.
Pearson, K. 1957. *Tables of the incomplete Γ-function.* London: Cambridge University Press.
Technical and Engineering Aids for Management (TEAM). Box 25, Tamworth, NH 03886.

Part III

Reliability-Confidence Assessment

Part III contains chapters 6, 7, 8, and 9, respectively titled "Statistical Assessment: The Bernoulli and Poisson Processes," "Bayesian Reliability Analysis: Attributes and Life Testing," "Normal Theory Tolerance Analysis," and "Hazard Analysis and the Weibull Distribution." Each chapter has a different approach, but all of them have in common that the principal concern is reliability-confidence measurement from real data for one or more of the principal distributions used in reliability: binomial, exponential, normal, or Weibull. Other topics are covered in each chapter as well, depending on the context.

Both chapters 6 and 7 deal with assessment for both the binomial-attributes case and the exponential-time-to-failure case. In chapter 6, assessment is covered in non-Bayesian terms from the viewpoint of a statistical technique called *maximum likelihood*. In chapter 7, the same types of problems are solved in Bayesian terms. It turns out that there is a close mathematical relationship between the non-Bayesian and the Bayesian approaches, and that the differences are insignificant for large-sample data but very noticeable for small-sample data or for data where there are relatively few failures. Chapter 8 discusses normal-theory reliability-confidence assessment from the viewpoint of either one-sided or two-sided tolerancing. Chapter 9 covers Weibull reliability-confidence assessment for life-test data.

Other topics covered in these chapters include the following:

Chapter 6: Minimum variance unbiased estimation, the two-parameter exponential distribution

Chapter 7: Prior and posterior distributions, the negative-log gamma distribution, the relationship between the Bayesian treatment of time-to-failure data and the Epstein–Sobel technique

Chapter 8: Tables of tolerance-limit factors

Chapter 9: Weibull and/or nonparametric hazard analysis, Weibull probability plotting and Weibull hazard plotting, the extreme-value distribution, Weibull estimation and reliability-confidence assessment for truncated life tests by exact linearized methods

6

Statistical Assessment: The Bernoulli and Poisson Processes

Introduction

In this chapter, certain aspects of reliability assessment are discussed for both the Bernoulli and Poisson processes, in greater detail than in chapters 3 and 5. These are statistical procedures for calculating not only reliability values, but also associated confidence levels. The maximum-likelihood method is used to obtain estimates of parameters of probability distributions for both processes. Unbiased linear estimation and the two-parameter exponential distribution are also introduced.

The Bernoulli parameter is p

One of the similarities between the Bernoulli and Poisson processes is that both processes are governed by a single parameter—p for the Bernoulli and λ for the Poisson process. There are other similarities that will be discussed in greater detail in this chapter and chapter 7, and there are also differences. The Bernoulli parameter, the probability of success p, is also the reliability, and the maximum likelihood "best" estimate (m.l.e.) of p is unbiased. The Poisson-process m.l.e. of λ is unbiased, but the corresponding estimate of the reliability is biased.

The Poisson parameter is λ

The Poisson-process parameter is the failure rate λ. The reliability R is calculated only if a specified time period of successful operation t_o is known, by the well-known formula

$$R = \exp(-\lambda t_o). \tag{6.1}$$

Possibly because of the extra operation of calculating an intermediate exponential function before obtaining the reliability, there is some loss of information. The unbiased m.l.e. of λ, total failures divided by total testing time, does not result in an unbiased estimate of R by Equation 6.1.

Complete data sets

The Bernoulli and Poisson processes also differ by the interpretation of what a complete data set is. A complete Bernoulli data set is usually thought of as being a specified number n of trials; there are n random events, successes or failures, with the result of each trial governed by a binary distribution. A complete Poisson data set is typically n identical items, all tested to end of life; as in the Bernoulli case, there are n random events, but all of them are failures governed by the exponential distribution.

Censored data sets

A censored data set is one for which the data collection process is terminated before all potential observations have been made. In the Bernoulli case, a data set is censored when either r failures or r successes are observed, and the number of trials n at the time of censoring is the r.v. In the Poisson case, a data set, such as a parallel life test of n items, is censored when $r < n$ failures have occurred, and the total test time T is the r.v. A type of premature termination similar to censoring is *truncation*, termination when a specified amount of testing time T has elapsed. A conservative way of handling truncated data is to assume that the last failure was at the time of truncation, regardless of whether a failure was actually observed. The techniques for processing incomplete data sets described in this chapter apply only to *single censoring*, censoring that cuts off the longer-lived items based on one dimension only, such as end of life.

The maximum-likelihood method

In the early 1920s R. A. Fisher introduced maximum likelihood as a means of calculating good estimates of the parameters of probability distributions from observed data. The technique can be described briefly in two paragraphs.

The likelihood function

Suppose the raw data consist of n random observations $x_1, \ldots, x_n$ of a random variable, all from the same population with probabilities governed by a single unknown parameter Θ. The p.d.f. for each of the n observations is

$$P(X_i = x_i) = f(x_i|\Theta), i = 1, \ldots, n.$$

Because of randomness the observations are independent and the joint probability is the product of the p.d.f.'s for all n observations. This joint probability is called the *likelihood function*

$$L = f(x_1|\Theta) \ldots f(x_n|\Theta). \qquad (6.2)$$

The idea of maximum likelihood is that by maximizing the natural logarithm of L and solving for Θ, we obtain an m.l.e. estimate $\hat{\Theta}$ most likely to be the value of the parameter for which the given data have occurred. To do so, take the derivative of $\ell n\,(L)$ with respect to Θ, and set this derivative to zero; then solve for Θ. The value obtained is the m.l.e.

Properties of maximum-likelihood estimates

It frequently turns out that an m.l.e. is intuitively quite reasonable. For example, with binomial data, the m.l.e. of the probability of success p is the number of successes divided by the number of trials. Also, under certain reasonable conditions, the m.l.e. of the Poisson-process failure rate λ is the number of failures divided by total testing time.

A number of statistics texts have extended discussions of the properties of m.l.e.'s. Among these properties are *efficiency*, comparability to a "best" estimate with minimum variance, and *sufficiency*, a concept also proposed by R. A. Fisher, a rather strong characteristic that the summary statistics upon which the estimate is based contain all of the relevant information in the sample. Maximum-likelihood estimates, however, are not always unbiased; for example, it is shown in chapter 2 that the sum of squares of the differences from the sample mean, divided by n, which is an m.l.e. of the variance, is biased; the correction factor $n/(n-1)$ corrects for the bias.

Attributes data

The m.l.e. of the reliability

The result of each trial is either a success "1" or a failure "0"; a "1" occurs with probability p and a "0" with probability $1-p$. The r.v. can be either the number of successes r in n trials, or else the number n of trials to attain a specified number of failures; and the reliability, the probability of success, is p. Since the trials are all independent by assumption, the likelihood L is the joint probability, the product of the probabilities, of the n trials,

$$L = p^r (1-p)^{n-r},$$

also called the *kernel* of the Bernoulli process in chapter 2. The m.l.e. results from maximizing $\ell n(L)$ by taking the derivative with respect to p, setting this derivative to zero, and solving for p.

$$\ell n(L) = r \, \ell n(p) + (n-r) \, \ell n(1-p).$$

The derivative of $\ell n(L)$ is

$$\frac{r}{p} - \frac{n-r}{1-p}.$$

After setting the derivative to zero the solution is

$$\hat{p} = \frac{r}{n}, \tag{6.3}$$

successes divided by trials.

Confidence level for the reliability

If the m.l.e. is used as the estimate of the reliability without specifying a confidence level, the resulting inference might be unacceptable. For example, if the device, system, or software being assessed is highly reliable, all trials could result in successes; then $n = r$ and $\hat{p} = 1$. Especially with small samples, this is not a reasonable estimate, because everything operates less than perfectly at some time or other.

In chapter 3 it is shown that with r successes and n trials, the LCL at any specified confidence level γ is that value p_L of p for which

$$\gamma = F_b(r|p_L, n+1), \qquad (6.4)$$

or alternatively,

$$\gamma = 1 - F_\beta(p_L|r+1, n-r+1), \qquad (6.5)$$

where F_b and F_β respectively denote the binomial and beta distribution functions. If there are no failures, then $r = n$ and $n-r = 0$ and γ and p_L are obtained from either Equation 6.4 or Equation 6.5 as

$$\gamma = 1 - p_L^{n+1}, \text{ and}$$

$$p_L = \sqrt[n+1]{1 - \gamma}. \qquad (6.6)$$

Censored attributes data

There is no distinction between complete and incomplete data sets in attribute maximum-likelihood analysis. If the test is terminated when r successes have occurred and r is the r.v., or, alternatively, $n-r$ failures with n the r.v., the m.l.e. is obtained from Equation 6.3, and the LCL is given by Equation 6.6.

EXAMPLE 6.1.

LCL with all successes: What is the LCL on the reliability at the 50 percent confidence level if there are no failures in 68 trials? From Equation 6.6

$$\ell n(p_L) = \frac{\ell n(1-\gamma)}{n+1} = \frac{\ell n\, 0.5}{69} = -0.01;$$

the answer is $p_L = 0.99$.

From Equation 6.5, p_L is the $100 \cdot (1-\gamma)$th percentage point of a beta distribution with parameters $r+1$ and $n-r+1$. From chapter 3 this distribution has the p.d.f.

$$f_\beta(p|r+1, n-r+1) = \frac{(n+1)!}{r!\,(n-r)!} p^r (1-p)^{n-r}. \qquad (6.7)$$

The mode of this beta distribution, the value of p with the highest probability, is obtained by differentiating Equation 6.7 and setting the derivative to zero; this results in r/n, identical to the m.l.e.

EXAMPLE 6.2.

Graphing the p.d.f.'s of beta distributions (Figure 6.1): Graph the probability density functions of the beta distributions on p for $n = 15$ and

1. $r = 7$
2. $r = 12$
3. $r = 14$
4. $r = 15$

The closer r, the number of successes, is to n, the number of trials, the greater the skewness. For each of the four cases r/n is the mode. With all successes ($r=n$), the distribution is J-shaped with a finite maximum, 16, at $p = 1$.

Exponential distributions

Maximum-likelihood assessment: The Epstein–Sobel technique

This section describes the Epstein–Sobel technique, also known as the chi-square method for assessing censored time-to-failure data. We assume that

1. The distribution governing failures is exponential with failure rate λ.
2. n items under life test are not repaired during the test.
3. The test is terminated when a fixed number r, $r \leq n$, failures occur.
4. The failure times are independent.

There are three steps in this process.

1. Define a distribution on λ.
2. Calculate the m.l.e. for λ.
3. Obtain confidence limits on λ and the reliability R.

The distribution on λ is gamma with parameters r and total testing time T. Because of the close relationship between the gamma and chi-square distributions, assessment is performed with chi-square tables.

Ordered failure times: The maximum-likelihood estimate

n identical units are placed under life test, r units fail at times $t_1, \ldots, t_r$, respectively, and the test is terminated at t_r. For each nonfailed item,

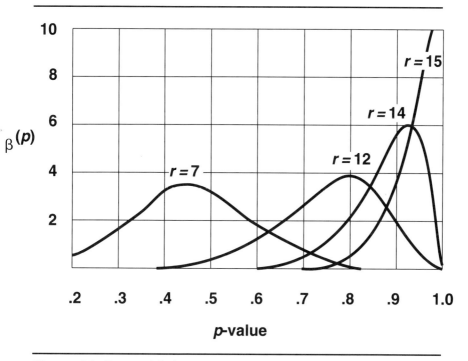

Figure 6.1. Beta p.d.f. for $n = 15$.

the probability of surviving to t_r is

$$1 - F(t_r) = \exp(-\lambda t_r).$$

Because of the independence assumption, the joint probability or likelihood L of failure of the first r items at times $t_1, \ldots, t_r$, respectively, and survival of the remaining $n-r$ items to t_r is the product of the probabilities

$$L = [1 - F(t_r)]^{n-r} \prod_{i=1}^{r} f(t_i)$$

$$= \lambda^r \exp\{-\lambda(\sum_{i=1}^{r} t_i + (n-r)\, t_r)\}.$$

The total testing time is the sum of two components, the total operating time for the r components that failed and also for the $n-r$ that did not fail:

$$T = \sum_{i=1}^{r} t_i + (n-r)t_r.$$

The m.l.e. is the value of λ that maximizes the natural log of L. Since

$$\ln(L) = r \ln(\lambda) - \lambda T,$$

the partial derivative is

$$\frac{\partial \ln(L)}{\partial \lambda} = \frac{r}{\lambda} - T;$$

setting this partial derivative to zero, the m.l.e. is

$$\hat{\lambda} = \frac{r}{T}. \tag{6.8}$$

This agrees with our intuition; the estimate of the failure rate is the number of failures r divided by the total accumulated testing time T. If the component under test is required to operate for t time, the corresponding m.l.e. estimate of the reliability R is

$$\hat{R} = \exp(-\hat{\lambda}t),$$
$$= \exp\left\{-\left(\frac{rt}{T}\right)\right\}. \tag{6.9}$$

Epstein–Sobel confidence limits

Epstein and Sobel showed that total testing time T is a random variable with a gamma-r distribution. In chapter 5 it is shown that the relationship between the chi-square and gamma-r distribution functions is

$$F_r(T) = F_{\chi^2(2r)}(2\lambda T).$$

After the test is completed, T is known, but λ is unknown. In order to find the confidence limits on λ and the reliability R, the reasoning, which is similar to the Bayesian argument that is developed more fully in chapter 7, is that the roles of T and λ are reversed, with T a parameter and λ as the r.v. Furthermore, the percentage points are obtained from the $\chi^2(2r)$ line of Table A.4.

The upper confidence limit on λ at confidence-level γ, λ_γ, is derived from

$$\gamma = P(\lambda \leq \lambda_\gamma)$$
$$= F_{\chi^2(2r)}(2\lambda T);$$

therefore, the γ percent upper confidence limit on λ is

$$\lambda_\gamma = \frac{\chi^2_\gamma(2r)}{2T} \tag{6.10}$$

where $\chi^2_\gamma(2r)$ is the γth percentile of the chi-square distribution with $2r$ degrees of freedom; and the γ percent lower confidence limit on R, the LCL at confidence level γ for a required operating time of t, is

$$R_\gamma = \exp\{-\lambda_\gamma t\}. \tag{6.11}$$

EXAMPLE 6.3.

m.l.e. and LCL with censored data: In Example 5.1, suppose that the test had been terminated after the fourth failure. What would be the m.l.e. of λ and of R, the UCL, for λ at the 95 percent confidence level, and the LCL for R at the 95 percent confidence level if the required operating time is 25 hours?

In this case the total operating time is

$$T = 35 + 100 + 240 + 440 + 440 = 1255 \text{ hours.}$$

The m.l.e.'s are, respectively,

$$\hat{\lambda} = \frac{4}{1255} = 0.003187$$

$$\hat{R} = \exp\{-(0.003187)\,25\} = 0.9234;$$

The 95 percent upper confidence limit for λ is

$$\lambda_{0.95} = \frac{15.507}{2\,(1255)} = 0.006178;$$

and the LCL is

$$R_{0.95} = \exp\{-(0.006178)\,25\} = 0.857.$$

Minimum variance unbiased estimation

Poisson process

Because every observation in a data sample from a population governed by a parent distribution is a realization of a random variable, an estimate of the parameter of this distribution based upon the data is also a random variable subject to a probability distribution that is different from the parent distribution. There are a number of ways of defining a good estimate; two characteristics are desirable: (1) minimum variance: smaller variance than any other estimate; and (2) unbiasedness: the expected value is equal to the parameter. An estimate that satisfies both of these properties is called a *minimum variance unbiased estimate* (m.v.u.e.).

When, as in previous sections of this chapter, the reliability of a Poisson process based upon time-to-failure data was estimated, the m.l.e. $\hat{\lambda}$ of the failure rate is an m.v.u.e., but the corresponding estimate of the reliability is biased and does not have minimum variance. Basu and Pugh have shown that the m.v.u.e. of the reliability is

$$R^*(t) = \left(1 - \frac{t}{T}\right)^{r-1}, \qquad (6.12)$$

where, as before, r is the accumulated number of failures, t is the required operating time, and T is the accumulated testing time.

EXAMPLE 6.4.

m.v.u.e. of R with censored data: For the data of Example 6.3, what is the m.v.u.e. of the reliability?

$$\text{Answer: } \left(1 - \frac{25}{1255}\right)^3 = 0.941;$$

this is higher than the m.l.e., 0.9234.

The two-parameter exponential distribution

General considerations

As a general rule, failure rates of equipment are not constant over an entire lifetime, but vary according to some probability law, sometimes increasing, sometimes constant, and sometimes decreasing. There are available a number of techniques for analyzing lifetime distributions from life-test data when the failure rates are variable. Some of these techniques are parametric, associated with a family such as normal, lognormal, Weibull, gamma, or Rayleigh. Others are nonparametric, meaning that the shape of the distribution and of the failure-rate function can be found, but no particular family is identified.

The simplest variable failure-rate family is the two-parameter exponential distribution. There are just two values for the failure rate, zero from the beginning of life up to some intermediate point of time, μ, and a constant λ thereafter. The p.d.f. is

$$f(t) = 0, \ t \le \mu,$$
$$= \lambda \exp\{-\lambda(t-\mu)\}, \ t > \mu.$$

The constant μ is known as a location parameter, because it is a point of time for a discrete change in the failure rate, and λ is a scale parameter, because it is a multiple of the r.v., $(t-\mu)$.

If μ is known in advance, estimating λ by graphical techniques or by computation is no different than for the one-parameter exponential case; the r.v. is $t-\mu$ instead of t. If μ is estimated graphically, some trial and error is necessary.

Best linear unbiased estimates

Sarhan and Greenberg developed a procedure for computing estimates of both parameters of the two-parameter exponential distribution using only the statistics $t_1 \leq t_2 \leq \ldots \leq t_n$ of the n ordered times to failure. The estimates obtained, called *best linear unbiased estimates* (BLUE) are weighted averages of the order statistics; the weights are based upon n and the order numbers. The technique can be used even for censored samples, with either the upper order statistics deleted, the lower order statistics deleted, or both the upper and lower order statistics deleted.

The formulas for the estimators of μ and λ for complete samples only (no censoring) are

$$\mu^* = \frac{n}{n-1} t_1 - \sum_{i=1}^{n} \frac{t_i}{n(n-1)}$$

$$\lambda^* = \frac{n-1}{\sum_{i=1}^{n} t_i - n t_1}. \tag{6.13}$$

The formulas for censored samples are quite complex and are not included in this text; they may be found in the Sarhan and Greenberg article.

EXAMPLE 6.5.

Given the eight data points representing field time-to-failure data in hours, fit a two-parameter exponential distribution.

45, 47, 50, 62, 64, 64, 80, 80

The plotting points for $n = 8$ are obtained from Table A.3. For the two sets of ties at 64 and 80, in Figure 6.2 the average plotting points, 0.612 and 0.834, respectively, were used. This plot would not fit an ordinary one-parameter exponential distribution because the location parameter is nonzero; however, with an assumed location parameter μ of

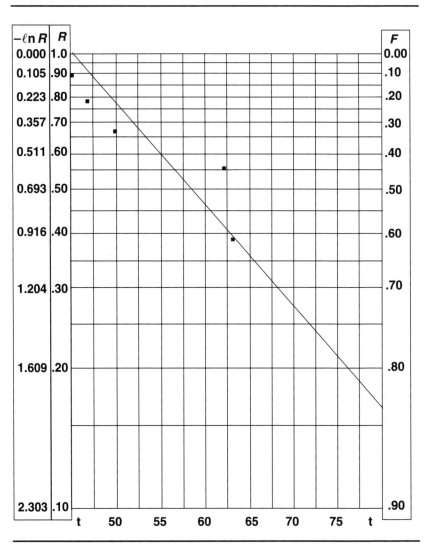

Figure 6.2. Two-parameter exponential probability plot.

40 hours, there is a reasonable visual fit to the two-parameter exponential with a failure rate λ of 0.047 failures per hour, starting from 40 hours. The BLUE estimates are

$$\mu^* = \frac{8}{7} \cdot 45 - \frac{45 + 47 + 50 + 62 + 64 + 64 + 80 + 80}{56} = 42.643$$

$$\lambda^* = \frac{7}{492 - 360} = 0.053.$$

Problems

6.1. The required lifetime under a producer's warranty is 10 months. In a life test, these times to failure, in months, were obtained.

$$2, 4, 7, 19, 21, 21, 37, 37$$

a. What is the unbiased estimate of the reliability by the Basu–Pugh formula Equation 6.12?
b. What is the m.l.e. of the reliability?
c. What is the upper confidence limit on the failure rate λ at confidence level 0.5, by the Epstein–Sobel formula?
d. What is the corresponding LCL for the reliability at the 50 percent confidence level?
e. Graph this problem on exponential graph paper and perform a Lilliefors K–S goodness-of-fit test. What is the value of the D-statistic?
f. How is this problem similar to Example 6.5?

6.2. For the stress-corrosion failure data of Problem 5.6,

a. What are the parameters μ^* and λ^* of the two-parameter exponential distribution, by Equation 6.13?
b. Assuming that the required lifetime free of stress corrosion is 100 hours, what is the unbiased estimate of the reliability by Equation 6.12?
c. Under the same assumption as 6.2b., what is the m.l.e. of the reliability?

6.3. For the hydraulic pump failure data of Problem 5.8,

a. What are the values of μ^* and λ^* for the two-parameter exponential distribution?

b. Graph these data on exponential graph paper after subtracting the calculated value of μ^* obtained from the original values.
c. What is the value of the Lilliefors D-statistic? Is the fit to the two-parameter exponential better than the one-parameter exponential in Problem 5.8?

6.4. Local area network: A local area network (LAN) consists of several personal-computer terminals interconnected through a file server that manages traffic between the terminals and controls the word-processing and printing functions. A failure of the LAN occurs when the system freezes and is nonresponsive to keyed input coming from a terminal. If a freeze occurs at a terminal, it must be shut off and rebooted to continue the work. A new installation was tested for a period of one week. There were 18 shutdowns, including five shutdowns that were not failures, but system cutoffs at the end of a working day. The following times between shutdowns, in minutes, were observed.

22, 30, 35, 38, 45, 50, 75, 85, 115, 120, 145, 163, 180, 185, 205, 210, 260, 327

Note: There are problems using this particular data set in both chapters 7 and 9. In chapter 9, the five end-of-day shutdowns are handled differently than the failures; however, for purposes of Problem 6.4, treat all 18 shutdowns as if they were all system freeze-ups.

a. Graph the distribution of shutdown times on exponential paper, Figure 5.4.
b. Perform a Lilliefors goodness-of-fit test for an exponential distribution. What is the calculated value of the D-statistic? Does this show a good fit?
c. Provide the calculated estimates of the parameters μ^* and λ^* of the two-parameter exponential distribution by Equation 6.13 and graph the data on exponential graph paper after subtracting μ^* from each value.
d. Perform a Lilliefors goodness-of-fit test on the two-parameter exponential. What is the value of the D-statistic, and does it indicate good fit?
e. Compare the results of the one-parameter versus the two-parameter fits.

References

Basu, A. P. 1964. Estimates of reliability for some distributions useful in life testing. *Technometrics* 6:215–49.

Cramer, Harald. 1946. *Mathematical methods of statistics.* Princeton, N.J.: Princeton University Press.

Epstein, B. 1960. Tests for the validity of the assumption that the underlying distribution of life is exponential. *Technometrics* 1:83–101.

———. 1960. Tests for the validity of the assumption that the underlying distribution of life is exponential. *Technomectrics* 2:167–84.

Epstein, B., and M. Sobel. 1953. Life testing. *Journal of the American Statistical Association* 48:486–502.

Fisher, R. A. [1922] 1950. Mathematical foundations of theoretical statistics. Reprinted in *Contributions to mathematical statistics.* Edited by R. A. Fisher. New York: Wiley.

Hogg, R. V., and A. T. Craig. 1959. *Introduction to mathematical statistics.* New York: Macmillan.

Pugh, E. L. 1963. The best estimate of the reliability in the exponential case. *Operations Research* 11:57–61.

Sarhan, A. E., and B. G. Greenberg. 1962. *Contributions to order statistics.* New York: Wiley.

———. 1971. The effect of two extremes in estimating the best linear unbaised estimate (BLUE) of the parameters of the exponential distribution. *Technometrics* 13:113–25.

7

Bayesian Reliability Analysis: Attributes and Life Testing

Introduction

In chapter 1 Bayes theorem was introduced, to calculate the conditional probability of an event, an element of a partition, given that some other event with known probability has occurred. This chapter introduces Bayesian techniques for setting confidence limits on the reliability for both the Bernoulli and Poisson processes, based on observed test data.

Lower confidence limits

In contrast to the discrete Bayesian model of chapter 1, for a Bayesian reliability model the partitioning is on the set of values of a parameter with continuous measure, the reliability p, a number in the range zero to one. The lower confidence limit p_L at confidence level γ partitions this range into two parts: the upper part from p_L to 1 with probability γ, and the lower part from zero to p_L with probability $1-\gamma$. Practically the same type of analysis is performed in chapter 6, but not in the Bayesian framework. The difference between the Bayesian and non-Bayesian approaches is that the former relies upon both current and prior data, whereas the non-Bayesian methodology requires only the current data. Under certain reasonable assumptions, however, both the Bayesian and non-Bayesian approaches to the reliability of either the Bernoulli or Poisson process will yield the same confidence limits.

Continuous prior and posterior distributions: Sufficient statistics

The type of Bayesian analysis illustrated in this chapter starts from a continuous prior distribution on the reliability p, over the range zero to one. For the Bernoulli process, the prior distribution is on p; for the Poisson process, the prior distribution could be either on p, the failure rate λ, or even the MTBF. The parameters of this prior distribution are sufficient statistics representing all of the accumulated prior data. Examples of prior data include testing time, number of units under test, and number of failures.

Treating the parameter of a process as a random variable, which is characteristic of the modern Bayesian approach to statistics, contrasts with the conventional approaches, such as those in the earlier chapters of this book, where a parameter is always a fixed number. There is a reversal of roles: the data, the prior sufficient statistics are the parameters of a prior distribution, and the parameters of the Bernoulli or Poisson process are the random variables. With new data, there is a linkage from the prior to a posterior distribution, with the posterior sufficient statistics being both the old and new data.

Current data

New data are incorporated into a Bayesian reliability model as observations of an event governed by a parameter which is the r.v. for a prior distribution. For the cases discussed in this chapter, the prior and data distributions are conjugate; this means that there are convenient mathematical similarities that make it possible to mix the available information, via Bayes theorem, so as to form a posterior distribution, with the same form as the prior and with the same r.v. and range, but with the parameters being additive sufficient statistics for both the prior and the current data.

Conjugate distributions

For the Bernoulli process, the prior distribution on p is beta, described in chapter 3, with a kernel of the form

$$p^{a-1}(1-p)^{b-1},$$

the same as the kernel of the probability density functions for both the binomial and negative binomial probability distributions. For the Poisson process, the prior distribution on the failure rate λ is gamma with the kernel

$$\lambda^a \exp(-\lambda b),$$

which is conjugate to the p.d.f. of either an exponential, Poisson, or gamma distribution, all of which are described in chapter 5.

In a broader sense, both the Bernoulli and Poisson processes are members of the Koopman–Pitman–Darmois class, also known as exponential families of distributions, with the following properties:

1. The likelihood function, described in chapter 6, for repeated independent trials is a function of additive sufficient statistics for all of the prior data.
2. If the prior distribution is conjugate, the posterior distribution is of the same mathematical form as the prior, with the parameters being the sums of the sufficient statistics for both the prior and current data.

Varieties of Poisson-process prior distributions

There are different types of gamma prior distributions for the Poisson process. If the failure rate λ is the parameter of the Poisson process, the usual form of the gamma distribution described in the appendix to chapter 5 is conjugate prior, with λ, instead of t, being the r.v. with a range of zero to infinity, and t the fixed number or parameter of the distribution. By a change of variables based on a redefinition of the time units, the gamma prior on λ becomes a negative-log gamma (n.l.g.) prior distribution on the reliability p, with a range of zero to one, like that of the beta distribution for the Bernoulli attribute process.

Structure of a Bayesian analysis

The discrete case

A Bayesian analysis proceeds from a two-way partitioning of the universal set. If the partitioning is discrete, there is one partition A with m events

$$A = \{A_1, \ldots, A_m\},$$

and a different partition, B, of the same universal set, with n elements

$$B = \{B_1, \ldots, B_n\},$$

and with the prior probability distribution

$$P(B_1), P(B_2), \ldots, P(B_n).$$

The joint probability or likelihood of A_j and B_i is

$$P(A_j \cap B_i) = P(B_i) P(A_j|B_i).$$

Given that event A_j occurs, the conditional probability of B_i is obtained by Bayes theorem

$$P(B_i|A_j) = \frac{P(A_j \cap B_i)}{P(A_j)}, \qquad (7.1)$$

known as the posterior probability of B_i given that event A_j occurs. The totality of all n terms of the form of Equation 7.1 is the posterior probability distribution. The denominator of Equation 7.1,

$$P(A_j) = \sum_{i=1}^{n} P(B_i) P(A_j|B_i), \qquad (7.2)$$

is the marginal probability of A_j. The set of all m terms of the form of Equation 7.2 is the marginal probability distribution, over all of the subsets in B.

EXAMPLE 7.1.

Prior probability distribution for the discrete case: In Example 1.13, the thrower continues to roll the dice; therefore, the first roll was a point, either 4, 5, 6, 8, 9, or 10. The prior probability distribution consists of the set of normalized probabilities for these six values.

value	prior probability
4	$\frac{1}{8}$
5	$\frac{1}{6}$
6	$\frac{5}{24}$
8	$\frac{5}{24}$
9	$\frac{1}{6}$
10	$\frac{1}{8}$

EXAMPLE 7.2.

Marginal and posterior distributions: The marginal probability distribution for Example 7.1 has two events.

A_1: makes the point, with probability $\frac{3216}{7920}$

A_2: does not make the point, with probability $\frac{4704}{7920}$

If it is known that the thrower of the dice won by making the point (and if the point is unknown), the posterior distribution is obtained by dividing the prior probabilities by 3216/7920.

value	posterior probability
4	$\dfrac{330}{3216}$
5	$\dfrac{528}{3216}$
6	$\dfrac{750}{3216}$
8	$\dfrac{750}{3216}$
9	$\dfrac{528}{3216}$
10	$\dfrac{330}{3216}$

Bayesian reliability

Bayesian reliability analysis relates to a continuous valued r.v. of the prior and posterior distributions, the reliability p, which can assume any one of an infinite set of possible values between zero and one. The marginal distribution can be either discrete or continuous; discrete if the data are governed by a discrete probability distribution such as the binomial or Poisson distribution, or continuous if the governing probability distribution is continuous, such as the gamma. The marginal or summation probabilities are obtained by mathematical integration.

Attributes data

The beta probability distribution

For a Bernoulli process, the same two cases are covered in this chapter as were discussed in chapter 3: fixed sample sizes n and fixed number of occurrences r. For fixed n, the distribution on the data is binomial, and for fixed r, the distribution is negative binomial. In both cases, the prior distribution on p is beta.

The parameters of a beta prior are sufficient statistics representing all of the relevant prior data; n' is the number of prior tests and r' is the number of previous occurrences. Following Raiffa and Schlaifer, primes (') are used in order to facilitate combining the prior sufficient

Chapter 7 ■ Bayesian Reliability Analysis

statistics with the current data in the Bayesian mixing operation. The prior p.d.f. is

$$f(p) = f_\beta(p|r', n')$$
$$= \frac{p^{r'-1}(1-p)^{n'-r'-1}}{B(r', n'-r')}; \qquad (7.3)$$

where the beta function is

$$B(r', n'-r') = \frac{(r'-1)!\,(n'-r'-1)!}{(n'-1)!}.$$

Fixed number of tests: Binomial sampling

The binomial probability of exactly r successes in n trials is

$$f(r|p) = f_b(r|p, n)$$
$$= \binom{n}{r} p^r (1-p)^{n-r}. \qquad (7.4)$$

The likelihood or joint probability of p and r is the product of Equation 7.3 by Equation 7.4.

$$f(r, p) = f(p)\,f(r|p)$$
$$= \frac{\binom{n}{r}}{B(r', n'-r')} p^{r+r'-1}(1-p)^{n+n'-r-r'-1} \qquad (7.5)$$

The marginal probability of r, the probability of r successes, including all possible values of p, is the integral of Equation 7.5 over the range zero to one.

$$f(r) = \int_0^1 f(r,p)\,dp$$
$$= \frac{\binom{n}{r}}{B(r', n'-r')} \int_0^1 p^{r+r'-1}(1-p)^{n+n'-r-r'-1}\,dp$$
$$= \frac{\binom{n}{r}}{B(r', n'-r')} B(r+r', n+n'-r-r') \qquad (7.6)$$

132   Chapter 7 ■ Bayesian Reliability Analysis

The posterior p.d.f. is obtained by dividing Equation 7.5 by Equation 7.6.

$$f(p|r) = \frac{f(r, p)}{f(r)}$$
$$= \frac{p^{r+r'-1}(1-p)^{n+n'-r-r'-1}}{B(r+r', n+n'-r-r')} \quad (7.7)$$

Equation 7.7 is a beta p.d.f., of the same mathematical form as the prior p.d.f., Equation 7.3, with $r+r'$ instead of r' and $n+n'-r-r'$ instead of $n'-r'$, as the new sufficient statistics for the combined prior and current data.

Fixed number of occurrences: Negative binomial sampling

If the testing is terminated after r successes have been attained, the number of trials n is the r.v. and the negative binomial distribution governs the data. Starting from the beta prior, Equation 7.3, the Bayesian-mixing operation results in a beta posterior distribution which is identical to Equation 7.7. The intermediate steps to generate this result are the same as those of the binomial process, Equations 7.4–7.6, except that there is a different combinatorial coefficient, $\binom{n-1}{r-1}$ instead of $\binom{n}{r}$. Since this coefficient cancels out when the likelihood is divided by the marginal probability, the same posterior is derived as in binomial sampling. The proof of this is left as a problem at the end of the chapter.

Bayesian reliability assessment for attributes data

Since the beta family of prior and posterior distributions is on the reliability p, the Bayesian attributes model leads directly to an interpretation of the percentage points of a beta distribution as confidence limits on the reliability. Given that the prior parameters are r' and $n'-r'$, respectively, the lower confidence limit p_L at confidence level γ is the value of p such that

$$\gamma = P(p > p_L)$$
$$= 1 - F_\beta(p_L|r', n'-r').$$

This means that p_L is the $(1-\gamma)$th percentage point of the beta distribution with parameters r' and $n'-r'$. Table A.1, which may be viewed as a table of the percentage points of selected beta distributions, is used for Bayesian reliability assessment in the same way described in chapters 3 and 6, with r' substituted for r and $n'-r'$ instead of $n-r$.

Chapter 7 ■ *Bayesian Reliability Analysis* 133

EXAMPLE 7.3.

Bayesian assessment with attributes data: Given the prior distribution with $n' = 25$ and $r' = 23$, a new test of 25 items of the same type is performed, and all are successes. What is the LCL at the 75 percent confidence level

1. Based upon the prior distribution
2. Based upon the posterior distribution

From the prior distribution data, the LCL is 0.855 ($\gamma=0.75$, $n'=25$, $r'=23$); for the posterior distribution, the LCL is 0.925 (0.75, 50, and 48). The LCL for the posterior distribution is the same as if there had been 50 prior tests with 48 successes.

Time-to-failure data

Gamma prior distribution

For Poisson-process time-to-failure testing, two cases are considered in this section, where testing time t is specified, or testing is performed until r failures occur. If t is fixed, the number of failures r is the r.v., and a Poisson distribution governs the probabilities for different values of r; if r is fixed, t is the r.v., and a gamma distribution governs the amount of time required for r failures to occur. The first case involves a discrete valued r.v. and the second a continuous valued r.v. The Bayesian approach is to use a continuous gamma prior distribution on the failure rate λ for both cases.

The sufficient statistics for the gamma prior are the available data about the Poisson process, the number of prior equivalent failures r' and the prior equivalent testing time t'. The p.d.f. is

$$f(\lambda) = f(\lambda|t', r') = \frac{(t')^{r'} \lambda^{r'-1} \exp(-\lambda t')}{(r'-1)!}. \qquad (7.8)$$

Fixed r

Parallel life testing is sometimes performed until a fixed number r of failures occur. This could be under either of two circumstances:

1. A complete life test of r items, all tested to failure: t is the total testing time for all r items.

2. A censored life test of n items, with the test terminated after r failures have occurred: in this case, t is the total testing time for the r items that failed plus the successful operating time for the $n-r$ items that did not fail.

The gamma distribution on the time to failure has the p.d.f.

$$f(t|\lambda) = f_r(t|\lambda) = \frac{(t)^{r-1} \lambda^r \exp(-\lambda t)}{(r-1)!}.$$

The joint probability is the product of Equation 7.8 and $f(t|\lambda)$

$$f(t, \lambda) = \frac{(t')^{r'} (t)^{r-1} \lambda^{r+r'-1} \exp\{-\lambda(t+t')\}}{(r-1)! (r'-1)!}.$$

The marginal p.d.f. on t is obtained by integrating the joint probability over λ from zero to infinity.

$$f(t) = \frac{(t')^{r'} t^{r-1}}{(r-1)! (r'-1)!} \int_0^\infty \lambda^{r+r'-1} \exp\{-\lambda(t+t')\} d\lambda$$

$$= \frac{(t')^{r'} t^{r-1}}{(t+t')^{r+r'}} \frac{(r+r'-1)!}{(r-1)! (r'-1)!}$$

Note that the integral is

$$\int_0^\infty \lambda^{r+r'-1} \exp\{-\lambda(t+t')\} d\lambda = \frac{(r+r'-1)!}{(r-1)! (r'-1)!}.$$

The posterior p.d.f. obtained by dividing the joint density by the marginal density is of the same form as the p.d.f. of the gamma prior, Equation 7.8, with $t+t'$ instead of t' and $r+r'$ instead of r'.

$$f(\lambda|t) = \frac{f(t, \lambda)}{f(t)}$$

$$= \frac{(t+t')^{r+r'} \lambda^{r+r'-1} \exp\{-\lambda(t+t')\}}{(r+r'-1)!} \quad (7.9)$$

Fixed t

A life test may be terminated when a specified amount of testing time t has been completed. In this case, the Poisson distribution governs the probabilities for different values of the number of failures r. Similar to what was observed for the Bayesian model of the Bernoulli process, starting from the gamma prior, Equation 7.8, the Bayesian-mixing operation results in the same gamma posterior, Equation 7.9, that was obtained with fixed r.

The Poisson probability function for r failures, given t testing time and failure rate λ is

$$f(r|\lambda) = \frac{(\lambda t)^r \exp(-\lambda t)}{r!}.$$

The joint probability is the product of $f(r|\lambda)$ by Equation 7.8,

$$f(r,\lambda) = f(\lambda) f(r|\lambda) = \frac{(t')^{r'} t^r \lambda^{r+r'-1} \exp\{-\lambda(t+t')\}}{r!(r'-1)!}.$$

The discrete marginal probability density on r is obtained by integrating the joint probability over λ.

$$f(r) = \frac{(t')^{r'} t^r}{r!(r'-1)!} \int_0^\infty \lambda^{r+r'-1} \exp\{-\lambda(t+t')\} \, d\lambda$$

$$= \frac{(t')^{r'} t^r}{(t+t')^{r+r'}} \frac{(r+r'-1)!}{r! \, (r'-1)!}$$

Dividing the joint density by the marginal density, the gamma posterior p.d.f., Equation 7.9, is obtained.

Bayesian reliability assessment: The Poisson process

Bayesian confidence limits for the failure rate λ are percentage points of a gamma prior or posterior distribution such as Equation 7.8 or Equation 7.9. As with the Epstein–Sobel technique discussed in chapter 6, the table used is a χ^2 table such as Table A.4.

Given the prior sufficient statistics t' and r', the γ upper confidence limit UCL on the failure rate λ_γ is the γ percentage point of the prior distribution

$$\gamma = P(\lambda \leq \lambda_\gamma)$$
$$= \frac{(t')^{r'}}{(r'-1)!} \int_0^{\lambda_\gamma} \lambda^{r'-1} \exp(-\lambda t') \, d\lambda = F(\lambda_\gamma).$$

Because of the relationship between gamma and chi-square, $2\lambda_\gamma t'$ is the

136 Chapter 7 ■ *Bayesian Reliability Analysis*

γ percentage point of the chi-square distribution with $2r'$ degrees of freedom.

$$\gamma = F_{\chi^2(2r')}(2\lambda_\gamma t')$$

In order to find the UCL for λ, find the chi-square value and divide by $2t'$. If the required mission time is t, the corresponding γ lower confidence limit on the reliability p is

$$p_L = \exp(-\lambda_\gamma t).$$

EXAMPLE 7.4.

Bayesian confidence limits on λ and p: Given that $r' = 4$, $t' = 50.2$, and the required mission time t is 1, what are the following confidence limits?

1. the 95% UCL for λ?
2. the 95% LCL for p?

The value of chi-square is 15.507. The solution to question 1 is

$$\lambda_{.95} = \frac{15.507}{100.4} = 0.15445.$$

The solution to question 2 is

$$p_L = \exp(-0.15445) = 0.857.$$

This LCL on the reliability p is identical to Example 6.3. The reason for this is that these are actually the same data: $r' = 4$ means that there were four failures, as in Example 6.3; likewise, t' was scaled in 25-hour units, so that $t' = 50.2$ corresponds to $(50.2)25 = 1255$ hours, the number of successful operating hours in Example 6.3.

Negative-log gamma prior distribution on the reliability

By rescaling the units for measuring time, after a simple change of variables from the gamma prior, the negative-log gamma (n.l.g.) distribution is derived, with a range for the r.v. from zero to one. The n.l.g. is useful as a prior distribution on the reliability for the Poisson process and the percentage points are interpreted as confidence limits on the reliability p.

Suppose that time t is measured in units of required operating time t_o, such as performance life or mission time, instead of the customary

Chapter 7 ■ **Bayesian Reliability Analysis** 137

units of hours, minutes, and seconds. For this scale, $t_o = 1$, so that the reliability formula is a 1:1 relationship between p and λ

$$p = \exp(-\lambda).$$

By taking the natural logarithm of both sides

$$\lambda = -\ell n(p) = \ell n(p^{-1}). \tag{7.10}$$

By a change of variables for the gamma prior distribution on λ, Equation 7.8, using Equation 7.10, the p.d.f. of the n.l.g. prior distribution on p is

$$f(p|r', t') = \frac{(t')^{r'} [\ell n(p^{-1})] p^{r'-1}}{(r'-1)!}, \quad 0 < p < 1. \tag{7.11}$$

If $r' = t' = 1$, Eq. 7.11 is a uniform distribution with $f(p) = 1$.

Similarities to beta distributions

Because both the n.l.g. and beta distributions have the same range for the r.v., zero to one, there are some interesting similarities. Frequently a member of the n.l.g. family will closely approximate a member of the beta family, and vice versa. This is shown in Example 7.5 where four different n.l.g. prior distributions are graphed, and each one of them is very close in shape to one of the four beta priors discussed in Example 6.2 and graphed in Figure 6.1.

EXAMPLE 7.5.

Graphing the n.l.g. prior: Graph the n.l.g. p.d.f. for each of the following four cases.

$$r' = 12.431, \; t' = 16$$
$$r' = 4.3471, \; t' = 16$$
$$r' = 2.0353, \; t' = 16$$
$$r' = 1, \qquad t' = 16$$

The graph is Figure 7.1; although it is not identical to Figure 6.1, the shapes of the four p.d.f.'s in both graphs are similar, with the same modes (values of the r.v. at maximum height of the graph), 0.4667, 0.8, 0.9333, and 1, respectively. In particular, the cases $r = n = 15$ in Example 6.2 and $r' = 1$, $t' = 16$ are identical with the p.d.f.

$$f(p) = 16p^{15}.$$

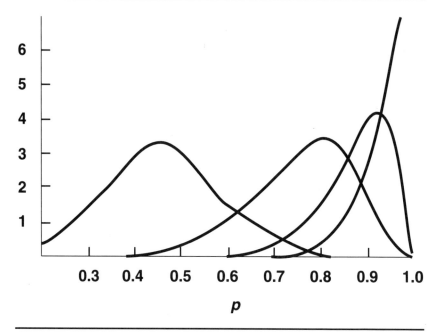

Figure 7.1. The p.d.f. of the n.l.g. distribution.

The diffuse n.l.g. initial uniform prior

In his famous 230-year-old essay on predicting the position of a billiard ball thrown on a table, Thomas Bayes assumed a uniform prior distribution on p, the probability of success on a single trial. Since the publication of that essay, the acceptability of this historic controversial assumption would determine whether or not the prominent mathematical scholars of that generation would find Bayes' work acceptable. It is common for the modern Bayesian school of statisticians to initiate an analysis with a *diffuse* prior distribution: the uniform n.l.g. is an example of such a diffuse prior.

There are several different ways of explaining why a uniform initial prior on p is a satisfactory, if not a correct, assumption. For example, suppose that at some stage in the process of testing and data accumulation r' failures had occurred in t' (in units of mission lifetimes) testing time. Then by the maximum-likelihood method of chapter 6 (Equation

Chapter 7 ■ Bayesian Reliability Analysis

6.8 and Equation 6.9), the m.l.e. of the failure rate λ and of the reliability p are, respectively,

$$\hat{\lambda} = \frac{r'}{t'}$$

$$\hat{p} = \exp\left(-\frac{r'}{t'}\right).$$

Because the concept of maximum likelihood is closely related to the mode, we would expect there to be a relationship between the m.l.e. of the reliability and the mode of the n.l.g. prior, Equation 7.11. Differentiating Equation 7.11 with respect to p and setting this derivative to zero yields the mode

$$\exp\left(-\frac{r'-1}{t'-1}\right).$$

If it is assumed that the initial n.l.g. prior (with no data) is uniform with $r'=t'=1$, both the m.l.e. and the Bayesian estimates of p are identical.

Another argument for using a uniform initial prior distribution on p is that without it, some failures are required to make confidence statements. For example, if $r'=t'=0$ instead of 1 at the initiation of the data-collection process, then the initial prior on p is indeterminate. If testing continues without any failures, then $r'=0$ implies that $\hat{p}=1$; this interpretation is not acceptable because it would be possible to take the results of a few tests that are all successes and claim perfect reliability.

The same problem is discussed in chapters 3 and 6 in connection with the Bernoulli process. The same solution is offered for that case: a uniform initial beta prior if there is no data.

Reliability assessment with the n.l.g. prior distribution

The n.l.g. distributions are used to set confidence limits for p based upon the results of time-to-failure testing of a Poisson process, in the same way that the beta distributions set confidence limits for p with attributes data of a Bernoulli process. The LCL at confidence level λ is the value of p such that

$$\begin{aligned}\gamma &= P(p > p_L) \\ &= 1 - F(p_L | r', t').\end{aligned} \quad (7.12)$$

140 Chapter 7 ■ Bayesian Reliability Analysis

The most convenient way of finding p_L is with the tables of the χ^2-distribution function. It can be shown by the change of variables, Equation 7.10, that Equation 7.12 can be restated as a chi-square probability

$$\gamma = F_{\chi^2(2r')}(2t'\, \ell n\{(p_L)^{-1}\}). \tag{7.13}$$

To find the LCL, look up the γ percentage point of the chi-square distribution with $2r'$ degrees of freedom, and then use the following formula to calculate the LCL.

$$p_L = \exp\left(-\frac{\chi^2(2r', \gamma)}{2t'}\right).$$

EXAMPLE 7.6.

Bayesian LCL for time-to-failure data (*see also* Example 6.3): Given a required operating life of 25 hours and a test history of 1255 successful operating hours with four failures, what is the LCL for the reliability at the 95 percent confidence level?

If the initial prior is uniform, r' is one greater than the number of failures and t' is one greater than the number of equivalent lifetimes successfully completed. Thus,

$$r' = 4 + 1 = 5$$
$$t' = \frac{1255}{25} + 1 = 51.2.$$

For 10 degrees of freedom, the chi-square value at the 0.95 level is 18.307; therefore the LCL is

$$p_L = \exp\left(-\frac{18.307}{102.4}\right) = 0.837.$$

In this case, the Bayesian 95 percent LCL, starting from a uniform prior distribution with no data, is lower than the conventional one (0.857 from Example 6.3). It is not always like this: sometimes the Bayesian LCL is higher and sometimes it is lower.

EXAMPLE 7.7.

Similarity of Bayesian binomial and exponential models: Interpret Example 7.6 as an attributes test (viz: four failures in approximately 50 tests). What is the 95 percent LCL on the reliability?

From Table A.1, for $\gamma=0.95$, $n=50$, $n-r=46$ results in $p_L = 0.829$, compared to 0.837 in Example 7.6. Since the n.l.g. and beta prior distributions are similar but not identical, it is to be expected that these two results would be close, but not the same.

Chapter 7 ■ *Bayesian Reliability Analysis* 141

Appendix to Chapter 7
The Bayesian and the Epstein–Sobel Models for Reliability Assessment of the Poisson Process: Interval Estimation

The purposes of this section are

1. To summarize the relevant portions of the discussions on the m.l.e. or classical Epstein–Sobel (ES) model of reliability assessment for the Poisson process covered in chapter 6 and the Bayesian version in this chapter for both censored and uncensored samples
2. To show the similarities and differences of these two alternative ways of interpreting the same set of data
3. To provide a simplified procedure with examples covering both methods for obtaining confidence limits for the reliability, the failure rate, and the MTBF or MTTF for a Poisson process, bypassing the technical discussion in both chapters 6 and 7 that explains the derivation of these methods
4. To demonstrate interval estimation of the reliability, failure rate, and MTTF or MTBF of a Poisson process

Similarities and differences of the ES and Bayesian models

Briefly, the ES and Bayesian techniques are similar in that they have the same types of mathematical formulas, functions, and tables (for example, the chi-square distribution), for calculating confidence levels. The differences between the results obtained by the two alternatives are potentially significant when the data set is relatively small, with few or no failures and/or small amounts of successful testing time T, because the classical usage can lead to indeterminate results with small data sets; however, when the data sets are large and there are significant numbers of failures, there is little difference in the resulting calculations. Because of the uniform initial prior distribution on the reliability in the Bayesian model, the formulas for assessment computations for the reliability, failure rate, and mean-time-between-failures MTBF (or else mean-time-to-failure MTTF) are modified by augmenting T by one mission time and also increasing the number of chi-square degrees of freedom by 2, for both complete and censored data sets.

Summary of assessment formulas

As noted in both chapters 6 and 7, the available data consist of T accumulated total successful testing time for a Poisson process, such as may be obtained from a parallel life test of a finite number n of identical items, r failures that occurred in the course of that testing, $r \leq n$, and a defined mission completion time t_o. For confidence level γ, let $\chi^2(2r+2, 1-\gamma)$ be the $(1-\gamma) \cdot 100$ percentage point of the chi-square distribution with $2r+2$ degrees of freedom. The Bayesian $\gamma \cdot 100$ percent UCL for the failure rate λ is

$$\lambda(\gamma) = \frac{\chi^2(2r+2, \gamma)}{2(T+t_o)}. \qquad (7.14)$$

By contrast, the formula for the classical m.l.e. $\gamma \cdot 100$ percent UCL is similar to Equation 6.14, but with different parameters.

$$\lambda(\gamma) = \frac{\chi^2(2r, \gamma)}{2T}$$

For either the Bayesian method described in this chapter or the m.l.e. method of chapter 6 the corresponding $\gamma \cdot 100$ percent LCL for the MTBF (or MTTF) is

$$\frac{1}{\lambda(\gamma)},$$

and the $\gamma \cdot 100$ percent LCL for the reliability is

$$\exp\{-\lambda(\gamma) \cdot t_o\}.$$

EXAMPLE 7.8.

For the data of Problem 5.7, six units on parallel life test with failures at 425, 1000, 1650, 2400, 3600, and 6500 hours, respectively, and a required mission time of 1000 hours, give both the m.l.e. and Bayesian 50 percent UCL for the failure rate and the 50 percent LCL's for both the MTTF and the reliability. The total successful testing time is 15,575 hours. The Bayesian 50 percent UCL for λ is

$$\frac{\chi^2(14, 0.50)}{2(15575+1000)} = 0.0004024 \text{ failures per hour.}$$

Chapter 7 ■ Bayesian Reliability Analysis

The corresponding Bayesian 50 percent LCL for the MTTF is

$$\frac{1}{(0.0004024)} = 2485 \text{ hours},$$

and the corresponding 50 percent LCL for the reliability is

$$\exp\{-(0.0004024)(1000)\} = 0.669.$$

The m.l.e. 50 percent UCL for λ is

$$\frac{\chi^2(12, 0.50)}{2(15575)} = 0.0003640 \text{ failures per hour};$$

the corresponding m.l.e. 50 percent LCL for the MTTF is

$$\frac{1}{(0.0003640)} = 2747 \text{ hours};$$

and the corresponding 50 percent LCL for the reliability is

$$\exp\{-(0.0003640)(1000)\} = 0.695.$$

In Example 7.8, a complete life test with all units tested to failure, the differences between the m.l.e. and the Bayesian confidence limits were small. Now consider some hypothetical cases where there are significant differences.

EXAMPLE 7.9.

No failures: Twenty units were life tested for 50 hours apiece, and there were no failures. If the required lifetime is 50 hours, what are the Bayesian 70 percent UCL for the failure rate, the Bayesian 70 percent LCL for the MTTF, the Bayesian 70 percent LCL for the reliability and the corresponding statistics for the m.l.e. model? The total successful test time in this case is $T = 1000$ hours. The Bayesian 70 percent UCL for the failure rate is

$$\frac{\chi^2(2, 0.70)}{2(1000 + 50)} = 0.001147 \text{ failures per hour};$$

the Bayesian 70 percent LCL for the MTTF is

$$\frac{1}{(0.001147)} = 872 \text{ hours};$$

and the Bayesian 70 percent LCL for the reliability is

$$\exp\{-(0.001147)(50)\} = 0.945.$$

In this case, the m.l.e. confidence limits for the failure rate, the MTTF, and the reliability do not exist because there is no chi-square distribution for zero degrees of freedom.

EXAMPLE 7.10.

One failure: In Example 7.9, suppose there had been one failure after 25 hours of testing. Then $T = 975$ hours, because 19 units are tested for 50 hours and the single failed unit for 25 hours. Do both the Bayesian and m.l.e. 70 percent confidence-limit calculations, as in Example 7.9. The Bayesian 70 percent UCL for the failure rate is

$$\frac{\chi^2(4, 0.70)}{2(975 + 50)} = 0.00238 \text{ failures per hour;}$$

the Bayesian 70 percent LCL for the MTTF is

$$\frac{1}{(0.00238)} = 420 \text{ hours;}$$

and the Bayesian 70 percent LCL for the reliability is

$$\exp\{-(0.00238)(50)\} = 0.888.$$

The m.l.e. 70 percent UCL for the failure rate is

$$\frac{\chi^2(2, 0.70)}{2(975)} = 0.001235 \text{ failures per hour;}$$

the m.l.e. 50 percent LCL for the MTTF is

$$\frac{1}{(0.001235)} = 806 \text{ hours;}$$

and the corresponding 50 percent LCL for the reliability is

$$\exp\{-(0.001235)(50)\} = 0.940.$$

Interval estimation

This section describes interval estimation of the reliability, the failure rate, and the MTTF for a Poisson process with the Bayesian model of the foregoing discussion. It is customary in interval estimation to compute central confidence intervals. A central $\gamma \cdot 100$ percent confidence interval is the range between the $\frac{(1-\gamma)}{2}$ and the $\frac{(1+\gamma)}{2}$ percentage points of the relevant distribution of the estimator or parameter. For example, the 90 percent confidence interval is the range between the 5th percentage point and the 95th percentage point.

EXAMPLE 7.11.

90 percent confidence intervals with censored data sets: For the data of Example 7.8, suppose that the test had been terminated at 3000 hours. Then the total amount of testing is

$$T = 425 + 1000 + 1650 + 2400 + 3000 + 3000 = 11475 \text{ hours}$$

and the total number of failures is $r = 4$. To find the 90 percent confidence intervals, similar to what was done in the foregoing, first find the UCL for λ, then the corresponding LCL for both the MTTF and the reliability. The order in which this is done, however, is immaterial. The 5th percentage point is

$$\lambda(0.05) = \frac{\chi^2(10, 0.05)}{2(11475 + 1000)} = 0.000158 \text{ failures per hour,}$$

and the 95th percentage point is

$$\lambda(0.95) = \frac{\chi^2(10, 0.95)}{2(11475 + 1000)} = 0.000734 \text{ failures per hour.}$$

The 95th percentage point for the MTTF is

$$\frac{1}{\lambda(0.05)} = 6329 \text{ hours,}$$

and the 5th percentage point is

$$\frac{1}{\lambda(0.95)} = 1362 \text{ hours.}$$

The 95th percentage point for the reliability is

$$\exp\{-(0.000158)(1000)\} = 0.854.$$

and the 5th percentage point is

$$\exp\{-(0.000734)(1000)\} = 0.480.$$

Problems

7.1. For the data of Problem 5.8 (twelve pumps failing at

830, 1173, 1304, 1737, 2304, 3047, 4018, 5189, 6912, 9029, 11,278, and 18,102

hours, respectively), the manufacturer's warranty guarantees 4000 hours with no failures.

  a. What are the calculated MTTF and failure rate?
  b. What is the classical (Epstein–Sobel) LCL for the reliability at the 90 percent confidence level based on a requirement of 4000 hours with no failures?
  c. What are the Bayesian UCL for the failure rate, the LCL for the MTTF, and the LCL for the reliability at the 90 percent confidence level, based on two assumptions: (1) the required lifetime of 4000 hours, and (2) an initial uniform prior distribution on the reliability?

7.2. For the data of Problem 6.4 (18 shutdowns with times between shutdown

22, 30, 35, 38, 45, 50, 75, 85, 115, 120, 145, 163, 180, 185, 205, 210, 260, and 327

minutes, respectively), what are the Bayesian LCLs for the reliability at the 50 percent and 90 percent confidence levels, if the standard time between system shutdowns is 150 minutes?

7.3. Prove that the mode of the n.l.g. prior distribution with p.d.f. Equation 7.11 is

$$\exp\left(-\frac{r'-1}{t'-1}\right).$$

7.4. Prove that the mean of the n.l.g. prior distribution is

$$\left(\frac{t'}{t'+1}\right)^{r'}.$$

7.5. Prove that the variance of the n.l.g. prior distribution is

$$\left(\frac{t'}{t'+2}\right)^{r'} - \left(\frac{t'}{t'+1}\right)^{2r'}.$$

7.6. Prove that the beta posterior p.d.f., Equation 7.7, obtained in binomial sampling with a fixed number of trials, n, is identical to the one obtained in negative binomial sampling where trials are conducted until there are a fixed number, r, of occurrences of the desired event.

7.7. Prove that the gamma posterior p.d.f., Equation 7.9, obtained in testing until a fixed number, r, of failures occurs is identical to the one obtained by testing for a fixed amount of time t.

7.8. For the data of Problem 7.1, assume that the data collection for each pump was terminated at 5000 hours, even if the pump was still operating at that time.
Note that the manufacturer's warranty guarantees 4000 hours.
   a. What is the total good operating time?
   b. How many failures were there?
   c. How many degrees of freedom for chi-square with the Bayesian model?
   d. With the Bayesian model, give the central 90 percent confidence intervals for λ, the MTTF, and the reliability.

References

Bonis, A. J. 1966. Bayesian reliability demonstration plan. In *Proceedings of the 1966 Annual Reliability and Maintainability Symposium,* 861–73.

———. 1975. Why Bayes is better. In *Proceedings of the 1975 Annual Reliability and Maintainability Symposium,* 340–48.

Box, G. E. P., and G. C. Tiao. 1973. *Bayesian inference in statistical analysis.* New York: Addison-Wesley.

Epstein, B., and M. Sobel. 1953. Life testing. *Journal of the American Statistical Association* 48:486–502.

Fisher, R. A. 1953. Chapter 1 in *The design of experiments.* 6th ed. New York: Hafner.

———. 1956. Chapter 1 in *Statistical methods and scientific inference.* New York: Hafner.

Grobowski, G., W. C. Hausman, and R. Lamberson. 1977. A Bayesian statistical inference approach to automotive reliability estimation. *Journal of Quality Technology* 8:197–208.

Lehmann, E. T. 1959. *Testing statistical hypotheses.* New York: Wiley.

Martz, H. F., and R. A. Waller. 1982. *Bayesian reliability analysis.* New York: Wiley.

Raiffa, H., and R. Schlaifer. 1961. *Applied statistical decision theory.* Cambridge, Mass.: Harvard University Press.

Schafer, R. E. 1973. The role of Bayesian methods in reliability. In *Proceedings of the 1973 Annual Reliability and Maintainability Symposium,* 290–91.

8

Normal Theory Tolerance Analysis

Introduction

This chapter is a continuation of chapter 4 on the normal distribution in reliability-confidence assessment, using test data for a characteristic of performance that is required to be within specified limits. The limits could be either upper limits only, lower limits only, or both upper and lower limits.

The typical scenario is that a set of data is observed, a random sample of measurements of the performance characteristic. The coverage or reliability, R, is the probability of an observation being inside the limits. The confidence level, γ, is a figure of merit in the range zero to one, associated with R; for a given set of data, the higher R is, the lower γ, and vice versa.

The objective of reliability assessment in this chapter is similar to that of the previous chapters, but there is a difference in the interpretation of results. In those chapters, the confidence level, γ, is viewed as a percentage point of a probability distribution; in this chapter it is a statistical function of the data, the limits and the value of R.

Tolerance limits and tolerance-limit factors

The operating limits for the equipment or components being tested are known as tolerance limits. Confidence assessment is performed with tolerance-limit factors, statistically derived multiples of the sample standard deviation, representing the allowable difference between the sample mean and the limit or limits for specified confidence levels and sample sizes. Tolerance limit factors for both one-sided and two-sided limits are given in Tables A.5 and A.6; these factors are inversely related to the sample size, and for very large samples, they are identical to the corresponding deviates of the standard normal distribution.

Appropriateness of normal theory

It is not always feasible to use normal theory in tolerance analysis, because the unreliability (estimate of the proportion defective) is sensitive to the normality assumption. For that reason, other forms of analysis that treat the problem of changing failure rates in a more flexible way, such as Weibull and nonparametric techniques, are frequently more attractive. D. B. Owen recommends that normal plans can be made more useful, even for nonnormal populations, if appropriate adjustments are made to account for the lack of normality; and he provides some examples of how to do so.

Historical note

Normal-theory tolerance analysis was originated by Shewhart and was an important aspect of the development of reliability and quality control. Some significant contributions were made by the World War II Statistical Research Group (SRG) at Columbia University, which developed both one- and two-sided tolerancing techniques, sequential analysis, and other statistical methods for analyzing data from the standpoint of quality assurance. The principal report of their work on tolerancing is included in sections of the book *Techniques of Statistical Analysis* by C. Eisenhart, M. Hastay, and W. A. Wallis; for the most part, the contents of this chapter are based on chapters 1 and 2 of that book, plus some refinements by D. B. Owen.

One-sided analysis

Upper specification limit

The normally distributed random variable $X : N(\mu, \sigma^2)$ (μ and σ^2 are the mean and variance, respectively, both unknown), denotes a performance characteristic; let U be the upper limit for X. The data consist of a random sample of n observations, $X_1, X_2, \ldots, X_n$. The sample mean is

$$\bar{X} = \sum_{i=1}^{n} \frac{X_i}{n}$$

and the unbiased estimate of σ^2 is the sample variance

$$s^2 = \sum_{i=1}^{n} \frac{(X_i - \bar{X})^2}{n-1}.$$

Since U is an upper limit, the reliability R or probability of successful performance is

$$R = P(X \le U) = \sigma^{-1}(2\pi)^{-1/2} \int_{-\infty}^{Y} \exp\left\{-\frac{(x-\mu)^2}{2\sigma^2}\right\} dx.$$

Because μ and σ are both unknown, R is unknown. The connection between R and U is derived statistically.

Let $Z : N(0, 1)$ denote the standard normal r.v. with mean zero and variance one, and let z_R be the value of Z such that

$$P(Z \le z_R) = (2\pi)^{-1/2} \int_{-\infty}^{z_R} \exp\left(-\frac{z^2}{2}\right) dz = R.$$

A value of X can be expressed as a linear function of the sample mean and standard deviation by including a constant multiplier or scale factor k into the definition

$$X = \bar{X} + ks.$$

A coverage-and-confidence assessment problem may be stated as: Given the observed data and R, what is the value of k such that the confidence is at least γ that

$$\frac{\bar{X} + ks - \mu}{\sigma} \ge z_R ? \tag{8.1}$$

Alternatively, given k, R, and the data, what is γ?

Relationship to noncentral-t tolerance-limit factors

After some algebraic manipulation, Equation 8.1 can be rewritten

$$\frac{\mu - \bar{X}}{\sigma} + z_R \ge \frac{ks}{\sigma}.$$

Multiplying through by the factor,

$$\frac{\sqrt{n}}{s/\sigma},$$

Equation 8.1 is rewritten again as

$$\text{confidence} \left(\frac{\frac{\sqrt{n}\,(\mu-\bar{X})}{\sigma} + \sqrt{n}\,z_R}{s/\sigma} \leq \sqrt{n}\,k \right) \geq \gamma. \qquad (8.2)$$

From chapter 4, note that

$$\frac{\sqrt{n}\,(\bar{X}-\mu)}{\sigma}$$

has the standard normal $N(0, 1)$ distribution, and that s/σ has the

$$\left(\frac{\chi^2}{n-1}\right)^{1/2}$$

(square root of chi-square divided by $n-1$) distribution with $n-1$ degrees of freedom. Therefore,

$$\frac{\frac{\sqrt{n}\,(\mu-\bar{X})}{\sigma} + \sqrt{n}\,z_R}{s/\sigma}$$

has the noncentral-t distribution with $n-1$ degrees of freedom and noncentrality parameter

$$\delta = \sqrt{n}\,z_R.$$

Corresponding to U as the upper limited for the r.v. X, the upper limit for the tolerance factor k is

$$\frac{U - \bar{X}}{s}. \qquad (8.3)$$

Let $t(n-1, \sqrt{n}\,z_R, \gamma)$ denote the γ percentage point of the noncentral-t distribution with noncentrality parameter $\sqrt{n}\,z_R$ and $n-1$ degrees of freedom. Then Equation 8.2 can be restated again

$$t(n-1, \sqrt{n}\,z_R, \gamma) \leq \sqrt{n}\left(\frac{U-\bar{X}}{s}\right). \qquad (8.4)$$

Equation 8.4 is in the appropriate form for a reliability-and-confidence statement that can be answered by consulting a table of the noncentral-t. Because the noncentral-t tables are bulky, a table with an abbreviated set of choices can be used instead, such as the table of one-sided tolerance limit factors in Table A.5. The tolerance-limit factor k is defined as

$$k(R, \gamma) = \frac{t(n-1, \sqrt{n}\, z_R, \gamma)}{\sqrt{n}}.$$

The one-sided tolerance-limit factors in Table A.5 are tabled for $\gamma = 0.9$ and 0.95, and for $R = 0.9, 0.95, 0.975, 0.99,$ and 0.999. If Equation 8.3 exceeds k, R reliability is assured with γ confidence level; otherwise the R reliability is obtained only at a lower level of confidence.

EXAMPLE 8.1.

One-sided analysis with an upper limit (based on MIL-STD 414): The maximum temperature of operation for a certain device is specified at 209° F. A sample with five units is obtained, with measurements of 197, 188, 184, 205, and 201° F, respectively. The sample mean is 195° F and the sample standard deviation is 8.81° F. What is the confidence that the unreliability is 1 percent?

Because the unreliability is 1 percent, the 0.99 probability column in Table A.5 applies. The computed tolerance factor is

$$\frac{209 - 195}{8.81} = 1.59.$$

This is less than the tolerance factor 4.668 in Table A.5 for $\gamma = 0.90$, $R = 0.990$, and $n = 5$; the confidence is less than 90 percent that the unreliability is no more than 1 percent. From the table of the noncentral-t, it was verified that the 1 percent unreliability requirement is satisfied only at about 10 percent confidence.

Lower specification limit

When a performance specification has a lower limit L (and no upper limit U), the procedure is the same as with a one-sided upper limit. The computed tolerance factor is

$$\frac{\overline{X} - L}{s},$$

which is compared to the tolerance limit factors in Table A.5.

EXAMPLE 8.2.

One-sided tolerance analysis with a lower limit L: It is required that the burst pressure of a tank shall be at least 420 pounds per square inch. A sample of six observations had burst pressures of 441, 452, 435, 455, 436, and 440 p.s.i., respectively. Assuming normality, what is the reliability

1. at the 90 percent confidence level?
2. at the 95 percent confidence level?

The mean burst pressure is 443.17 p.s.i., and the standard deviation is 7.637 p.s.i. The computed tolerance factor is

$$k = \frac{443.17 - 420}{7.637} = 3.02.$$

From Table A.5, for $n = 6$, $\gamma = 0.9$, and $R = 0.95$, $k = 3.093$; and for $\gamma = 0.95$ and $R = 0.90$, $k = 3.008$. Therefore the tank is 95 percent reliable at 90 percent confidence or 90 percent reliable at 95 percent confidence. A more favorable assessment is obtained by normal theory than by interpreting the test results as binary success-or-failure data. If the test results are treated as attributes, there would be six successes out of six trials; from Table A.1, the reliability is 0.72 at 90 percent confidence, or 0.652 at 95 percent confidence.

Two-sided tolerance

Tolerance-limit factors

Two-sided tolerance analysis is concerned with characteristics that are required to stay inside of both upper and lower specification limits. As with one-sided tolerance, two-sided normal-theory tolerance analysis employs tabled tolerance limit factors k that give the minimum number of standard deviations of difference between the sample mean and the limits, to guarantee R reliability or coverage at γ confidence.

The size of a two-sided k-factor depends not only upon γ, R, and the sample size, but also upon how the interval for exclusion is formed. The most widely used factors are those of Wald and Wolfowitz, which were tabled extensively by Bowker in *Techniques of Statistical Analysis*. In those tables, approximately $1-R$ probability is excluded from coverage, not necessarily split evenly between both tails of the distribution. The other type of exclusion is a more conservative one by D. B. Owen, in which at least $(1-R)/2$ probability is excluded from each tail separately. The Owen factors, therefore, are larger than the Wald–Wolfowitz factors.

Problems in developing two-sided tolerance-limit factors

It is a more difficult job to calculate two-sided tolerance-limit factors than one-sided factors, because the problem cannot be reduced to a probability statement about a distribution of known form. By contrast, a one-sided factor is derived from a noncentral-t percentage point.

The original Wald–Wolfowitz factors are approximations which are quite accurate. In 1968 Webb and Friedman produced exact two-sided limits using the same distributional assumptions as Wald and Wolfowitz, and found that the error in the approximation was insignificant for 10 degrees of freedom or more, and that the error was no more than 3 percent for as few as 2 degrees of freedom. A widely used table which is a subset of the Bowker tables is reproduced as Table A.6.

EXAMPLE 8.3.

(*See* Example 8.1). The minimum temperature for operation of a certain device is 180° F, and the maximum temperature is 209° F. A sample of five devices is tested, with measurements of 197, 188, 184, 205, and 201° F, respectively. As in Example 8.1, the mean temperature is 195° F and the standard deviation is 8.81° F. Is the reliability at the 75 percent confidence level at least 90 percent?

In Table A.6, for $N = 5$, $\gamma = 0.75$, and $R = 0.9$, the tolerance limit factor is 2.599. In order to have 90 percent reliability at a 75 percent confidence level, $195 \pm (2.599)(8.81)$ would have to be inside both the upper and lower limits of 209 and 180. Since this is not the case, the answer is negative. With a tolerance-limit factor of 1.825, at 75 percent confidence, R is slightly less than 75 percent.

Hald's approximate two-sided tolerance-limit factors

A useful formula for approximate two-sided factors for $n \geq 5$ was given by Hald. Let $z_{(1+R)/2}$ denote the standard normal deviate with $(1-R)/2$ probability excluded from each tail, and $\chi^2_{n-1}(1-\gamma)$ the $(1-\gamma)$th percentage point of the chi-square distribution with $n-1$ degrees of freedom. The tolerance factor is

$$k = z_{(1+R)/2} \left(\frac{n-1}{\chi^2_{n-1}(1-\gamma)} \right)^{1/2} \left(1 + \frac{1}{2n} \right).$$

Problems

8.1. A torsion spring is required to have a torque in the range of 60 ± 10 grams per centimeter. Since overnight production runs are made with less supervision than daytime production runs, and quality control inspection is not made on overnight production until the following morning, the problem of assuring that the overnight runs are within tolerance is a significant one. On a production run of 20 items, the following torque measurements were obtained:

54, 56, 56, 57, 57, 58, 58, 59, 60, 61, 61, 61, 62, 62, 63, 63, 63, 63, 65, 67.

a. Provide a normal probability graph for these data, and also perform a Lilliefors g.o.f. test for a fit to an unspecified normal population. What are the mean and standard deviation of these values? Do these results indicate a good fit?
b. For the results of 8.1a, what is the computed two-sided tolerance factor?
c. For the data of 8.1a and 8.1b, what is the confidence level for a coverage (reliability or probability of success) of 0.90? for a coverage of 0.95?
d. Provide a lognormal probability graph for these data, and also perform of g.o.f. test for a fit to an unspecified lognormal population. What are the mean and standard deviation of the logarithms of the values? Do these results indicate a good lognormal fit?
e. For the data of 8.1d, what is the computed two-sided tolerance factor?
f. For the data of 8.1d and 8.1e, what is the confidence level for a coverage of 0.90? and for a coverage of 0.95?

8.2. The thicknesses of 10 mica washers were sampled at random from a normally distributed population with unknown mean and unknown variance. The thicknesses obtained, are respectively, in inches,

0.123, 0.124, 0.126, 0.129, 0.120, 0.132, 0.123, 0.126, 0.129, 0.128

   a. What is the sample mean?
   b. What is the sample standard deviation?
   c. For a one-sided upper specification limit U of 0.136 inches, what is the computed tolerance factor $(U-\bar{X})/s$?
   d. What is the highest confidence level (γ) at which we can guarantee 95 percent reliability or population coverage below the specification limit?
   e. If the specifications are changed to include not only an upper limit of 0.136 inches, but also a lower limit of 0.116 inches, what is the relevant tolerance-limit factor in Table A.5 for 95 percent reliability or coverage with 75 percent confidence?
   f. For the two-sided case in 8.2 e, does the data show 95 percent reliability with 75 percent confidence?

8.3. An adult volume ventilator (AVV) facilitates breathing for many types of surgical and critically ill patients. Because a ventilator is a critical item of health care, log books of operation of AVV ventilators are maintained by the State of California Department of Health. Assume that there is a required industry standard of 1 year (4380 hours round-the-clock for 365 days) of continued operation of an AVV without failure. The log books of eight units of the same type show failures at the following times, in hours,

8972, 11,079, 12,044, 19,880, 19,918, 25,683, 27,255, and 28,957.

   a. Graph these eight failure times on normal probability paper. Does the graphical fit appear to be good?
   b. What are the calculated mean and standard deviation?
   c. Perform a Lilliefors K–S goodness-of-fit test to an unspecified normal distribution; what is the calculated value of the D-statistic? Does this result in acceptance of the null hypothesis of a good fit at the 10 percent level?
   d. Let the industry-standard 4380 hours of continuous operation without failure be the lower tolerance limit L. What are the computed tolerance factor based on L and your answers to 8.3 b? Is there 90 percent coverage inside the tolerance limits at 90 percent confidence?

8.4. Two-sided tolerance: The diameters of castings produced by a numerically controlled lathe are required to be within the tolerance range of 0.5000 ± 0.002 inches. Ten castings yield the following measurements.

0.5006, 0.5009, 0.5008, 0.5005, 0.5006, 0.5006, 0.5000, 0.5003, 0.5001, and 5.0004

a. What are the calculated mean and standard deviation of the deviations from the 0.5000 inches requirement?
b. Show both graphically and by a Lilliefors goodness-of-fit test that there is no evidence to disprove that these data fit a normal distribution.
c. From Table A.6, what is the minimum number of (computed) standard deviations above the (computed) mean for the upper tolerance limit 0.502 to have 95 percent coverage within the tolerance limits with 95 percent confidence? How many standard deviations above the mean is 0.502? Does this show 95 percent coverage with 95 percent confidence above the mean?
d. What can you say about coverage below the mean?

8.5. A power supply is required to have an output voltage of 500 ± 12 volts. In a test of 25 new power supplies, the mean output voltage was 502.5 volts and the standard deviation of output was 4.1 volts; also a normal probability graph and a statistical test of goodness-of-fit to a normal distribution showed that there was sufficient reason to believe that the data fit a normal distribution.

a. What is the tolerance limit factor (from Table A.6) for 95 percent coverage (that is, reliability) at 75 percent confidence?
b. Do the data show 95 percent reliability at 75 percent confidence?
c. Do the data show 95 percent reliability at 90 percent confidence?
d. Do the data show 95 percent reliability at 95 percent confidence?

References

Dixon, W. J., and F. J. Massey, Jr. 1957. *Introduction to statistical analysis*. New York: McGraw-Hill.

Eisenhart, C., M. Hastay, and W. A. Wallis, eds. 1947. *Techniques of statistical analysis*. Especially chapters 1 and 2. New York: McGraw-Hill.

Hahn, G. J. 1970. Statistical intervals for a normal population. *Journal of Quality Technology* 2:126–49, 195–206.

Hald, A. 1952. *Statistical theory with engineering applications.* New York: Wiley.

Locks, M. O., M. J. Alexander, and B. J. Byars. 1963. *New tables of the noncentral-t distribution.* U.S. Office of Aerospace Research, ARL 63-19.

MIL-STD 414. 1957. *Sampling procedures and tables for inspection by variables for percent defective.* U.S. Department of Defense.

Owen, D. B. 1958. Tables of factors for one-sided tolerance limits for a normal distribution. Albuquerque, N. Mex.: Sandia Corporation Monograph SCR-13; Washington, D.C.: Office of Technical Services, U.S. Department of Commerce.

———. 1964. Control of percentages in both tails of the normal distribution. *Technometrics* 6:377–87.

———. 1966. Errata. *Technometrics* 8:570.

———. 1967. Variables sampling plans based on the normal distribution. *Technometrics* 9:417–23.

———. 1969. Summary of recent work on variables acceptance sampling with emphasis on nonnormality. *Technometrics* 11:631–37.

Owen, D. B., and W. H. Frawley. 1971. Factors for tolerance limits which control both tails of the normal distribution. *Journal of Quality Technology* 3:69–79.

Resnikoff, G. J., and G. J. Lieberman. 1957. *Tables of the noncentral-t distribution.* Stanford, Calif.: Stanford University Press.

Shewhart, W. A. 1939. *Statistical method from the viewpoint of quality control.* Washington, D.C.: Graduate School Press, U.S. Department of Agriculture.

Wald, A., and J. Wolfowitz. 1946. Tolerance limits for a normal distribution. *Annals of Mathematical Statistics* 17:208–15.

Webb, S. A., and H. Friedman. 1968. Exact two-sided tolerance limits under normal theory. Rocketdyne Division of Rockwell International Corporation NAR 54416. NASA Tech Brief.

9

Hazard Analysis and the Weibull Distribution

Introduction

Life testing, hazard, mortality tables, and the Weibull distribution

The term *hazard*, as it is used in quality assurance and reliability, is related to mortality in vital statistics and insurance. Hazard is a generalization of the fact that the failure rate of objects or the death rate of people varies over a lifetime. Hazard analysis provides a way of measuring a changing failure rate.

The first recorded attempt to use mortality data for hazard analysis was made 300 years ago by Halley, widely known as the discoverer of a famous comet. Halley's technique, which would be called a nonparametric method in modern statistical terminology, is still in use by actuaries who use vital statistics to construct mortality tables for setting insurance rates. Reliability engineers who have available either life-test or field-equipment mortality data are usually concerned with warranties for products with finite useful lives. Because the sample sizes are small, it is convenient to employ mathematical functions derived from time-to-failure probability distributions. The Weibull distribution is preferred for this purpose, because there is a convenient mathematical form for making inferences and predictions about the variable failure rates and length of life, based on the values of the parameters, and for plotting life-test data.

This chapter has four major parts. The first part is an overview of hazard analysis, including both nonparametric and simple parametric approaches based on the exponential distribution. The second part is on the Weibull distribution, including the mathematical form of the probability functions, and probability plotting, for both complete samples and censored data. The third part deals with Weibull hazard functions and hazard plotting. Finally, the fourth part covers the use of linear weighting techniques for estimating the parameters of the Weibull distribution and for reliability-and-confidence assessment.

Applications of the Weibull distribution

The Weibull distribution is related to the exponential, but with two additional parameters, a location parameter and a shape parameter. Instead of a single constant failure rate λ, as for the exponential case, a wide variety of types of hazards can be expressed mathematically. For example, the failure rate can be continually increasing, continually decreasing, or constant. Major shifts in the pattern of changing hazards at different stages of a lifetime are accommodated by plotting or estimating a different Weibull distribution for each time period; an example of this is the so named *bathtub curve*, with a decreasing failure rate for the first period, a constant failure rate for the second period, and increasing failure rate for the third period. A further convenience is the relative ease of probability plotting to estimate parameters, detect outliers, and perform crude goodness-of-fit analysis.

Historical note

The distribution is named after the Swedish scientist Waloddi Weibull, who published a paper in 1939 giving some of its uses for the analysis of the breaking strength of solids. It was also known prior to that by R. A. Fisher and L. H. C. Tippett, who published a paper in 1928 presenting it as the asymptotic form of the distribution of the smallest values for a sample of size n. The family of Weibull distributions is also a subfamily within the family of extreme-value distributions discussed extensively by E. J. Gumbel.

Hazard

Constant failure rate: The exponential distribution

Hazard measurement is based upon the failure intensity, the ratio of the concentration of failures to the probability of survival. A simple example of hazard is provided by the exponential distribution. With a constant failure rate λ, the concentration of failures at any given time t is the exponential p.d.f.

$$f(t) = \lambda \exp(-\lambda t),$$

and the probability of survival up to t is the reliability

$$R(t) = \exp(-\lambda t).$$

There are two types of hazard functions, instantaneous and cumulative. The instantaneous hazard $h(t)$ is the failure intensity ratio

$$\frac{f(t)}{R(t)} = \lambda,$$

identical to the constant failure rate; and the cumulative hazard up to and including t, $H(t)$, is

$$H(t) = \int_0^t h(x)\, dx = \lambda t.$$

If the cohort of data cases, such as n identical items all put on life test at the same time, is large enough, the cumulative hazard is proportional to t, with λ as the factor of proportionality.

Discrete hazard analysis: Halley's method

We shall illustrate a discrete version of hazard analysis, with an example that is similar in principle to Halley's method.

EXAMPLE 9.1.

Discrete hazard analysis: Ninety identical items were put onto an overstress life test at the same time; four items failed in the first hour, 21 items in the second hour, 30 in the third hour, 25 in the fourth hour, 8 in the fifth hour, and the 2 remaining items failed in the sixth and final hour.

A hazard analysis for this case must take account of the fact that failure observations are made once every hour; thus the frequency of failures $f(t)$ is the proportion of the original 90 items that failed during the tth hour, while the reliability $R(t)$ is the proportion remaining at the end of the $(t-1)$st hour. Table 9.1 supplies the hazard calculations for Example 9.1; the $h(t)$ percent column shows that the hazard rate is constantly increasing.

Table 9.1. Hazard calculations by Halley's method.

t	Failed	$R(t)\%$	$f(t)\%$	$h(t)\%$	$H(t)\%$
1	4	100.0	4.4	4.4	4.4
2	21	95.6	23.3	24.4	28.8
3	30	72.2	33.3	46.1	74.9
4	25	38.9	27.8	71.5	146.4
5	8	11.1	8.9	80.2	226.6
6	2	2.2	2.2	100.0	326.6

Probability functions and probability plotting

The exponential distribution

The exponential reliability formula is

$$R(t) = \exp(-\lambda t). \tag{9.1}$$

Equation 9.1 is linearized by taking the logarithms of both sides

$$-\ln R(t) = \lambda t. \tag{9.2}$$

Based upon Equation 9.2, the exponential reliability, or its complement, the distribution function, is graphed as a straight line on semilogarithmic paper, with the intercept at the origin and the slope of λ.

Elements of Weibull probability plotting

Modify the right-hand sides of both Equations 9.1 and 9.2 as follows. Subtract the quantity μ from t, then define the scale factor $\delta = 1/\lambda$. The right-hand side of Eq. 9.2 is now

$$\frac{t-\mu}{\delta}. \tag{9.3}$$

Note that t can be no less than μ, because the numerator must be nonnegative. Then raise Equation 9.3 to the power β. Instead of Equation 9.1, this gives

$$R(t) = \exp\left\{-\left(\frac{t-\mu}{\delta}\right)^{\beta}\right\}. \tag{9.4}$$

Equation 9.4 is the reliability function for the Weibull distribution with scale parameter δ, location parameter μ, and shape parameter β. Taking the natural logarithms of both sides of Equation 9.4 yields

$$-\ell n\, R(t) = \left(\frac{t-\mu}{\delta}\right)^{\beta}. \tag{9.5}$$

By taking the logarithms of both sides of Equation 9.5, we have

$$\ell n(-\ell n\, R(t)) = \beta\, \ell n\left(\frac{t-\mu}{\delta}\right)$$
$$= \beta\, \ell n(t-\mu) - \beta\, \ell n\delta. \tag{9.6}$$

On log-log vs. log graph paper, Equation 9.6 graphs as a straight line. The plot is for the line

$$y = \beta x + c,$$

with a slope of β and with the variables y and x and the constant c, respectively defined as follows:

$$y = \ell n(-\ell n\, R(t))$$
$$x = \ell n(t-\mu)$$
$$c = -\beta\, \ell n\delta.$$

The values of two of the Weibull parameters, β and δ, can be read from a straight line of the form of Equation 9.6 fitted to Weibull log-log vs. log graph paper, such as Figure 9.1. The location parameter μ, the minimum time to failure, must be supplied by trial and error, guessing, or some external source of information. A useful first approximation could be the value of the location parameter μ^* (*See* Equation 6.13) of the two-parameter exponential distribution.

Weibull probability density function

The Weibull probability density function may be derived from Equation 9.4. Since the reliability R is the complement of the distribution function F,

$$F(t) = 1 - R(t),$$

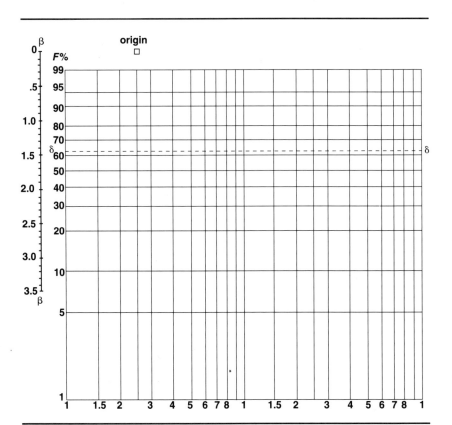

Figure 9.1. Weibull probability graph paper.

Chapter 9 ■ Hazard Analysis and the Weibull Distribution

and the p.d.f. is the derivative of $F(t)$

$$f(t) = -\frac{d}{dt} R(t) = \frac{\beta(t-\mu)^{\beta-1}}{\delta^\beta} \exp\left\{-\left(\frac{t-\mu}{\delta}\right)^\beta\right\}.$$

Two-parameter Weibull distribution

Since the location parameter μ is subtracted from every t, the analysis is simplified by redefining $t = t - \mu$. The resulting two-parameter p.d.f. is

$$f(t) = \frac{\beta\, t^{\beta-1}}{\delta^\beta} \exp\left\{-\left(\frac{t}{\delta}\right)^\beta\right\}, \tag{9.7}$$

and the reliability function is

$$R(t) = \exp\left\{-\left(\frac{t}{\delta}\right)^\beta\right\}. \tag{9.8}$$

If $\beta = 1$, Equation 9.7 simplifies further to

$$\frac{1}{\delta} \exp\left\{-\left(\frac{t}{\delta}\right)\right\},$$

and Equation 9.8 becomes

$$\exp\left(-\frac{t}{\delta}\right).$$

Thus if $\beta = 1$, the scale parameter is the MTTF, the reciprocal of the failure rate λ. This shows that the exponential distribution can be viewed as a special case of the Weibull with $\beta = 1$.

Weibull probability graph paper

Equation 9.6 explains why a Weibull distribution graphs as a straight line on log-log vs. log graph paper. Figure 9.1 is Weibull probability graph paper that was prepared especially for use with this text; reproduction is permitted. The vertical scale is accumulated probability, and the horizontal scale is time to failure. Life data are plotted using the plotting points of Table A.3 in the same way this is done for normal distributions in chapter 4 and exponential distributions in chapter 5.

Two parameters are obtained from a life-test data plot, β and δ. First, plot the points and then the straight line of apparent best fit. To obtain the value of β, the slope of the fitted line, draw a line from the origin

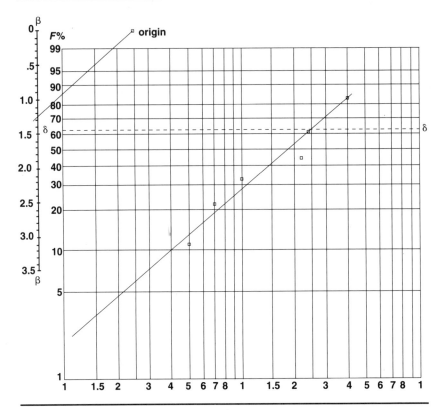

Figure 9.2. Probability plot of Example 9.2.

at the top of the page which is parallel to the fitted line. The estimate of β is at the point where the line from the origin crosses the β-scale at the left side of the graph. The estimate of δ is the horizontal value where the fitted line crosses the dotted horizontal line at the 63d percentile.

EXAMPLE 9.2.

Fitting a Weibull probability graph: In Example 6.5, eight identical items are all life tested to failure. The failure times in hours are

45, 47, 50, 62, 64, 64, 80, and 80.

This problem graphs well as a two-parameter exponential fit (see Figure 6.2) with an assumed location parameter of 40 hours and a failure rate of 0.0471 failures per hour, corresponding to an MTTF of 21.2 hours (after the first 40). Using 40 hours as the estimate of μ, fit a Weibull distribution to these data graphically and compare the results to the two-parameter exponential plot of Figure 6.2.

The adjusted failure times after subtracting 40 from each observation are 5, 7, 10, 22, 24, 24, 40, and 40. The plotting positions from Table A.3 are, respectively, 11.1, 22.2, 33.3, 44.4, 61.2, 61.2, 83.4, and 83.4 (the last four positions are averaged because of ties). The straight line that averages these plots has the shape parameter β of approximately 1.22 and the scale parameter δ of 25, corresponding to 65 hours after adding back the assumed location parameter of 40. With a two-parameter exponential location parameter of 40 as in Figure 6.5, the shape parameter would be 1 (why?), and the scale parameter would be 21.2.

This problem is sensitive to the assumed value of the location parameter; for example, with $\mu = 0$, β is more than 4.0—with $\mu = 43$, it is approximately 0.9. More stable results would be obtained with a larger data set.

EXAMPLE 9.3.

Ten identical items are put onto a parallel life test, and are all tested to failure. The failure times in hours are, respectively,

37, 58, 72, 88, 115, 136, 152, 165, 185, and 213

and are plotted in Figure 9.3. Estimate the Weibull parameters graphically. What is the reliability if the item should operate in the field for at least 25 hours without a failure?

Assuming that $\mu = 0$, the estimated parameters are $\beta = 1.6$ and $\delta = 140$. The estimated reliability is

$$R = \exp\left\{-\left(\frac{25}{140}\right)^{1.6}\right\} = 0.938.$$

Hazard functions and hazard plotting

An alternative to Weibull probability plotting described in the preceding section is Weibull cumulative hazard plotting. Conventional wisdom has it that hazard plotting is preferable to probability plotting for two types of situations: first, for fitting field-life data that is *multiply censored*, meaning that the data is an ordered mixture of times to failure and times to removal from service because of aging; second, for mixing failure data of different types or due to different causes, such as dif-

170   Chapter 9 ■ *Hazard Analysis and the Weibull Distribution*

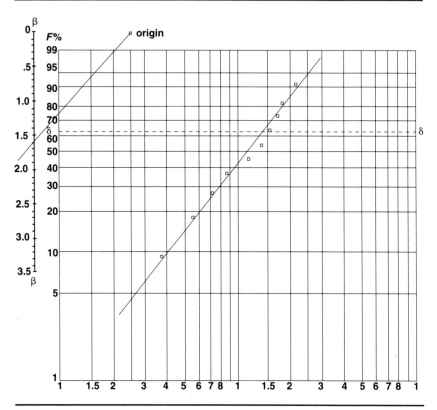

Figure 9.3. Probability plot of Example 9.3.

ferent failure modes or failures of different types of related equipment in a common data set. Data of both the first and second types are often obtained for the power generating and transmitting equipment of public utilities.

A feature of hazard plotting is that the mathematical structure of a Weibull cumulative hazard plot is simple and graphs as a straight line on log-vs.-log graph paper without using standardized plotting positions, such as those in Table A.3.

Weibull hazard and cumulative hazard

The instantaneous hazard function $h(t)$ is the ratio of the p.d.f. $f(t)$ to the reliability $R(t)$. For a two-parameter Weibull, dividing Equation 9.7 by Equation 9.8,

Chapter 9 ■ Hazard Analysis and the Weibull Distribution 171

$$h(t) = \frac{f(t)}{R(t)} = \frac{\beta\, t^{\beta-1}}{\delta^\beta}. \tag{9.9}$$

The cumulative hazard $H(t)$ is the integral of Equation 9.9,

$$H(t) = \int_0^t \frac{\beta\, x^{\beta-1}}{\delta^\beta}\, dx = \left(\frac{t}{\delta}\right)^\beta. \tag{9.10}$$

Taking the logarithms of both sides of Equation 9.10,

$$\log H(t) = \beta \log t - \beta \log \delta, \tag{9.11}$$

a linear function in $\log H(t)$ vs. $\log t$. Weibull hazard paper, such as Figure 9.4 and other commercially available hazard graph paper, have $\log t$ on the vertical scale and $\log H(t)$ on the horizontal scale. By algebraically rearranging Equation 9.11, it becomes

$$\log t = \left(\frac{1}{\beta}\right) \log H + \log \delta, \tag{9.11a}$$

with a slope of $\left(\frac{1}{\beta}\right)$.

Plotting and fitting Weibull hazard paper

In a data set consisting of n ordered times to removal from service, m items failed and the remaining $n-m$ items were removed because of aging. The ordering of the n items is based solely on the time of removal, regardless of the cause. Let $j = 1, \ldots, n$, respectively, be the ranks of the n ordered times; the corresponding reverse ranks are $n-j+1$. For the m failure times only, let j_i, $i = 1, \ldots, m$, be the rank of the ith failure and calculate the inverted reverse rank as an estimate of the instantaneous hazard for the ith failure

$$h_i = \left(\frac{1}{n-j_i+1}\right), \quad i = 1, \ldots, m.$$

This follows from the definition of $h(t)$ as the ratio of the probability density of a failure at t to the probability of survival up to t. The cumulative hazard is

$$H_i = \sum_{k=1}^{i} h_k.$$

In Figure 9.4, the ith point has coordinates H_i percent on the logarithmic horizontal scale and t_i on the logarithmic vertical scale. After

172   Chapter 9 ■ *Hazard Analysis and the Weibull Distribution*

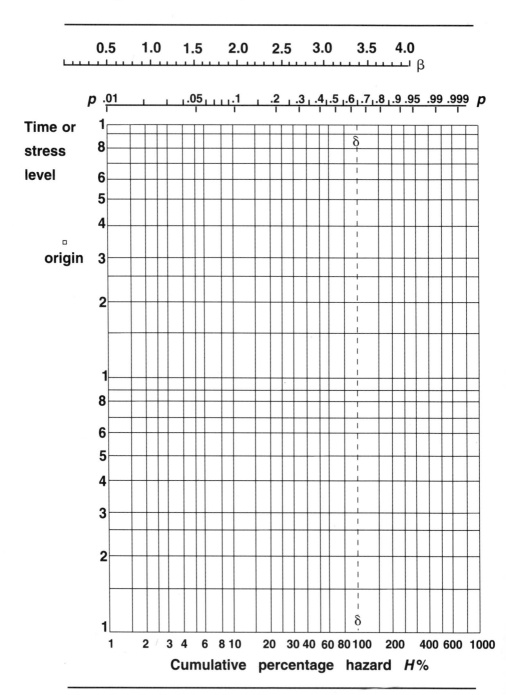

Figure 9.4. Weibull hazard graph paper.

drawing the line of visual best fit, draw a parallel line through the origin. The estimate of β is the point at which the line through the origin crosses the β-scale at the top of the page. The estimate of δ is the value of t where the fitted line crosses the vertical dotted line corresponding to 100 percent on the H scale.

EXAMPLE 9.4.

Hazard analysis of a complete data set: Fit a Weibull cumulative hazard plot to the data of Example 9.2 and estimate β and δ. Discuss differences, if any, between the parameter estimates obtained from the probability plot of Example 9.2 in Figure 9.3 and the hazard plot.

The hazard calculations are given in the following table and the hazard plot is graphed in Figure 9.5. The estimates of the parameters are about the same as for the probability plot: 1.20 for the shape parameter compared to 1.22; and the scale parameter is 23 compared to 25.

i	$t_i - 40$	$h_i\%$	$H_i\%$
1	5	12.5	12.5
2	7	14.3	26.8
3	10	16.7	43.5
4	22	20.0	63.5
5	24	25.0	88.5
6	24	33.3	121.8
7	40	50.0	171.8
8	40	100.0	271.8

EXAMPLE 9.5.

Fit a Weibull hazard plot to the data of Example 9.3: For the 10 data points in this problem, the corresponding cumulative hazard percentages are, respectively,

10.00, 21.11, 33.61, 47.90, 64.56, 84.56, 109.56, 142.90, 192.90, and 292.90.

The hazard plot in Figure 9.6 has a shape parameter of 1.7 compared to 1.6 for the probability plot, and a scale parameter of 140 hours, the same as that for the probability plot. With 25 hours as the required operating time, the calculated reliability is

$$R = \exp\left\{-\left(\frac{25}{140}\right)^{1.7}\right\} = 0.948.$$

174 Chapter 9 ■ *Hazard Analysis and the Weibull Distribution*

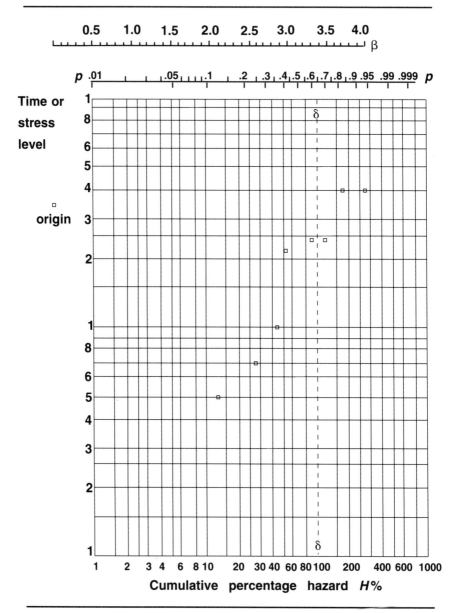

Figure 9.5. Weibull hazard plot of Example 9.4.

Chapter 9 ■ *Hazard Analysis and the Weibull Distribution* 175

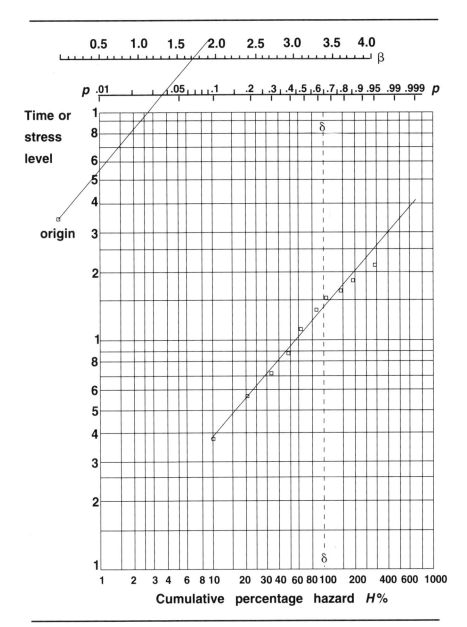

Figure 9.6. Weibull hazard plot of Example 9.5.

Precise Weibull estimation and confidence assessment

Purpose of this section

Because it requires judgment, plotting on special graph paper is a crude way to estimate the parameters of a probability distribution, because different answers are obtained depending both on the choice of graph paper and person to do the estimating. Precision is important, however, to enhance the results and to increase the acceptability of the analysis and conclusions by the clients or colleagues of the analyst. Therefore, techniques based on statistical criteria that are independent of the judgment of any individual and that would provide the same answers for any given data set should be used.

Mann best linear invariant estimation

One of the ways to perform precise estimation is *linearizing*; this means that the estimates are linear combinations of functions of the observed data points. In chapter 6, for example, formulas are presented for best linear unbiased estimates (BLUE) of the parameters of the two-parameter exponential. A similar technique is available for the two-parameter Weibull distribution, called best linear invariant (BLI) estimates, created by Nancy Mann.

With BLI, estimates of functions of both the shape and scale parameters are linear combinations of the natural logarithms of the observed failure times; that is, the natural log of the ith failure time, ℓn t_i, is multiplied by a specified weight, and the linear combinations are sums of the products of the weights by the ℓn t_i's. This technique can be used for either complete or censored sets of life data, but not for multiple censoring, for samples of up to $n = 25$ observations and for any censoring number m from 2 to n. Because a different set of weights is needed for each different (m,n) pair, the table of weights is bulky and takes up many pages in Table A.7. The bulkiness of the overall table, however, does not cause a problem in analyzing a data set because each different set of weights requires just a few lines.

Linear estimation of a location and scale parameter

The reduced form

Let us suppose there is a random variable $X(u, b)$, with the location parameter u and the scale parameter b. X has a reduced form

$$Z = \frac{X-u}{b}, \tag{9.12}$$

which is in the standardized form of an r.v. discussed in chapter 2, and also is related to the standard normal $Z: N(0, 1)$ distribution covered in chapter 4.

Order statistics

The observed data consist of n observations of the r.v. $X(u, b)$. The ordered observed data points or order statistics,

$$X_1 \leq X_2 \leq \ldots \leq X_n$$

constitute a set of sufficient statistics for the distribution of X. This means that this ordered set of observations contains all of the information available from the data as to the form of the distribution. Another way of saying this is that with good technique, reasonably good estimates of the parameters u and b can be calculated from functions of the order statistics.

Linear weighted estimates of u *and* b

The estimates of u and b are weighted sums of the natural logarithms of the order statistics, with a different set of weights for each parameter, sample size n, and censoring number m. For every (n, m) pair, the weights for the location parameter u sum to 1 and the weights for the scale parameter b sum to 0.

A similar example of linear estimation is the normal distribution. For a random sample from a normal population $X: N(\mu, \sigma^2)$, the sample mean

$$\bar{X} = \frac{X_1 + X_2 + \ldots + X_n}{n}$$

is the best estimate of the location parameter μ, the population mean. This computes a weighted sum of the n order statistics, assigning a weight of $1/n$ to each one, so that the sum of the n equal weights is unity. Likewise, a good estimate of the scale parameter σ, the population standard deviation, can be obtained by a weighted average of the order statistics with the sum of the weights being zero; numerically equal, but opposite in sign weights are assigned to the order statistics (X_1 has the same weight as X_n but opposite in sign, X_2 has the same weight as X_{n-1}, and so on). This is shown in appendix Table A.8b(3)

178 Chapter 9 ■ *Hazard Analysis and the Weibull Distribution*

of the popular statistics text by Dixon and Massey, where linear estimates of the standard deviation with efficiencies of 0.97 or better for small samples are obtained using a weighting scheme of this type with the order statistics.

The extreme-value distribution

In order to derive a reduced distribution of the form of Equation 9.12 from the Weibull, with both a location parameter u and a scale parameter b, a change of variables and a redefinition of parameters is necessary. Let $x = \ln t$, $b = 1/\beta$, and $u = \ln \delta$. Then Equation 9.8 is changed to

$$R(x) = \exp\left\{-\exp\left(\frac{x-u}{b}\right)\right\}, \qquad (9.13)$$

with the d.f. $F(x) = 1 - R(x)$. The resulting distribution, known as the smallest extreme value or Fisher–Tippett type I distribution (also sometimes called the log-Weibull distribution), depends upon the parameters u and b. The r.v. inside the parentheses of Equation 9.13 has the reduced form of Equation 9.12. For every sample size n up to 25 and for every censoring number m, $2 \leq m \leq n$, the BLI method calculates linearized estimates of both u and b, where these estimates are linear combinations of the $\ln t_i$'s, $i = 1, \ldots, m$. The set of weights for every (n, m) pair are in Table A.7. The estimates of the Weibull parameters are obtained by reversing the parameter changes:

$$\delta = \exp u \text{ and } \beta = 1/b.$$

A technique is also provided for computing estimates for sample sizes greater than 25.

EXAMPLE 9.6.

Linearized estimates of Weibull parameters for a complete data set: What are the BLI estimates of β and δ for the 10 ordered failure times in Example 9.3?

In Table 9.2, there are three columns of data: t_i, the ordered times to failure; a_i, the weights for computing $u = \ln \delta$ for $m = n = 10$; and c_i, the weights for computing $b = 1/\beta$ for $m = n = 10$. The columns of weights for a_i and c_i are from Table A.7.

Chapter 9 ■ Hazard Analysis and the Weibull Distribution

Table 9.2. Ordered failure times and weights for Example 9.6.

i	t_i	a_i	c_i
1	37	0.027331	−0.072734
2	58	0.040034	−0.077971
3	72	0.052496	−0.077242
4	88	0.065408	−0.071876
5	115	0.079263	−0.061652
6	136	0.094638	−0.045420
7	152	0.112414	−0.020698
8	165	0.134239	0.017927
9	185	0.164178	0.085070
10	213	0.230001	0.324597

The BLI estimates of Example 9.6 are

$$\hat{u} = \sum_{i=1}^{10} a_i \ln t_i = 4.96; \quad \hat{\delta} = \exp(\hat{u}) = 142.590;$$

$$\hat{b} = \sum_{i=1}^{10} c_i \ln t_i = 0.425; \quad \hat{\beta} = 1/\hat{b} = 2.353.$$

EXAMPLE 9.7.

Weighted estimates of Weibull parameters for censored data sets: What are the estimates of the Weibull parameters if the data in Example 9.6 are censored at the fifth failure?

In Table 9.3 the columns for a_i and c_i are taken from Table A.7 for the sample size $n = 10$ and censoring number $m = 5$. The estimates for $n = 10$ censored at $m = 5$ are

$$\hat{u} = \sum_{i=1}^{5} a_i \ln t_i = 4.962; \quad \hat{\delta} = \exp(\hat{u}) = 142.870;$$

$$\hat{b} = \sum_{i=1}^{5} c_i \ln t_i = 0.443; \quad \hat{\beta} = 1/\hat{b} = 2.256.$$

The estimates of δ and β in Examples 9.6 and 9.7 differ from those obtained by probability plotting in Examples 9.3, 9.4, and 9.5, but are remarkably close to one another. Since nearly the same estimates were obtained with or without censoring, this illustrates the superiority of precise computed methods over graphical approximations for estimating parameters of distributions. It also shows that there could be substantial

Table 9.3. Ordered failure times and weights for Example 9.7.

i	t_i	a_i	c_i
1	37	−0.115524	−0.185169
2	58	−0.090868	−0.181821
3	72	−0.051341	−0.160697
4	88	0.000925	−0.125311
5	115	1.256809	0.652997

economies in censoring data, because it is not necessary to life-test all n items to the end of lifetime in order to obtain statistically justifiable estimates of the parameters.

Confidence bounds for the reliable lifetime and the reliability

Mann, Fertig, and Scheuer (MFS) developed a method and a set of tables for obtaining confidence bounds for the reliability, in conjunction with the Mann BLI point estimates, for a wide range of sample sizes and censoring numbers. Given the estimates $\hat{u}$ and $\hat{b}$ of the location and scale parameters for a sample of size n with censoring number m, the MFS method obtains $(1-\gamma)$ percentage points of the distribution of t_R, the reliable lifetime associated with the reliability or probability of survival R; from these percentage points, bounds are obtained for the LCL R_γ on R with confidence γ.

For the values of u and b and for any specified R, the reliable lifetime is

$$t_R = \exp(u) \, [\ell n \, (1/R)]^b.$$

The natural logarithm of t_R is

$$\ell n \, t_R = u + b \, \ell n \, [\ell n \, (1/R)].$$

The MFS tables provide percentage points of the standardized statistic

$$V_R = \frac{\hat{u} - \ell n(t_R)}{\hat{b}}, \quad (9.14)$$

that were obtained by Monte Carlo simulation; these are tabulated in Table A.8 at probabilities 0.02, 0.05, 0.10, 0.25, 0.40, 0.50, 0.60, 0.75, 0.90, 0.95, and 0.99, for all sample sizes n, $3 \le n \le 25$, for three dif-

Chapter 9 ▪ Hazard Analysis and the Weibull Distribution 181

ferent values of R, 0.90, 0.95, and 0.99, and for all Type-II (upper) censorings of size m, $3 \leq m \leq n$.

EXAMPLE 9.8.

For the data in Example 9.3, an uncensored sample of $n = 10$, and the values of u and b calculated in Example 9.6, given that the required operating time is 25 hours, what are the confidence levels γ, or bounds for the confidence levels, for reliabilities R of 0.90, 0.95, and 0.99?

From Equation 9.14, for any R

$$V_R = \frac{4.959 - \ln 25}{0.425} = 4.09;$$

this is compared to the closest entries in the rows for $m = n = 10$ in Table A.8 for $V_{.90}$, $V_{.95}$, and $V_{.99}$.

For $R = 0.90$: $0.90 \leq \gamma \leq 0.95$;
For $R = 0.95$: $\gamma \approx 0.75$;
For $R = 0.99$: $0.10 \leq \gamma \leq 0.25$.

EXAMPLE 9.9.

For the same data and calculated values of u and b as Example 9.7, a sample of $n = 10$, censored at $m = 5$, and, as in Example 9.8, a required operating time of 25 hours, what are the confidence levels γ, or bounds for the confidence levels, for reliabilities R of 0.90, 0.95, and 0.99?

From Equation 9.14,

$$V_R = \frac{4.962 - \ln 25}{0.443} = 3.93;$$

this is compared to the closest entries in the rows for $n = 10$ and $m = 5$ in Table A.8 for $V_{.90}$, $V_{.95}$, and $V_{.99}$.

For $R = 0.90$: $0.75 \leq \gamma \leq 0.90$;
For $R = 0.95$: $0.50 \leq \gamma \leq 0.60$;
For $R = 0.99$: $0.10 \leq \gamma \leq 0.25$.

The fact that the confidence levels for specified values of R are lower in Example 9.9 than in Example 9.8 seems to be a reflection of the fact that censoring reduces the amount of available data and, therefore, also the confidence for specified values of the reliability.

Problems

9.1. Modify the terminal failure data for a local area network (LAN) of Problems 6.4 and 7.2, so as to account for the five end-of-day shutdowns out of 18 that were not failures,

22, 30, 35, 38, (45), 50, 75, 85, 115, 120, (145), 163, (180), 185, (205), 210, 260, and (327)

minutes, respectively. The five parenthesized times—45, 145, 180, 205, and 327 minutes—all represent end-of-day shutdowns and not failures; the other 13 observations are all shutdowns due to failure.

  a. Prepare a table of hazard calculations using the procedure from the discussion of Example 9.4, giving $h(i)$ percent based on inverted reverse ranks and cumulative $H(i)$ percent for each observation.
  b. Graph the cumulative hazard data on Weibull hazard graph paper.
  c. What are the estimated values of β and δ?
  d. Does the hazard plot show a constant, increasing, or decreasing failure rate?
  e. Does this modify any conclusions you might have drawn from the analysis of Problem 6.2?

9.2. A cardiac surgery unit at a medical center has performed 100 prosthetic aortic valve replacements during a specified year. We shall define this group of 100 patients as a *cohort*. The following table gives the annual failure and survival information for this cohort for the ensuing 11 years. Failure means either that another replacement was performed or that the patient died; *survive* means that a patient survived that many full years without requiring a valve replacement.

year t	survive	$f(t)$	$F(t)$	$R(t)$	$h(t)\%$	$H(t)\%$
1	90					
2	83					
3	78					
4	73					
5	67					
6	63					
7	58					
8	52					
9	46					
10	39					
11	32					

Chapter 9 ▪ Hazard Analysis and the Weibull Distribution 183

a. Fill in the remainder of this table for calculating both the instantaneous hazard $h(t)$ percent and the cumulative hazard $H(t)$ percent. The columns $f(t)$, $F(t)$, $R(t)$, $h(t)$, and $H(t)$ have the usual meaning: $f(t)$ is the proportion of the cohort of 100 aortic valves that failed during year t; $F(t)$ is the cumulative proportion of failures up to and including t; and $R(t) = 1 - F(t-1)$ is the proportion surviving to the end of the previous period. The hazard formulas are

$$h(t) = \frac{f(t)}{R(t)} \times 100\%, \text{ and } H(t) = \sum_{i=1}^{t} h(i).$$

b. What pattern of a failure rate curve does this show?
c. What is the characteristic lifetime for this surgery?

9.3. A family of 40-foot utility poles have had the following lifetimes (that is, years of service) (F denotes failure; those that did not fail were usually removed from service because of the rerouting of lines).

years (t)	$h(t)\%$	$H(t)\%$
6		
(F) 6		
8		
9		
(F) 10		
(F) 12		
12		
13		
(F) 16		
(F) 17		
(F) 19		
20		
25		
(F) 28		
31		

a. Calculate the instantaneous hazard values $h(t)$ percent and the cumulative hazard values $H(t)$ percent and put these numbers into the spaces provided for that purpose.
b. Graph the results on Weibull hazard graph paper.

c. What are the estimated values of
The Weibull shape parameter;
The characteristic lifetime?

9.4. Twenty-five units of the ML-XXX transistor were life tested for 5000 hours at a temperature of 200° C. There were seven failures at times $t = $ 370, 1167, 1167, 1798, 2429, 2429, and 3715 hours, respectively. Provide a Weibull probability graph of the failure-history data.

   a. From the graph, what are the Weibull shape parameter and the scale parameter or characteristic life?
   b. Calculate the values of the parameters of the Weibull and extreme-value lifetime distributions at 200° C by the Mann B.L.I. method. Determine the following values.
      i. The extreme-value location parameter u;
      ii. The extreme-value scale parameter b;
      iii. The Weibull shape parameter;
      iv. The characteristic lifetime.
   c. If specifications call for the transistor to have a lifetime of 500 hours at 200° C, what is the value of the V-statistic used for evaluating the Mann–Fertig–Scheuer confidence bounds on the reliability?
   d. What are the confidence levels or confidence bounds for
      i. $R = .90$;
      ii. $R = .95$;
      iii. $R = .99$?

9.5. A certain type of gyroscope has a planned operational lifetime of 25 hours with the motor spinning at synchronous speed at the correct voltage output. In an automatic laboratory test, the output of each gyroscope on test is connected to a voltmeter. When the output goes to zero volts, the time of failure is recorded. In an automatic laboratory test of nine gyroscopes from the same supplier, the following failure times, in hours, were recorded.

        13, 41, 50, 71, 87, 169, 189, 226, 329

   a. Graph these failure times on a Weibull probability plot, Figure 9.1, and provide the estimated values of the shape parameter β and the scale parameter δ.
   b. Based on the complete sample of nine gyros on test, using the Mann BLI method, what are the calculated values of

Chapter 9 ■ *Hazard Analysis and the Weibull Distribution* 185

   i. The extreme-value location parameter $\hat{u}$;
   ii. The extreme-value scale parameter $\hat{b}$;
   iii. The Weibull scale parameter $\hat{\delta}$;
   iv. The Weibull shape parameter $\hat{\beta}$?
c. Suppose the test had been terminated after the sixth failure at 169 hours, instead of being carried out to the very last failure at 329 hours. What would have been the calculated values of the BLI estimates for
   i. The extreme-value location parameter $\hat{u}$;
   ii. The extreme-value scale parameter $\hat{b}$;
   iii. The Weibull scale parameter $\hat{\delta}$;
   iv. The Weibull shape parameter $\hat{\beta}$?
d. For the complete-sample calculated values of the parameters for Problem 9.5b, with a planned lifetime of 25 hours, what is the value of the statistic V_R used for confidence-bound calculations? Provide the confidence bounds for the reliability values
   i. $R = 0.90$;
   ii. $R = 0.95$;
   iii. $R = 0.99$?
e. For the truncated-sample calculated values of the parameters for Problem 9.5c, with a planned lifetime of 25 hours, what is the value of the statistic V_R for confidence-bound calculations? Provide the confidence bounds
   i. $R = 0.90$;
   ii. $R = 0.95$;
   iii. $R = 0.99$?

9.6.

a. For the data of Problem 6.1, draw a Weibull probability graph and provide the graphical estimates of β and δ.
b. Provide estimates of the following parameters of the Weibull distribution of lifetimes by the Mann BLI method.
   i. The extreme-value location parameter $\hat{u}$;
   ii. The extreme-value scale parameter $\hat{b}$;
   iii. The Weibull shape parameter $\hat{\beta}$;
   iv. The characteristic lifetime.

c. Suppose that the product being tested has a warranty of successful operation for 3 months. What is the value of the V_R statistic? What are the confidence levels or confidence bounds for
   i. $R = 0.90$;
   ii. $R = 0.95$;
   iii. $R = 0.99$?
d. Assume that the life test had been censored after the 4th failure, at 19 months. By the Mann BLI method, what are
   i. The extreme-value location parameter $\hat{u}$;
   ii. The extreme-value scale parameter $\hat{b}$;
   iii. The Weibull shape parameter $\hat{\beta}$;
   iv. The characteristic lifetime?
What is the value of the V_R statistic?
What are the confidence levels or confidence bounds for
   v. $R = 0.90$;
   vi. $R = 0.95$;
   vii. $R = 0.99$?

9.7. For the same data used in Problems 5.8 and 6.3 (12 identical hydraulic pumps tested to failure), the estimated value of the location parameter μ^* for the two-parameter exponential distribution by the Sarhan–Greenberg BLUE method is 414 hours. Use 414 hours as an estimate of the value of the location parameter μ for the Weibull distribution. This means that no failures are assumed to occur before 414 hours, but that the time for purposes of estimating parameters or for confidence assessment is not the actual time t, but $t - 414$; for example, the first time to failure is $830 - 414 = 416$. This also implies that the reliable lifetime for purposes of confidence assessment is not t_R, but $t_R - 414$.

a. What are the estimated values of the extreme-value parameters u and b after subtracting 414 from each observation?
b. What is the characteristic lifetime?
c. For a reliable lifetime t_R of 500 hours, what is V_R? This means that the lifetime for this purpose is not 500, but $500 - 414 = 86$.
   What are the confidence bounds for
   i. $R = 0.90$;
   ii. $R = 0.95$;
   iii. $R = 0.99$?

Chapter 9 ▪ *Hazard Analysis and the Weibull Distribution* 187

9.8. Failure mode hazard analysis: Twenty-four production units of a personal computer system were life tested for 24 hours per day for different periods of time, up to two months. All subsystems were exercised, including random access memory, power supply, hard disk, disk drive, keyboard, and monitor. Five failures occurred during testing; in each case, the failing subsystem was replaced or repaired, and the testing of that unit continued until it was terminated. Thus, these data can be treated as if there were 29 different lifetimes. For these five failures, three are due to failure of the hard disk.

a. Perform a Weibull hazard analysis of the lifetime data for the complete system, based on the data in the following table, including the instantaneous hazard values $h(t)$ percent, the cumulative hazard values $H(t)$ percent, graphing on Weibull hazard paper, and estimating β and δ. What is the estimated probability that a system will last six months with no failures?

b. Perform a Weibull hazard analysis similar to Problem 9.8a for the three disk failures (column labeled *h.d.*).

c. Discuss any possible discrepancies between the results of Problems 9.8a and 9.8b.

i	$t(i)$	h.d.	Overall failures, including hard disk			Hard disk failures	
			Failed	$h(i)\%$	$H(i)\%$	$h(i)\%$	$H(i)\%$
1	96		96				
2	297						
3	336	y	336				
4	336	y	336				
5	480	y	480				
6	504						
7	504						
8	528						
9	648						
10	672		672				
11	744						
12	744						
28	1320						
29	1320						

All data are in hours at removal from service

References

Bain, L. J. 1978. *Statistical analysis of reliability and life-testing models.* New York: Marcel Dekker.
Crowder, M. J. 1991. *Statistical analysis of reliability data.* New York: Chapman Hall.
Dreyfogle, F. W., III. 1992. *Statistical methods for testing, development, and manufacture.* New York: Wiley.
Fertig, K. W., and N. R. Mann. 1980. Life-test sampling plans for two-parameter Weibull populations. *Technometrics* 22:165–77.
Finklestein, J. M. 1976. Confidence bounds on the parameters of the Weibull process. *Technometrics* 18:115–17.
Fisher, R. A., and L. H. C. Tippett. [1928] 1950. Limiting forms of the frequency distribution of the smallest member of a sample. Reprinted in R. A. Fisher. *Contributions to mathematical statistics.* New York: Wiley.
Gumbel, E. J. 1968. *Statistics of extremes.* New York: Columbia University Press.
Ireson, W. G., and C. F. Coombs. 1988. *Handbook of reliability engineering and management.* New York: McGraw-Hill.
Johns, M. V., Jr., and G. J. Lieberman. 1966. An exact asymptotically efficient confidence bound for reliability in the case of the Weibull distribution. *Technometrics* 8:135–76.
Kao, J. H. K. 1958. Computer methods for estimating Weibull parameters in reliability studies. *Institute of Radio Engineers Transactions on Reliability and Quality Control* 13:15–22.
Lawless, J. F. 1982. *Statistical models and methods for lifetime data.* New York: Wiley.
Leiblein, E. J. 1954. A new method for analyzing extreme data. *National Advisory Committee for Aeronautics Technical Note 3053.*
Leiblein, E. J., and M. Zelen. 1956. Statistical investigation of the fatigue life of deep groove ball bearings. *Journal of Research of the National Bureau of Standards* 57:273–316.
Locks, M. O. 1971. Analysis of stress corrosion testing of 2014-T651 aluminum, using Weibull techniques. *Corrosion* 27:386–89.
Mann, N. R. 1967. *Results on location and scale parameter estimation with application to the extreme-value distribution.* Dayton, Ohio: Office of Aerospace Research, U.S. Air Force, ARL 67-0023.
―――. 1967. Tables for obtaining the best linear invariant estimates of parameters of the Weibull distribution. *Technometrics* 9:629–46.

———. 1968. Point and interval estimation procedures for the two-parameter Weibull and extreme-value distributions. *Technometrics* 10:231–56.
Mann, N. R., K. W. Fertig, and E. M. Scheuer. 1971. *Confidence and tolerance bounds and a new goodness-of-fit test for two-parameter Weibull or extreme-value distributions.* Dayton, Ohio: Aerospace Research Laboratories, U.S. Air Force ARL 71-0077.
Mann, N. R., R. E. Schafer, and N. D. Singpurwalla. 1974. *Methods for statistical analysis of reliability and life data.* New York: Wiley.
Martz, H. F., and R. A. Waller. 1982. *Bayesian reliability analysis.* New York: Wiley.
Misra, K. B. 1992. *Reliability analysis and prediction.* Amsterdam and New York: Elsevier.
Nelson, W. 1969. Hazard plotting for incomplete failure data. *Journal of Quality Technology* 1:27–52.
———. 1970. Hazard plotting methods for analysis of life data with different failure modes. *Journal of Quality Technology* 2:126–49.
———. 1982. *Applied life data analysis.* New York: Wiley.
———. 1990. *Accelerated testing.* New York: Wiley.
Nelson, W., and V. C. Thompson. 1971. Weibull probability paper. *Journal of Quality Technology* 3:45–50.
O'Connor, P. D. T. 1991. *Practical reliability engineering.* 3d ed. New York: Wiley.
Office of Assistant Secretary of Defense. 1961–62. *Sampling procedures and tables for life and reliability testing based on the Weibull distribution.* TR-3 and TR-4. Washington, D.C.: Office of Assistant Secretary of Defense.
Rigdon, S. E., and A. P. Basu. 1989. The power law process: A model for the reliability of repairable systems. *Journal of Quality Technology* 21:251–60.
Technical and Engineering Aids for Management (TEAM). Box 25, Tamworth, NH 03886.
Thoman, D. R., L. J. Bain, and C. E. Antle. 1970. Reliability and tolerance limits in the Weibull distribution. *Technometrics* 12:363–71.
Villemeur, A. 1992. *Reliability, availability, maintainability, and safety assessment.* 2 vols. New York: Wiley.
Weibull, W. A. 1939. A statistical theory of the strength of material. *Ing Vetenskaps Akad Handl* 151:15 ff.

Part IV

Maintainability and Availability

Part IV of this book consists of a single chapter, chapter 10, that deals with measuring the maintainability or availability of a repairable system that is not subject to reliability growth or improvement, using some of the statistical techniques developed in the earlier chapters for time-to-failure and/or time-to-repair distributions. Assessing the maintainability (that is, maintainability-confidence assessment) with time-to-repair data, as it is defined herein, is analogous to assessing the reliability for a nonrepairable system with time-to-failure data. Assessing the availability, however, requires Monte Carlo simulation. Other topics covered in this chapter include the two types of availability, inherent and observed, and the results of several simulation programs prepared especially for this text.

10

Maintainability and Availability

Introduction

The foregoing chapters are all concerned with a common problem, confidence levels for the reliability of a nonrepairable component or system, based upon the failure histories of components or systems of that type. This chapter deals with repairable systems that can be restored to service after a failure; it covers both the maintainability M, the probability of restoration within a specified downtime, and the availability A, the probability of being in service during a scheduled operating period.

Since assessing M or A is a form of applied probability analysis that is analogous to assessing the reliability, the contents of this chapter, for the most part, are adapted from the discussions of time-dependent phenomena in earlier chapters, including Bayesian methods for a Poisson process, and Weibull and lognormal analysis. Monte Carlo simulation, which is used extensively in availability computations, is also included.

Relationship to queueing

Maintainability and availability analysis is similar to queueing. Measuring M or A may be compared to the problem of determining the occupancy, arrival, and service rates at a queue if the service performed is repair, the servers are the maintenance crew or repairers, and the customers or patrons are the machines, components, or systems that are repaired after the occurrence of random events called *failures*. A simple form of queueing model that describes all of the cases discussed in this chapter would be a single repairer and a single system; the repairer services the system when a failure occurs, and the system is in operating condition as soon as he or she is finished. The *operating time* is the period between completion of the repair and the occurrence of a subsequent failure. The *repair time* is the period between the failure and completion of the repair.

Maintainability

The maintainability $M(t)$ is the probability that a repair is completed in at most t time; thus t is the Mth percentage point of the time-to-repair (TTR) or unscheduled downtime distributions. This definition is complementary to that of the reliability $R(t)$, the probability that the system operates without failure for at least t time, so that t is the $(1-R)$th percentage point of the time to failure. Because M and R are similar in that both of them relate to the occurrence of a single type of event over time, the same families of techniques can be used for assessment.

The simplest type of parametric M analysis is performed with exponential TTR. Taking data such as downtimes, one can assess M for a specified confidence level γ, with the Bayesian version in chapter 7 or the Epstein–Sobel method described in chapter 6.

Availability

The availability A is the ratio of the actual operating time to the scheduled operating time excluding preventive maintenance or scheduled servicing. Since it is a kind of a probability of being in service when required, it has the same meaning as the reliability for a nonrepairable system. The difference, however, is that whereas R is a measure of system quality for only one event, failure, A requires two events to occur, failure and repair.

Assessing confidence levels for A is more complicated than it is for R or for M, because an extra probability distribution is involved, as well as another set of data, and closed-form formulas are not obtainable, even in the simplest cases when both events are exponential. Therefore, it is necessary to employ simulation.

Inherent and observed availability

Two types of availability are covered, inherent and observed (or actual) availability. Inherent availability A_I depends only on the mean TBF and mean TTR; actual availability A_O is a random variable. The difference between A_I and A_O can be illustrated in terms of an availability cycle.

Availability cycle

An availability cycle has two consecutive time periods: (1) operation, terminating with a failure; and (2) downtime, ending with a completed repair. The availability is the ratio of the length of the first period to the length of the cycle. For each cycle, A_O is the measured or observed value of that ratio. By contrast A_I is the ratio of the average length of the first period to the average length of the cycle. Thus, the mean of A_O, the mean of a distribution of ratios, is different from A_I, the ratio of the means of two different distributions.

Assessing A_I

The value of A_I, the ratio of the mean length of the operating period to that of the cycle, can easily be established for any known or estimated TTR and TBF distributions. Assessing A_I, however, in the sense of establishing confidence limits for different values of A_I, can be done only after first using observed data to establish distributions on the TTR and TBF parameters, similar to the Bayesian treatment of time-to-failure distributions in chapter 7. If both the TTR and the TBF are exponential, both the mean TBF and the mean TTR can be sampled from Bayesian prior distributions that are functions of the prior data. Beyond this simple case, assessing A_I is a difficult problem that is not considered in this text.

Assessing A_O

The observed availability A_O may be assessed either nonparametrically, using historical data, or parametrically, by simulation. A nonparametric technique involves using data from a series of availability cycles, each cycle supplying both an actual TBF and an actual TTR. In each cycle, the ratio of the TBF to the TTR is the r.v. A_O.

In a parametric approach, the parameters of the distribution governing both the TBF and the TTR are estimated either from historical data or from prior knowledge, and are taken as given. Monte Carlo simulation samples TTRs and TBFs from these distributions in order to furnish observed values of A_O.

Simulating A_O is not as restrictive as A_I, because it is possible to sample failure and repair times from any distribution, as long as the choice of distribution is justified by data, experience, or a thorough goodness-of-fit analysis. The results of simulations of this type are less sensitive to the amount of data available than to the values of the estimates of the parameters of the distributions. In other words, the choice of parameters is a more important consideration than the size of the sample, since valid techniques can usually be found for making inferences from small samples.

Three examples of ways to assess A_O are covered in this text: nonparametric, exponential TTR with exponential TBF, and lognormal TTR with Weibull TBF. The nonparametric approach is not a simulation, but an analysis based only upon the percentage points of an empirical probability distribution of values of A_O obtained from a historical sequence of availability cycles. The exponential TBF with exponential TTR case is a simulation, with the values of the parameters taken from Bayesian prior distributions, and the times to the events are then sampled from these distributions. The lognormal TTR with Weibull TBF analysis is a simulation derived from sampling of failure and repair times from empirically determined lognormal and Weibull TBF distributions.

Maintainability

Exponential TTR: Point estimates and confidence assessment

When the time to repair is exponential, the maintainability is assessed in a way that is similar to assessing the reliability with exponential time to failure. The fundamental parameter of the process, however, is the repair rate μ, the reciprocal of the mean TTR, instead of the failure rate λ, the reciprocal of the mean TBF. Also, the definition of success is reversed in the sense that it is desirable to have μ to be as high as possible, so that the repair is completed quickly, whereas λ should be as small as possible, so that the time between failures is long.

Let t denote the time to completion of the repair. If the repair rate is μ, the maintainability is

$$M(t) = 1 - \exp(-\mu t).$$

If t is specified as a particular or standard repair time t_O,

$$M(t_O) = 1 - \exp\{-\mu t_O\},$$

and an estimate of μ is all that is necessary to assess M.

Suppose that there are r repairs, $i = 1, \ldots, r$, with repair times $t_1, t_2, \ldots, t_r$, respectively. The total downtime is

$$T = \sum_{i=1}^{r} t_i.$$

By analogy to Equation 6.8, the maximum-likelihood point estimates are

$$\hat{\mu} = \frac{r}{T} \text{ and } \hat{M} = 1 - \exp(-\hat{\mu} t_O).$$

By analogy to chapter 7, the $(100\,\gamma)$th percentage lower confidence limit for μ is

$$\mu_\gamma = \frac{\chi^2(2r+2, 1-\gamma)}{2(T+t_O)},$$

where $\chi^2(2r+2, 1-\gamma)$ is the $(1-\gamma)$th percentage point of the χ^2 distribution with $2r+2$ degrees of freedom; and the corresponding LCL for M is

$$M_\gamma = 1 - \exp(-\mu_\gamma t_O).$$

EXAMPLE 10.1.

Exponential maintainability: For a mainframe computer installation, the maintenance logbook shows that over a period of a month there were 15 unscheduled maintenance actions or downtimes, with 1200 minutes in emergency maintenance status. The maintenance contract calls for a partial rebate if an unscheduled downtime exceeds 100 minutes; thus, 100 minutes is effectively the standard time t_O. Assuming that the TTR is distributed exponentially, what are the m.l.e.'s $\hat{\mu}$ and $\hat{M}$ and the LCLs for μ and M at the 10 percent, 50 percent, and 90 percent confidence levels?

The m.l.e. $\hat{\mu}$ of the repair rate is $15/1200 = 0.0125$ repairs per minute of downtime and the corresponding m.l.e. of the maintainability is

$$\hat{M} = 1 - \exp\{-(0.0125)(100)\} = 0.713.$$

In order to compute the LCLs, the chi-square values for 32 degrees of freedom at the 90, 50, and 10 percentage points are, respectively, 46.194, 31.336, and 22.271. The factor $2(T+t_O)$ is $2(1200+100) = 2600$. At the 10 percent confidence level

$$\hat{\mu} = \frac{46.194}{2600} = 0.0178, \text{ and}$$

$$\hat{M} = 1 - \exp\{-(0.0178)(100)\} = 0.831;$$

at the 50 percent confidence level

$$\hat{\mu} = \frac{31.336}{2600} = 0.0121, \text{ and}$$

$$\hat{M} = 1 - \exp\{-(0.0121)(100)\} = 0.700,$$

practically the same as the m.l.e.; and at the 90 percent confidence level

$$\hat{\mu} = \frac{22.271}{2600} = 0.0086$$

$$\hat{M} = 1 - \exp\{-(0.0086)(100)\} = 0.577.$$

This example shows that there is 80 percent confidence that the probability of successfully completing an emergency maintenance action in 100 minutes or less is between 0.577 and 0.831.

Lognormal TTR

There is some empirical evidence that the time required to perform service tasks of the same type but of different levels of difficulty is lognormally distributed. Because the normal distribution involves two parameters, Bayesian assessment of the maintainability is more difficult than for the exponential case, which has only one parameter, the repair rate μ. Lognormal-theory maintainability assessment, however, can employ methods analogous to normal-theory tolerancing, such as that described in chapter 8.

Suppose the TTR is lognormally distributed with unknown parameters: μ, the mean of the logarithms of the repair times; and σ^2, the variance of the logarithms of the repair times. Suppose that U is a specified standard or maximum time for a repair; U would have to be a rather liberal upper limit, because a lognormal distribution of the times to repair would be highly skewed.

The observed or sample data consist of n TTRs $t_1, t_2, \ldots, t_n$. The sample mean of the log TTRs is

$$\bar{X} = \sum_{i=1}^{n} \frac{\log t_i}{n},$$

and the sample variance is

$$s^2 = \sum_{i=1}^{n} \frac{(\log t_i - \bar{X})^2}{n-1}.$$

For each of the two choices of the confidence level, 90 and 95 percent, M can be approximated somewhere in the range between 0.90 and 0.999 by consulting Table A.5 for one-sided tolerance factors.

EXAMPLE 10.2.

Lognormal maintainability assessment: A maintenance logbook for a continuously operating numerically controlled milling machine for airplane parts shows both the operating time between failures and the time to repair. Table 10.1 is a record of 20 consecutive failure-repair, or availability, cycles, excluding preventive maintenance periods; for each cycle the table shows both the TTF, in hours, and the TTR, in tenths of an hour.

Only the column labelled TTR is needed for Example 10.2. The TTF column is also used in a subsequent example to illustrate availability. The base-10 logarithms of the 20 TTR data points are probability plotted on normal probability paper in Figure 10.1; the data show that there is a reasonably good fit to the normal distribution of the logarithms (that is, a lognormal distribution). The calculated mean of the logarithms is 0.160 and the standard deviation of the logarithms is 0.473. A Lilliefors K–S test also shows a reasonably good fit; the calculated value of the

Table 10.1. Availability data for Example 10.2.

Cycle	TTF	TTR	Cycle	TTF	TTR	Cycle	TTF	TTR
1	125	1.0	2	44	1.0	3	27	9.8
4	53	1.0	5	8	1.2	6	46	0.2
7	5	3.0	8	20	0.3	9	15	3.1
10	12	1.5	11	58	1.0	12	53	0.8
13	36	0.5	14	25	1.7	15	106	3.6
16	200	6.0	17	159	1.5	18	4	2.5
19	79	0.3	20	27	9.8			

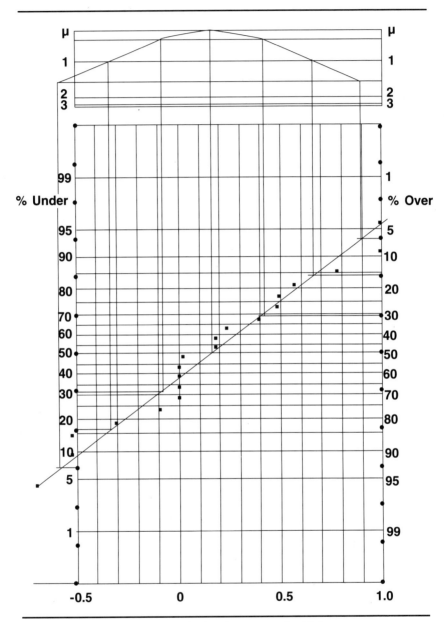

Figure 10.1. TTR lognormal plot for Example 10.2.

D-statistic is 0.118, compared to 0.174 for a sample of size 20 at the 10 percent level of significance, from Figure 4.6.

Based upon a standard time limit for the repair of 10 hours, what is the maintainability M at the 90 percent and 95 percent confidence levels? The computed tolerance factor is

$$k = \frac{\log 10 - 0.160}{0.473} = 1.776.$$

From Table A.5, for $\gamma = 0.9$, M is between 0.9 and 0.95; for $\gamma = 0.95$, $M \leq 0.9$. This means that M is at least 0.9 at the 90 percent confidence level, but not at the 95 percent level.

Inherent availability A_I: Exponential TTR and TBF

This section describes a computerized procedure for obtaining confidence bounds on the inherent availability when both the failure rate λ and repair rate μ are exponential. The p.d.f. at failure time x is

$$f(x) = \lambda \exp(-\lambda x). \tag{10.1}$$

The p.d.f. for completion of the repair at a subsequent time t, the end of a cycle, is

$$f(t-x) = \mu \exp\{-\mu (t-x)\}. \tag{10.2}$$

The p.d.f. for a failure at x followed by a repair completed at t is the convolution or accumulated product of Equations 10.1 and 10.2.

$$f(t) = \int_0^t f(x) f(t-x) \, dx$$

$$= \frac{\lambda \mu}{\mu - \lambda} (\exp(-\lambda t) - \exp(-\mu t)).$$

The average length of an availability cycle is the mean of t

$$E(t) = \int_0^\infty t f(t) \, dt$$

$$= \frac{1}{\lambda} + \frac{1}{\mu} = \text{mean TBF} + \text{mean TTR}.$$

Therefore the average fraction of an availability cycle during which the system is available is

$$A_I = \frac{\text{mean TBF}}{\text{mean TBF} + \text{mean TTR}}. \tag{10.3}$$

Exponential TTR and exponential TBF: The prior distributions

If both the TTR and the TBF are exponential, A_I depends upon the values of both λ and μ. In chapter 7, it was shown that λ for an exponential TTF is governed by a gamma prior distribution and that the parameters of this prior distribution are additive sufficient statistics for the prior data—total test time and total failures. Likewise, if the TBF for a repairable system is exponential, the gamma prior distribution for λ is governed by additive sufficient statistics—total operating time and total failures. By the same reasoning, if the TTR is exponential, a gamma prior distribution on μ is governed by the additive sufficient statistics—total repair time and total repairs.

Assume that the data for λ consist of r_1 failures and T_1 operating time and a required or specified operating time between failures t_1; and that the data for μ are r_2 repair actions in T_2 total repair time, with a specified repair time of t_2. T_1 and t_1 and T_2 and t_2 are all measured in the same time units (seconds, hours, missions, and so on). Assuming that the number of failures is equal to the number of repairs, let $r_1 = r_2 = r$. The gamma prior distribution on λ has a p.d.f. similar to Equation 7.8

$$f(\lambda) = \frac{(T_1+t_1)^{r+1} \lambda^r \exp\{-\lambda(T_1+t_1)\}}{r!}; \qquad (10.4)$$

and the gamma prior distribution on μ has the p.d.f.

$$f(\mu) = \frac{(T_2+t_2)^{r+1} \mu^r \exp\{-\mu(T_2+t_2)\}}{r!}. \qquad (10.5)$$

Approximate confidence bounds for A_I by Monte Carlo simulation

Simulation of the inherent availability A_I requires a set of computerized repeated trials. If both the TBF and TTR are exponential, at each trial two random numbers are selected; one random number results in a value of λ from the gamma distribution defined by Equation 10.4, and the other in a value of μ from Equation 10.5. These are then substituted into Equation 10.3. From the set of repeated trials, an empirical distribution results, from which confidence bounds are obtained.

With the procedure described here, approximate confidence bounds for A_I can be obtained with a personal computer. For each trial and for each of the two distributions, Equations 10.4 and 10.5, the first step is obtaining a random number from a uniform distribution (refer to chapter 2 for a description of a uniform distribution) in the range zero to one. This random number becomes the complementary d.f. $1 - F(\lambda)$ for a value of λ to be found by an iterative solution. In the case of Equation 10.5, the random number is the d.f. $F(\mu)$, for a value of μ to be solved iteratively. Because r is an integer, the solution is based on a repeated application of a sum of Poisson distribution terms.

Pseudorandom number generator

As of this book's publication, all computer systems have random number generators. Because these random numbers are obtained mathematically, they are called *pseudorandom*. If a random generator is not available, the procedure described in this section generates a sequence of fractional pseudorandom numbers that are very nearly uniformly distributed. Initially, a seed number α_0 is placed in one of the storage registers of the computer; α_0 is the fractional portion of the absolute value of any number whatsoever and it is updated each time a new random number is needed by the recursive formula

$$\alpha_j = \text{fractional portion of } \{9821\, \alpha_{j-1} + .211327\},$$

where α_j is the pseudorandom number at the jth step.

The equation

For simplicity in notation when solving for λ, let $t = T_1 + t_1$. The equation to solve is

$$\alpha_j - \int_0^\infty \frac{(t)^{r+1}\, x^r\, \exp(-xt)}{r!}\, dx = 0. \qquad (10.6)$$

It is shown in the supplement to chapter 5 that the integral in Equation 10.6 is a discrete sum of Poisson terms

$$\int_0^\infty \frac{t^{r+1}\, x^r\, \exp(-xt)}{r!}\, dx = \sum_{i=0}^r \frac{(\lambda t)^i\, \exp(-\lambda t)}{i!}.$$

Because of this identity, Equation 10.6 is solved in this form.

$$g = \alpha_j - \sum_{i=0}^r \frac{(\lambda t)^i\, \exp(-\lambda t)}{i!} = 0 \qquad (10.6a)$$

Iterative solution for λ: The root finder

Since t, r, and α_n are known in advance, to solve Equation 10.6a means finding the root, the value of λ for which $g = 0$. The solution is obtained by Sir Isaac Newton's well-known iterative method for finding the root of a complex function. Starting from an initial guess λ_0, let g_n denote the value of Equation 10.6a at the nth iteration, λ_n the corresponding value of λ, and g'_n the derivative of g_n with respect to λ. By calculus, this derivative is simply

$$g'_n = \frac{(\lambda_n)^r \, t^{r+1} \, \exp(-\lambda_n t)}{r!}. \qquad (10.7)$$

The estimated root of Equation 10.6a, the value of λ that solves the function, is obtained when g_n is close enough to zero, for example 10^{-8} or less. At the $(n + 1)$st iteration

$$\lambda_{n+1} = \lambda_n - \frac{g_n}{g'_n}. \qquad (10.8)$$

Fortunately, the derivative g'_n, which is also the denominator of the second term on the right-hand side of Equation 10.8, is exact, as well as a positive number. The iterative process is simplified and converges rapidly in just two or three iterations for two reasons.

1. The computer cannot overflow because there would not be a division by zero in the quotient term of Equation 10.8.
2. No matter how small g_n is, the sign, $+$ or $-$, of the quotient term is identical to the sign of g_n.

Obtaining a value for μ and for A_I

The value of μ for each Monte Carlo trial is obtained by a process that mirrors the procedure for finding λ. First, obtain the next random number in the sequence, and then in Equation 10.5 and also in both Equations 10.6a and 10.7, let $t = t_2 + T_2$ and substitute μ for λ. Then after an initial guess μ_0, continue to solve for μ by Newton's method. The value of the inherent availability A_I for that trial is then obtained from Equation 10.9, which follows from Equation 10.3

$$A_I = \frac{1/\lambda}{1/\lambda + 1/\mu}. \qquad (10.9)$$

EXAMPLE 10.3.

The simulation procedure just described was programmed for operation on an Apple Macintosh™ computer using the Microsoft Quick-Basic™ language. For convenience in programming, we elected to use a standard

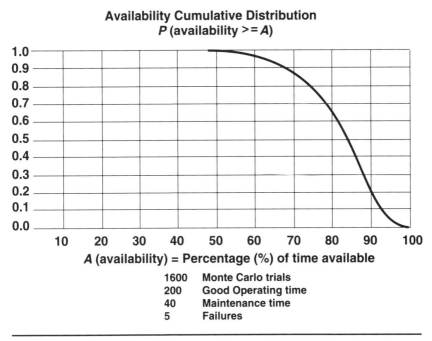

Figure 10.2. A_I distribution: Exponential TTF, exponential TTR.

number of trials, 1600, for each simulation run, regardless of the values of the parameters of the Bayesian prior distributions: r, t_1, T_1, t_2, and T_2. Figure 10.2 represents the results of the 1600 runs, with

$$t_1 + T_1 = 200, \quad t_2 + T_2 = 40, \text{ and } r = 5.$$

The results show that the 90 percent confidence bound is 68 percent availability; the 75 percent confidence bound is 78 percent availability; the 50 percent confidence bound is 83 percent availability; and the 25 percent confidence bound is 89 percent availability. With 1600 trials, the theoretical maximum error in a confidence bound is approximately 0.4 percent.

Observed availability A_o: Exponential failure and exponential repair

Differences between A_o and A_I

Observed availability A_o is the ratio of productive operating time to total available time in a single cycle. In order to establish confidence bounds for A_o, the simplest case to consider is exponential failure combined with exponential repair, with the time to failure governed by failure rate λ and the time to complete the repair governed by repair rate μ.

As in assessing the inherent availability A_I with both exponential failure and exponential repair, confidence bounds on A_o are derived from an empirical probability distribution obtained by repeated computerized Monte Carlo trials. The difference, however, is that the A_I parameters λ and μ are sampled from Bayesian prior gamma distributions; whereas for A_o the time to failure is obtained from an exponential distribution with parameter λ and the time to repair from an exponential distribution with parameter μ. As a result, the A_I input data consist of total failures, total good operating time, and total repair time; and the A_o input data are λ and μ. For A_I the outcome of each trial is the ratio of the MTTF to the average length of an availability cycle; the A_o outcome for each trial is the ratio of good operating time to total time in a single cycle.

The simulation program for A_o

The computer program for simulating A_o was prepared with a standard run of 1600 trials, the same as the A_I simulation program described previously. The data processing, sorting, and graphing subroutines that configure the empirical probability distribution with the results of the trials are identical to those of the program that obtains confidence bounds for A_I. However, with A_o each trial is simpler because the time to failure and the time to repair are both obtained algebraically from a formula for an exponential d.f. without requiring an indirect solution by Newton's method for the root of a complex function.

Let t_1 denote the time to failure, λ the failure rate, and α ($0 < \alpha < 1$) a random number representing the d.f. at a specified trial. The formula for t_1 is

$$t_1 = -\frac{\ln(1-\alpha)}{\lambda}.$$

Similarly, let t_2 denote the time to repair, μ the failure rate, and β, $0 < \beta < 1$, a random number. The formula for t_2 is

$$t_2 = -\frac{\ln(1-\beta)}{\mu}.$$

For each trial, the observed availability is

$$A_o = \frac{t_1}{t_1+t_2}.$$

This process is repeated 1600 times, with different random numbers at each trial, and the 1600 resulting values of A_o are sorted to obtain the empirical probability distribution.

EXAMPLE 10.4.

The results of a run with $\lambda = 0.025$ and $\mu = 0.125$ are displayed in the computer-generated graph in Figure 10.3. By comparing Figure 10.3 to Figure 10.2, the distribution of A_o is broader than A_I; this can be explained statistically by the fact that A_I is based on simulating averages, whereas A_o is obtained from the results of individual trials. The method of interpretation, however, is the same in both cases. The 75 percent confidence level is at the 25th percentile, or 61 percent availability; the 50 percent confidence level is at 84 percent availability; and the 25 percent confidence level is at 95 percent availability.

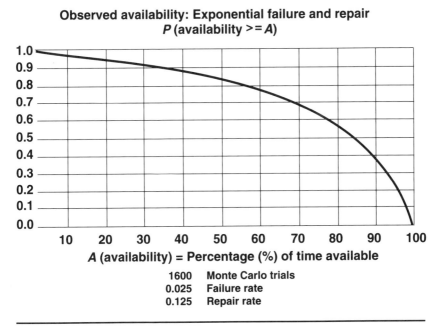

Figure 10.3. Distribution of A_o: Exponential TTF, exponential TTR.

Observed availability A_o: Weibull TTF and lognormal TTR

A_o can be simulated with any combination of a failure distribution and a repair distribution providing that for both distributions there is a functional relationship between a computer-generated random number and the value of the d.f. Exponential failure-exponential repair, described previously, is the simplest example, because the times to failure and repair are obtained easily. As a more complicated example, which is also more realistic, this section shall describe simulation with Weibull failure and lognormal repair.

Weibull time to failure

When the TTF is Weibull, the distribution function for the time t_1 to a failure is

$$F(t_1) = 1 - \exp\left\{-\left(\frac{t_1}{\delta}\right)^\beta\right\};$$

and the relationship between t_1 and the d.f. F is rewritten as

$$t_1 = \delta(-\ln(1-F))^{1/\beta}. \qquad (10.10)$$

In computerized simulation, if the parameters δ and β are known or specified, a random number α, $0 < \alpha < 1$, is substituted for F in Equation 10.10 to obtain t_1.

Lognormal time to repair

If the TTR is lognormal, the logarithm of the time t_2 to completion is normally distributed with mean μ and standard deviation σ. Let z represent a random selection from the standard normal distribution Z: $N(0, 1)$, discussed in chapter 4. Given z and the lognormal parameters μ and σ, t_2 is calculated by

$$t_2 = \exp(\mu + \sigma z). \qquad (10.11)$$

A random selection of z for a Monte Carlo trial is obtained by a widely used approximation due to Hastings. A uniform random number is generated in the usual way and used to approximate a value for z, which is then substituted into Equation 10.11 to calculate t_2. The approximating formula calculates positive values for z corresponding to the upper half of the distribution; negative values of z are obtained by symmetry.

Generate the random number α, $0 < \alpha < 1$. If $\alpha \leq 0.5$, let $q = \alpha$; otherwise $q = 1 - \alpha$. Then calculate

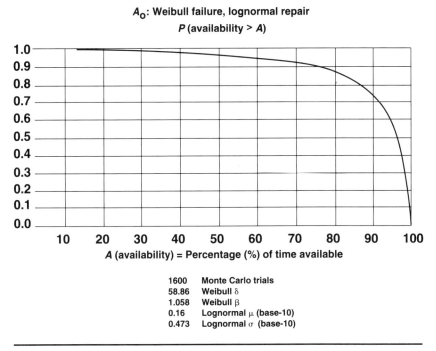

Figure 10.4. Distribution of A_O: Weibull TTF, lognormal TTR.

$$\eta = (-\ln q^2)^{1/2}.$$

The value of z is calculated by the approximating polynomial

$$z = \eta - \frac{a_0 + a_1\eta + a_2\eta^2}{1 + b_1\eta + b_2\eta^2 + b_3\eta^3},$$

where the constants have the following values:

$a_0 = 2.515517$ $b_1 = 1.432788$
$a_1 = 0.802853$ $b_2 = 0.189269$
$a_2 = 0.010328$ $b_3 = 0.001308.$

If $\alpha > 0.5$, the value of z would be taken as negative, corresponding to the lower half of the normal distribution.

EXAMPLE 10.5.

Weibull TBF-lognormal TTR simulation and confidence assessment for A_O: For the data of Example 10.2, what are the values of the observed availability A_O at the 10, 50, and 90 percent confidence levels, based upon Weibull TTF and lognormal TTR?

The base-10 lognormal distribution of repair times has a calculated mean of 0.160 and a standard deviation of 0.473. The Weibull parameters are, respectively, $\delta = 58.86$ and $\beta = 1.058$, calculated by the Mann BLI method. The computer printout of the results of this run is displayed in Figure 10.4. It shows that, at the 90 percent level of confidence, the availability is 75 percent; at the 50 percent level of confidence, it is 96 percent; and at the 10 percent level of confidence, it is 99 percent.

References

Ascher, H., and H. Feingold. 1984. *Repairable systems reliability.* New York: Marcel Dekker.

Bergmann, B. 1977. Some graphical methods for maintenance planning. *Proceedings of the 1977 Annual Reliability and Maintainability Symposium.*

Berman, M. B. 1971. Generating gamma distributed variates for computer simulation models. Report R-641. Santa Monica, Calif.: RAND Corporation.

Bily, M. 1989. *Dependability of mechanical systems.* Amsterdam: Elsevier.

Blanchard, B. S., Jr, and E. E. Lowery. 1969. *Maintainability, principles and practices.* New York: McGraw-Hill.

Bovaird, R. L., and H. I. Zagor. 1961. Lognormal distribution and maintainability in support system research. *Naval Logistics Research Quarterly* 8:343–56.

Calabro, S. R. 1962. *Reliability, principles, and practices.* New York: McGraw-Hill. (In particular, see Ch 9 on "Maintainability and Availability.")

Christian, N. L. 1989. *Fiber optic design, operation, testing, reliability, and maintainability.* Park Ridge, N.J.: Noyes Data Corp.

Dhillon, B. S. 1983. *Systems reliability, maintenance, and management.* New York: Petrocelli.

Dhillon, B. S., and H. Reiche. 1985. *Reliability and maintainability management.* New York: van Nostrand Reinhold.

Fox, B. L. 1963. Generation of random samples from the beta and F distributions. *Technometrics* 5:269–71.

Gertsbakh, I. B. 1977. *Models of preventive maintenance.* Amsterdam: North-Holland.

Goldman, A. S., and T. B. Slattery. 1964. *Maintainability: A major element of system effectiveness.* New York: Wiley.

Goyal, A., V. F. Nicola, A. N. Tantawi, and K. S. Trivedi. 1987. Reliability of systems with limited repairs. *IEEE Trans. Reliability* 36: 202–7.

Hastings, C. 1955. *Approximations for digital computers.* Princeton, N.J.: Princeton University Press.

Horvath, W. J. 1959. Application of the lognormal distribution to servicing times in congestion problems. *Operations Research* 7:126–28.

Howard, R. R., W. J. Howard, and F. A. Hadden. 1959. Study of the down time in military equipment. *Annals of Reliability and Maintainability.* 5th Annual Symposium on Reliability.

Institute of Electrical and Electronic Engineers. 1983. *IEEE guide to the collection and presentation of reliability data for nuclear-power generating stations.* New York: IEEE.

Ireson, W. G., and C. F. Coombs. 1988. *Handbook of reliability engineering and management.* New York: McGraw-Hill.

Knezevic, J. 1993. *Reliability, maintainability, and supportability.* Maidenhead, Berkshire: McGraw-Hill of Europe.

Morse, Philip M. 1957. *Queues, inventories, and maintenance.* New York: Wiley.

Moss, M. A. 1985. *Designing for minimal maintenance expense.* New York: Marcel Dekker.

Naylor, T. M., J. M. Balintfy, D. S. Burdick, and K. Chu. 1968. *Computer simulation techniques.* New York: Wiley.

Raiffa, H., and R. S. Schlaifer. 1961. *Applied statistical decision theory.* Cambridge, Mass.: Harvard University Press.

Rigdon, S. E., and A. P. Basu. 1989. The power law process: A model for the reliability of repairable systems. *Journal of Quality Technology* 21:251–60.

Thomson, J. R. 1987. *Engineering safety assessment.* New York: Wiley.

Villemeur, A. 1992. *Reliability, availability, maintainability, and safety assessment.* 2 vols. New York: Wiley.

Part V

System Reliability

Introduction to system reliability

Part V of this book, on system reliability, is organized differently than the first four parts. There are three chapters, each one covering a different technique for performing system reliability analysis and calculations. However, because the topics are interrelated such that all of the techniques can solve the same problems, instead of each chapter having a unique set of problems and references as in the first four parts, there is a common set of problems and a common set of references for all of the chapters in Part V.

The introduction to this book distinguished between reliability evaluation of a component and that of a system.

> *A component is assessed based upon testing it as a unit. For a system, on the other hand, a reliability evaluation is more likely to be a prediction based upon two types of data: failure data obtained by testing the components, and a system formula derived from the system logic.*

In the sense of this distinction, chapters 1–9 cover only the assessment of a single component. This group of three chapters, however, deals with the reliability of a system.

Part V ■ System Reliability

System reliability analysis introduces a new dimension that is not relevant to the evaluation of components, symbolic logic, based upon set-theoretic foundations originally set forth by Boole, and amplified by his contemporaries Venn and DeMorgan in the nineteenth century; as well as the work of the French mathematicians and probabilists Poincare and Frechet in the early part of the twentieth century; the philosopher Quine and the mathematicians Shannon, Birnbaum, and von Neumann in the 1950s and 1960s; and more recently by contemporary graph theorists. Many authors besides those mentioned above have contributed to this field, particularly during the 1980s.

There are three principal steps to obtain a numerical value for the system reliability, or the probability of success. First, identify the elements of the system and, with the help of graphical methods, classify these elements into the logical groupings or configurations necessary for success or failure. Second, employ an algebraic technique to interpret these configurations and create a mathematical formula giving the reliability of the system as a function of the reliabilities of its component elements. Third, substitute the component probabilities into the formula.

Each of the three following chapters describes a way to generate formulas. Chapter 11 discusses inclusion-exclusion (IE), also known as Poincare's theorem or the classical method; chapter 12 addresses sum of disjoint products (SDP), by a modification of a method originally developed by J. A. Abraham; and chapter 13 describes topological reliability (TR), a graph theoretical method which is closely related to IE.

For both IE and SDP, the starting point for implementing the technique is to identify the *minimal paths,* the smallest groupings of successfully operating components necessary to guarantee system success, or alternatively, the *minimal cuts,* the smallest groupings of failed components for failure. The system reliability formula, a sum of terms that are monomial products of component reliabilities, is developed iteratively (a monomial is a polynomial sum of terms which are products of variables, with every variable raised only to the power one). SDP and IE differ in that the terms of the SDP formula are all additive, with only plus signs (+) separating the terms, whereas the terms of an IE formula alternate between subtraction (−) and addition (+). Except for rare cases, SDP always results in a smaller formula than IE and with less total effort.

Topological reliability differs from both IE and SDP in that mathematical graph theory and search rules designed to prevent duplications of terms develop a factored system reliability formula in nested form, a product of terms that are sums of products of the reliabilities of the elements. In fully expanded form, the factored TR formula is identical to that of IE; however, it is not necessary to fully expand the TR formula in order to obtain a numerical value for the system reliability, because substituting the component reliability values into the shorter TR factored form will yield the correct result.

Because of the size and complexity of many systems, it is impractical to attempt to implement large-scale system reliability analysis without certain simplifications or requirements. It is almost always necessary to simplify the model by assuming that the component failure probabilities are independent; dependent probabilities can become entirely too complex to handle. Another requirement is computerization to implement the logical manipulations and numerical calculations. If just a few components are involved, the work often can be done easily by hand; with large systems, human algebraic manipulations and calculations invariably lead to errors. The third requirement is a database that provides failure history data or the reliability value for every component.

Chapter 11 introduces, in addition to IE, the concepts of system reliability: diagramming success or failure; definitions of minimal paths and minimal cuts; inversion of logical formulas by a two-stage application of DeMorgan's theorems; and approximations. Chapter 12, on SDP, further develops concepts of Boolean algebra, minimization, minimized inversion, and disjointness. In addition, it contains a discussion on the use of SDP formulas to evaluate the reliability importances of components. Chapter 13, on TR, describes the graph-theoretic concepts, search trees, and rules for efficiently deriving a factored system formula, as well as extensions of TR for more complex cases other than the simple source-to-terminal systems that are usually considered in the literature.

11

Inclusion-Exclusion and Related Matters

Introduction

A common way of viewing a system reliability problem is as a network of interrelated but possibly diverse elements organized to achieve a common objective, frequently called a *mission*. In some cases—examples are strings of Christmas-tree bulbs on a common line; or a flashlight with a bulb, switch, case, and batteries—all elements are necessary to achieve mission success; this is known as a *serial* system. In other cases, some of the elements are redundant, meaning that one element backs up another—such as when using a flashlight in case the power goes off in an outage caused by a catastrophe—this is known as a *parallel* system. More frequently, systems are mixtures of serial and parallel modules or groupings of elements, organized along lines that could be state-of-the-art within the range of engineering and economic feasibility.

In the most general sense, a system network is a binary (two-valued) set consisting of binary elements, with the value 1 for either a component or the system denoting success and the value 0 denoting failure. The arcs, lines, or boxes of the network graph represent components and the nodes joining the arcs are logical or physical connections between elements or they are representations of sequences of operations or of interrelationships between elements. Since both the components and the system are two-valued, it is appropriate to treat the network as a Boolean system.

Boolean source-to-terminal networks

For the discussion in this chapter as well as chapter 12 and a part of chapter 13, two unique nodes are common to every network: a source node S, initiation of the mission, and a terminal node T representing successful completion. The flow of information, instructions, sequences, and so on, and the direction of an arc is always outward from S and inward to T, from left to right; intermediate arcs between S and T, except those connected directly to S or to T, are either one-way or two-way. Networks of this type are identified as ST; the assumption that all networks are ST is modified in the discussion on overall reliability in chapter 13 on topological reliability (TR).

Network graphs: Minimal paths and minimal cuts

A *path* is an unbroken string of arcs from S to T, denoting system success if all arcs in the path are functional or 1-valued. If a path is just a single string with no redundancies or alternative ways to achieve success, it is a *minimal path*. A *cut* is a set of 0-valued elements such that if all of them fail, the system also fails. A *minimal cut* is a smallest set of 0-valued elements that would break all paths or cause system failure.

A system can have many minimal paths and many minimal cuts; the minimal paths are derived from the minimal cuts, or vice versa, by minimized inversion. For both IE and SDP, a system reliability formula is derived iteratively from either the minimal paths or the minimal cuts by augmenting the system probability formula at each stage successively with either the contribution of each minimal path to the system probability of success, or else the contribution of each minimal cut to the system probability of failure.

Inclusion-exclusion

Example: Bridge network reliability by IE

Bridge networks are used to measure electrical resistances, inductances, or capacitances. Because of the simplicity and symmetry of a bridge

Chapter 11 ■ Inclusion-Exclusion and Related Matters

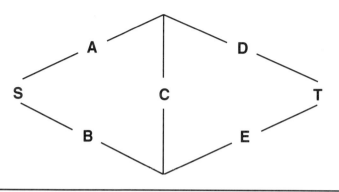

Figure 11.1. Idealized graph of a bridge network.

network and the fact that there are very few components, it is also convenient to represent reliability concepts. An idealized bridge network has five components A, B, C, D, and E, and the network graph is represented by Figure 11.1. In Figure 11.1, S and T are the source and terminal, respectively; A, B, D, and E are one-way elements; and C is a two-way element.

Minimal paths and the overlapping success states

In this and the subsequent discussion, a protocol of ordering lists of terms, paths, cuts, and so on by a combined size-and-alphabetical ordering is followed: smaller terms first, and within each size group, alphabetical ordering. The bridge circuit has four minimal paths, two with two elements each and two with three elements: AD, BE, ACE, and BCD. Each path covers a certain number of success states as shown in the following listing.

	AD					BE					ACE					BCD		
A	B	C	D	E	A	B	C	D	E	A	B	C	D	E	A	B	C	D	E
1	1	1	1	1	1	1	1	1	1	1	1	1	1	1	1	1	1	1	1
1	0	1	1	1	0	1	1	1	1	1	0	1	1	1	0	1	1	1	1
1	1	0	1	1	1	1	0	1	1	1	1	1	0	1	1	1	1	1	0
1	1	1	1	0	1	1	1	0	1	1	0	1	0	1	0	1	1	1	0
1	0	0	1	1	0	1	0	1	1										
1	0	1	1	0	0	1	1	0	1										
1	1	0	1	0	1	1	0	0	1										
1	0	0	1	0	0	1	0	0	1										

Although this table has 24 different items, there are just 16 distinct success states when overlapping items are accounted for. For example, the success state 1 1 1 1 1, all five components working, is included in all four paths; 1 0 1 1 1, four components working and B failed, is included in both paths AD and ACE; 0 1 1 1 1 is listed under both BE and BCD; and so on.

IE Reliablity formula algorithm

The IE reliability formula is built up one path at a time, with each successive step adding to the formula only the incremental probability for the incumbent path that is not also in a prior path. For example, the second path, BE, has two success states that are also in the first path AD, 1 1 1 1 1 and 1 1 0 1 1. When adding the probability of BE to the probability of AD, the probability of these two success states must be excluded from the sum, in order to prevent double counting of the same probabilities. Under the assumption that the component success or failure probabilities are independent, this probability is the product of the probabilities of the four one-valued elements A, B, D, and E that are common to both states.

For convenience in notation, let A be the probability of component A, B the probability of B, and so on. The formula for Figure 11.1 is built up in four steps, because there are four paths. At each step combine the components of the incumbent path with the variables of every term generated by all of the prior paths and change the sign of the resulting term. Let R_1 be the system formula at the first step, R_2 at the second step, and so on. By the algorithm, the steps are, respectively,

$R_1 = AD$;
$R_2 = R_1 + BE - ABDE$;
$R_3 = R_2 + ACE - ACDE - ABCE + ABCDE$; and
$R_4 = R_3 + BCD - ABCD - BCDE + ABCDE$.

Note that combining the letters BCD with the letters of the four terms listed in the R_3 line results in a net cancellation, since every resulting term would have all five letters in it, two of them with plus signs and two with minus signs. If we assume for simplicity that every component has the same reliability R, the system reliability R_s is

$$R_s = 2R^2 + 2R^3 - 5R^4 + 2R^5.$$

For example, if $R = 0.9$, $R_s = 0.97848$.

Generating an IE reliability formula with the minimal cuts

A minimal cut is a smallest set of components that will cause all paths to fail. A system formula can be generated using the minimal cuts with the same IE procedure employed with the minimal paths. However, since this formula builds up the system probability of failure rather than the probability of success, the letter Q is used to denote probability of failure to distinguish it from R, the probability of success.

The network in Figure 11.1 has four minimal cuts: AB, DE, ACE, and BCD. Let the lowercase letters a, b, c, d, and e respectively denote the probabilities of the corresponding components. Then by a procedure that is analogous to the one used for the paths, the probability of failure is built up in four steps, one step for each cut.

$$Q_1 = ab;$$
$$Q_2 = Q_1 + de - abde;$$
$$Q_3 = Q_2 + ace - abce - acde + abcde; \text{ and}$$
$$Q_4 = Q_3 + bcd - abcd - bcde + abcde.$$

If every component has the same probability of failure $Q = 0.1$, the system probability of failure Q_s is

$$Q_s = 2Q^2 + 2Q^3 - 5Q^4 + 2Q^5 = 0.02152;$$

and the system probability of success is

$$R_s = 1 - 0.02152 = 0.97848,$$

identical to the answer obtained with the minimal paths.

Minimized inversion of paths or cuts

The minimal cuts can be derived from the minimal paths, or vice versa, by a process of minimized inversion. This procedure is similar to the cycling used in deriving an IE system reliability formula; it also includes repeated use of the DeMorgan theorems, as well as *absorption*, whereby larger terms that contain all of the characters of smaller terms are absorbed into the smaller terms. For example, if a given success logic polynomial contains both ABC and AB, ABC is deleted and the polynomial is shortened because AB contains all of the Boolean sets represented by ABC.

The DeMorgan theorems

Inversion of Boolean forms is obtained by a two-stage application of the DeMorgan theorems. Since a Boolean form is a polynomial sum (inclusive-or union) of monomial terms, where each term is a product (intersection) of variables, both DeMorgan theorems are required for this purpose, inversion of a sum and inversion of a product.

Chapter 11 ■ Inclusion-Exclusion and Related Matters

In this section (and in the next wherever it is convenient, necessary, or advisable to do so, and it is clear from the context of the discussion why it is being done) instead of having 1's and 0's represent Boolean functions, alphabetic characters, such as A, B, C, and their respective inverses, $\overline{A}, \overline{B}, \overline{C}$, are used instead. A letter of the alphabet in its greatest generality could represent either a variable, a term, or an entire polynomial.

Inversion of a sum

The inverse of a Boolean sum of terms is the Boolean product of the inverses of the terms of the sum

$$\overline{A + B + C + \ldots} = \overline{A}\,\overline{B}\,\overline{C}.\ldots \tag{11.1}$$

For convenience, the multiplication sign is often deleted.

Inversion of a product

Inversion of a Boolean product is the dual of inversion of a Boolean sum. This means that addition is replaced by a multiplication, and multiplication is replaced by addition.

$$\overline{A\,B\,C} = \overline{A} + \overline{B} + \overline{C} \ldots \tag{11.2}$$

Corollaries to inversion of a sum or product

Since the inverse of $\overline{A}$ is A, Equations 11.3 and 11.4 are corollaries to Equations 11.1 and 11.2.

$$\overline{\overline{A} + \overline{B} + \overline{C}\ldots} = A\,B\,C\ldots \tag{11.3}$$

$$\overline{\overline{A}\,\overline{B}\,\overline{C}\ldots} = A + B + C + \ldots \tag{11.4}$$

Deriving the cuts from the paths: The bridge circuit

Inverting the paths to obtain the cuts is a stepwise repeated application of Equation 11.1 to the terms of a minimized logic polynomial. At each step, this is followed by multiplying the inverse obtained at the previous step by the inverse of the next minimal path derived by Equation 11.2, then minimizing and rearranging.

The minimized success logic polynomial, S, obtained from Figure 11.1 is simply the Boolean sum of the minimal paths

$$S = AD + BE + ACE + BCD.$$

The inverse of AD is $\bar{A}+\bar{D}$. At the second step, the inverse of $AD+BE$ is

$$(\bar{A}+\bar{D})(\bar{B}+\bar{E}) = \bar{A}\bar{B} + \bar{A}\bar{E} + \bar{B}\bar{D} + \bar{D}\bar{E}.$$

The minimized inverse of $AD + BE + ACE$ at the third step is

$$(\bar{A}\bar{B}+\bar{A}\bar{E}+\bar{B}\bar{D}+\bar{D}\bar{E})(\bar{A}+\bar{C}+\bar{E}) = \bar{A}\bar{B} + \bar{A}\bar{E} + \bar{D}\bar{E} + \bar{B}\bar{C}\bar{D}.$$

In this expansion, most of the terms vanish because of absorption. Examples are $\bar{A}\bar{B}\bar{E}$ is absorbed into $\bar{A}\bar{B}$ and $\bar{B}\bar{D}\bar{E}$ into $\bar{D}\bar{E}$. At the fourth and final step, the minimized inverse of the success polynomial S is the inverse of $AD + BE + ACE + BCD$

$$\bar{S} = (\bar{A}\bar{B}+\bar{A}\bar{E}+\bar{D}\bar{E}+\bar{B}\bar{C}\bar{D})(\bar{B}+\bar{C}+\bar{D})$$
$$= \bar{A}\bar{B} + \bar{D}\bar{E} + \bar{A}\bar{C}\bar{E} + \bar{B}\bar{C}\bar{D}.$$

These four terms are the minimal cuts. This same procedure can be used to find the minimal paths from the minimal cuts. The details are omitted and left as an exercise for the reader.

It is to the advantage of the analyst to employ inversion in procedures involving success or failure logic, because the size of a polynomial and the effort required to derive a system probability formula grows exponentially with the number of minimal states. Since the formula is obtained with either the minimal paths or the minimal cuts, the amount of work can be significantly reduced by taking the shorter of the two lists. Inversion is also employed with the sum-of-disjoint products technique described in chapter 12.

Series and parallel reductions

When the system can be reduced to a set of independent modules, where each module is either strictly serial or else strictly parallel, a simple and efficient alternative to IE is to develop a reliability formula for each module, as well as for the system as a configuration of subsystems or modules. For a strictly serial system having n independent elements 1, 2, ..., n, respectively, with corresponding reliabilities $p_1, p_2, \ldots, p_n$, the reliability R is the product of the probabilities of the elements

$$R = p_1 p_2 \ldots p_n.$$

For a strictly parallel system with n elements, since all elements must fail in order for the system to fail, the reliability is the complement of the probability that all n elements fail, or

$$R = 1 - (1-p_1)(1-p_2)\ldots(1-p_n). \tag{11.5}$$

EXAMPLE 11.1.

Develop the reliability formula for the success-logic diagram in Figure 11.2 and compute the reliability, assuming that the nine components A, B, ..., I all have success probabilities of 0.7.

Denote the reliabilities of components A, B, ..., I respectively by the corresponding lowercase letters $a, b, \ldots, i$, all numerically equal to 0.7. The subformula for the top branch is

$$1 - (1-a)(1-b) = 0.91.$$

The subformula for the bottom branch is the product of the probabilities of success for the three parallel structures in that branch is

$$[1 - (1-c)(1-d)] \cdot [1 - (1-e)(1-f)(1-g)] \cdot [1 - (1-h)(1-i)] = 0.805741.$$

The system formula is obtained by considering the two branches as two elements in parallel.

$$1 - [(1-a)(1-b)][1 - \{1 - (1-c)(1-d)\} \cdot \{1 - (1-e)(1-f)(1-g)\} \cdot \{1 - (1-h)(1-i)\}]$$

The system reliability is 0.982517. If all of the factors are multiplied out, the resulting system formula is identical to the one obtained by IE. Since there are 24 minimal paths, the maximum number of IE terms could be as large as $2^{24}-1 = 16,777,215$. Because of overlapping

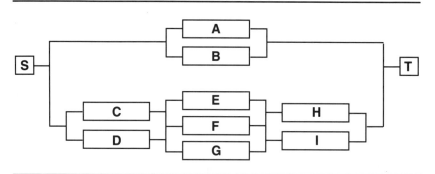

Figure 11.2. Success logic for Example 11.1.

elements between some of the paths, the actual size of the formula would be considerably less than the maximum.

Approximations

The Esary–Proschan lower bound

Exact methods, such as those described in Part V of this book, all have the property of *exponentiality*, also sometimes called *the curse of dimensionality*, the tendency for the formula to grow exponentially with the size of the system under worst-case conditions.

For IE, the worst case is deriving a success probability formula for a strictly parallel system with n minimal paths and no overlapping elements between the paths, where n can be any integer. By duality, the worst case for a serial failure probability formula is n minimal cuts of a serial system. In those cases, the maximum number of terms of the IE formula is $2^n - 1$. This can be a very large number; for example, if there are 10 minimal paths for a strictly parallel system, the IE success probability formula would have 1023 terms. Although this would be an unusual case, the problem of exponential growth of the size of the formula is serious enough that IE often is not a practical technique to employ to obtain exact success probability formulas.

The Esary–Proschan lower bound obtains an approximate lower-bound value of the reliability for relatively small (≤ 50 components) systems consisting only of highly reliable (≥ 0.99 reliability) components. It is derived from a fundamental assumption that all minimal paths fail and none of them have any elements in common, by a formula that looks like Equation 11.5. Assume that the system has k minimal paths $M_1, M_2, \ldots, M_k$. Define $m_1, m_2, \ldots, m_k$ as the corresponding products of the probabilities of failure for the components in the paths; for example, if the first path for the bridge circuit is $M_1 = AD$, then

$$m_1 = (1-a)(1-b),$$

where a and b are the probabilities of success for components A and B, respectively. The lower-bound formula is

$$R_L = (1-m_1)(1-m_2) \ldots (1-m_k). \tag{11.6}$$

EXAMPLE 11.2.

Lower bound for the bridge circuit: What is a lower-bound value for the reliability of the bridge circuit if every component has a reliability of 0.9?

In this case there are two minimal paths with two elements each and two paths with three components. Thus,

$$R_L = [1 - (0.1)^2]^2 \, [1 - (0.1)^3]^2 = 0.978141.$$

This is lower than the exact reliability of 0.97848, but it is correct to the first three significant figures.

Approximation with a subset of the minimal cuts

An upper-bound technique for approximating the reliability is derived using a subset of the minimal cuts of a system for which all components are highly reliable. Only those minimal cuts with few components are taken into account. The reasoning is that since the probability of failure of any individual component is small, say 0.1 or less, the probability of simultaneous failure of more than one component is an order of magnitude smaller yet; therefore, a minimal cut with more than three or four components makes no noticeable contribution to the calculation of the probability.

EXAMPLE 11.3.

Approximating the reliability of a double bridge circuit: A double bridge circuit is represented by Figure 11.3 with nine minimal cuts, three with two components each: *AB*, *DE*, and *GH*; four with three components

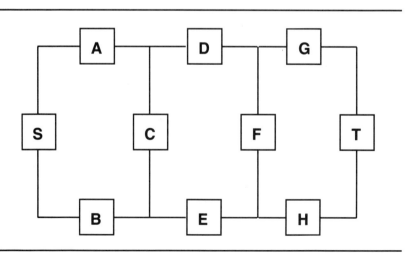

Figure 11.3. A double bridge circuit.

Table 11.1 System reliability for Example 11.3.

p	q	2s only	2s & 3s	Exact
0.99	0.01	0.999700	0.999696	0.999696
0.95	0.05	0.992519	0.992069	0.992059
0.90	0.10	0.970299	0.967106	0.966975
0.80	0.20	0.884736	0.865403	0.864092
0.50	0.50	0.421875	0.335938	0.328125

each: *ACE*, *BCD*, *DFH*, and *EFG*; and two with four components each: *ACFH* and *BCFG*.

Estimates of the reliability for five different values of the component probability p—0.99, 0.95, 0.9, 0.8, and 0.5—will be obtained in three steps, first considering only the three 2's, then the three 2's plus the four 3's, and then the exact reliability considering the 4's as well as the 2's and 3's. At each step the minimal cut formula is derived by accumulating the probabilities of failure with IE and subtracting the total from 1. For brevity, the derivation of these formulas is omitted; however, in chapter 12 the exact formula for this system is obtained by SDP.

Assume that every one of the eight components has the same reliability p; therefore, the unreliability is $q = 1-p$. Table 11.1 provides the system-reliability values for each of the three cases and for the five different sets of values for p and q.

Example 11.3 illustrates three features of this method.

1. The estimate obtained is an upper bound, because it is based on the minimal cuts. Since a subset of the minimal cuts is used, the probability of system failure is underestimated, and the probability of success is overestimated.
2. For systems with all components highly reliable, the error in deleting the higher-order minimal cuts can be negligible since most of the probability is carried by the smaller cuts. If there are many components (say 100 or more), the amount of error can be serious.
3. As more minimal cuts are used, the error in the approximation is reduced.

12

Sum of Disjoint Products

Introduction

This chapter describes sum of disjoint products (SDP) as an improved alternative to inclusion-exclusion (IE) for deriving a system-reliability formula. The advantages of SDP come largely from the fact that disjointness guarantees that the logic function of system success is 1:1, identical with the probability polynomial. In reading this chapter, the following should be noted.

1. Like IE, SDP is a stepwise procedure for deriving a formula. As in IE, at each stage the incremental probability accounted for by the incumbent minimal path that is not covered by any of the prior paths is added to the function.
2. Since the logic and probability expressions are 1:1, the terms of the probability polynomial in disjoint form are all additive, instead of alternately additive and subtractive as in IE.

The description of the ALR 6-step algorithm is from M. O. Locks, "A Minimizing Algorithm for the Sum of Disjoint Products," *IEEE Transactions on Reliability* (vol. 36, 1987, 445–53) with permission, © 1987 IEEE.

3. By comparison to IE, the number of terms in the probability polynomial is smaller with SDP, for very large systems possibly orders of magnitudes smaller. This means that there is much less effort and storage required for the tedious routine computations.
4. The mathematics of SDP facilitates the derivation of the formula for the *importance* of a component, the relative marginal contribution of the component to the reliability of the system. Because of the convenient way of deriving an importance formula by differentiating the system SDP formula, the subject of importance is introduced in this chapter.

The principal SDP computational technique described in this chapter is a modification, called ALR (Abraham–Locks revised), of an algorithm originally proposed by J. A. Abraham. Abraham–Locks revised introduces minimizing Boolean procedures for deriving the SDP formula, as well as rules for ordering paths and terms, that are not also included in the original Abraham version.

Logical and probabilistic aspects of disjointness

Venn diagrams, such as those in chapter 1, help explain the effects of disjointness in set formation and system-probability calculation. Suppose there are two minimal paths, 1 and 2; for example, 1 could be the set of success states subtended by minimal path AD in the bridge-circuit diagram Figure 11.1, and 2 the set of success states for path BE. In Figure 12.1 the universe of possibilities includes ellipses representing

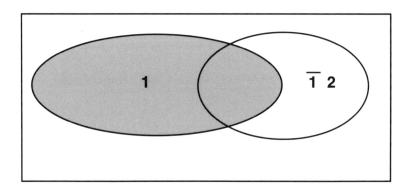

Figure 12.1. Venn diagram for two minimal paths.

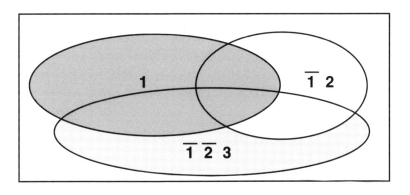

Figure 12.2. Venn diagram for three minimal paths.

the sets of success states subtended by these paths. As in the discussion on inversion in chapter 11, the notation $\bar{1}$ stands for the set of states not included in 1.

Since the two sets 1 and $\bar{1}\,2$ have no elements in common, the probability of the inclusive-or union $1 \cup 2$ is

$$P(1) + P(\bar{1}\,2).$$

Now let us add a third set 3, such as the success states for minimal path ACE in Figure 12.1, as is shown in Figure 12.2. Since the added set $\bar{1}\,\bar{2}\,3$ is disjoint from both 1 and $\bar{1}2$, the probability of the inclusive-or union of sets 1, 2, and 3 is

$$P(1+2+3) = P(1) + P(\bar{1}\,2) + P(\bar{1}\,\bar{2}\,3).$$

This equality illustrates the principal advantage of disjointness. The argument inside the parentheses on the left-hand side is a logical expression where the + sign between elements means *inclusive-or*. The + sign on the right-hand side, on the other hand, has the usual meaning of numerical addition. When the sets are disjoint, the logical inclusive-or union and numerical addition of probabilities are interchangeable.

Example: The bridge circuit

Like IE, SDP is a stepwise procedure for obtaining a system formula, with the number of steps equal to the number of minimal paths. With SDP, each step augments the system formula with a subformula of disjoint terms for the probability subtended by the incumbent path and not by any of the prior paths.

The bridge circuit of Figure 11.1 has four minimal paths AD, BE, ACE, and BCD; the formula is derived in four steps. At the first step the subformula is

$$R_1 = AD.$$

At the second step, note that path BE has different variables than AD. By the DeMorgan theorem, Equation 11.1, and by disjointness

$$\overline{A+D} = \bar{A} + \bar{D} = \bar{A} + A\bar{D}.$$

To obtain the added contribution of BE to the system disjoint formula, multiply both of these terms by BE and augment the system formula with the resulting sum of products

$$R_2 = R_1 + \bar{A}BE + AB\bar{D}E.$$

At the third step, since ACE does not contain D, as does AD, or B, as does BE, it is only necessary to invert both B and D to obtain the next disjoint term

$$R_3 = R_2 + A\bar{B}C\bar{D}E.$$

At the fourth step, since BCD does not contain either A or E,

$$R_4 = R_3 + \bar{A}BCD\bar{E}.$$

It can be verified that for a uniform component reliability of 0.9, this formula yields the same answer for the system reliability, 0.97848, that was obtained by IE in chapter 11.

The ALR algorithm

The starting point for implementing the algorithm is the minimal paths. As in chapter 11, the protocol is to order the paths by size, smaller paths first, and then alphabetically.

The outer loop

The procedure involves a succession of inner loops within an outer loop. There are as many stages of the outer loop as there are paths; at each stage, the system formula is augmented by a subformula of disjoint terms generated by the incumbent path that are also disjoint from any terms formed by any of the prior paths.

Chapter 12 ■ *Sum of Disjoint Products* 233

The fundamental disjointness equality is: Let there be n ordered minimal paths $A_1, A_2, \ldots, A_n$; for the jth path, $j \leq n$,

$$P(A_1 + A_2 + \ldots + A_j) = P(A_1) + P(\bar{A}_1 A_2) + \ldots + P(\bar{A}_1 \bar{A}_2 \ldots \bar{A}_{j-1} A_j). \quad (12.1)$$

The + sign inside the parentheses on the left-hand side of Equation 12.1 has the logical meaning of inclusive-or union; on the right-hand side the + sign means numerical addition.

The inner loop

The disjoint terms of a subformula for the incumbent minimal path are obtained in the inner loop, which is a six-step sequence of operations within the outer loop. The total number of operations is variable and depends primarily upon the orderings and sizes of the paths and terms and whether or not there are opportunities for minimizing Boolean expressions.

The six steps of the inner loop at the jth stage of the outer loop, corresponding to the incumbent minimal path A_j, are

1. Create a Boolean polynomial of $j-1$ terms, where each term is the product of all of the variables in a single prior path that are not also in the incumbent path. For example, with the bridge circuit, at the 4th outer-loop stage, this polynomial is $A+E+BE$.
2. Minimize the polynomial by absorbing larger terms into smaller terms. For example, by absorbing BE into E, BE is deleted and the minimized result is $A+E$.
3. Invert the minimized polynomial. The result for the fourth stage is $\bar{A}\bar{E}$—for the second stage, it is $\bar{A}+\bar{D}$.
4. Express the result of inner-loop step 3 in disjoint form. For the second stage, this is $\bar{A}+A\bar{D}$.
5. Multiply each disjoint term by every variable in the incumbent path.
6. Augment the system formula with the subpolynomial obtained at the previous step.

Features of the algorithm

Minimization

The most important labor-saving aspect of the ALR algorithm is the minimization of the polynomial of terms of variables in prior paths that are not in the incumbent path, at step 2, by absorbing larger terms into smaller terms with all of the same variables. This reduces the size of the polynomial enough to make the number of operations in subsequent steps manageable.

Continuous minimized reordering of terms

By continuously reordering terms in a polynomial at steps 1, 2, 3, or 4 in the inner loop and minimizing, the sizes of the expressions and the amount of effort is reduced. For example, steps 1 and 2 can be combined; this is illustrated in Example 12.2.

Rapid inversion at step 3

Rapid inversion at step 3 is illustrated in Example 12.2.

Inverse Boolean minimization at step 4

Inverse Boolean minimization at step 4 is illustrated in Example 12.2.

EXAMPLE 12.1.

Obtain the system-reliability polynomial for the double-bridge circuit of Figure 11.3 by ALR, and give the value of the system reliability if all components have 0.8 reliability.

The circuit has eight minimal paths: *ADG*, *BEH*, *ACEH*, *ADFH*, *BCDG*, *BEFG*, *ACEFG*, and *BCDFH*. The resulting SDP formula is

$$R_1 = A D G;$$
$$R_2 = R_1 + \overline{A} B E H + A B \overline{D} E H + A B D E \overline{G} H;$$
$$R_3 = R_2 + A \overline{B} C \overline{D} E H + A \overline{B} C D E \overline{G} H;$$
$$R_4 = R_3 + A D \overline{E} F \overline{G} H + A \overline{B} \overline{C} D E F \overline{G} H;$$
$$R_5 = R_4 + \overline{A} B C D \overline{E} G + \overline{A} B C D E G \overline{H};$$
$$R_6 = R_5 + B \overline{D} E F G \overline{H} + \overline{A} B \overline{C} D E F G \overline{H};$$
$$R_7 = R_6 + A \overline{B} C \overline{D} E F G \overline{H}; \text{ and}$$
$$R_8 = R_7 + \overline{A} B C D \overline{E} F \overline{G} H.$$

For a uniform component-reliability value of 0.8, the calculated exact system reliability is 0.864092, the same as the exact value obtained in Example 11.3.

EXAMPLE 12.2.

Sixty-one-term SDP formula for a 12-component, 24-path example: We use an example that is larger and more complex than the bridge and

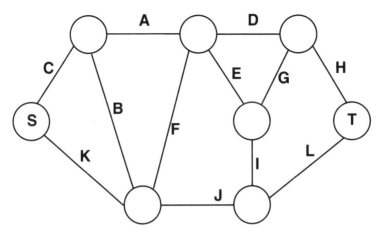

Source: M. O. Locks, "A Minimizing Algorithm for the Sum of Disjoint Products," *IEEE Transactions on Reliability* (vol. 36, 1987, 445–53), with permission, © 1987 IEEE.

Figure 12.3. Communications network with 24 minimal paths.

double-bridge circuits in the previous chapter to illustrate the SDP algorithm. These features include Boolean minimization, continuous reordering, rapid inversion at step 3 of the inner loop, and inverse minimization at step 4.

The example is taken from one that has been used in several articles that were published in *IEEE Transactions on Reliability*, starting from the original Abraham paper in 1979. The communications network on which this example is based is depicted in Figure 12.3.

The 61-term disjoint formula for this network is listed in Figure 12.4. The 24 minimal paths are listed in size and alphabetical order in the left-hand column; for each path the associated disjoint subformula of terms is shown to the right, with 1 representing success for the specified component and 0 denoting failure.

Since it would be a formidable task to explain the entire formula, three paths with fairly complex solutions were selected for detailed descriptions of the inner steps: *ACEIL, GHIJK,* and *BCGHIJ*.

ACEIL: At the first step, the polynomial of terms consisting of the variables in the six prior paths that are not also in *ACEIL* is

$$JK + DH + BJ + DFHK + BDHK + GH.$$

Path	A B C D E F G H I J K L	Path	A B C D E F G H I J K L
JKL	. . . . 1 1 1	EFIKL	0 . . . 1 1 . . 0 1 0 1 1
ACDH	1 . 1 1 . . . 1		1 . 0 . 1 1 . . 0 1 0 1 1
	1 . 1 1 . . . 1 . 1 0 .		0 . . 1 1 0 . . 1 1 0 1 1
	1 . 1 1 . . . 1 . 1 1 0		1 . 0 1 1 0 . . 1 1 0 1 1
BCJL	. 1 1 . . . . . . 1 0 1	GHIJK	0 . . . . . 0 1 1 1 1 1 0
	1 1 1 0 . . . . . 1 0 1		1 . . 0 0 . 1 1 1 1 1 0
	1 1 1 1 . . . . 0 1 0 1		0 . . 0 0 1 1 1 1 1 0
DFHK	0 . . 1 . 1 . 1 . . 0 1		1 0 0 1 . 0 1 1 1 1 1 0
	0 . . 1 . 1 . 1 . 1 1 0		1 0 0 0 1 0 1 1 1 1 1 0
	1 . 0 1 . 1 . 1 . . 0 1		1 1 0 0 1 0 1 1 1 1 1 0
	1 . 0 1 . 1 . 1 . 1 1 0	ABEGHK	1 1 0 0 1 0 1 1 . . 0 1
ABDHK	1 1 0 1 . 0 . 1 . . 0 1		1 1 0 0 1 0 1 1 0 1 1 0
	1 1 0 1 . 0 . 1 . 1 1 0	ABEIKL	1 1 0 . 1 0 . . 0 1 0 1 1
ACEGH	1 . 1 0 1 . 1 1 . . 0 .		1 1 0 0 1 0 0 1 1 0 1 1
	1 . 1 0 1 . 1 1 . 1 . 0	ACDGIL	1 . 1 1 0 . 1 0 1 0 . 1
	1 0 1 0 1 . 1 1 . 1 0 1		1 0 1 1 0 0 1 0 1 1 0 1
ACEIL	1 . 1 . 1 . . . 0 1 0 1	BCEFGH	0 1 1 0 1 1 1 1 . . 0 0
	1 0 1 . 1 . . . 0 1 1 0 1		0 1 1 0 1 1 1 1 . 1 0 0
	1 . 1 0 1 . 0 1 1 0 . 1	BCEFIL	0 1 1 . 1 1 . . 0 1 0 0 1
	1 0 1 0 1 . 0 1 1 1 0 1		0 1 1 0 1 1 0 1 1 0 0 1
ACFJL	1 0 1 0 0 1 . . . 1 0 1	BCGHIJ	0 1 1 . . . 0 1 1 1 0 0
	1 0 1 1 0 1 . 0 . 1 0 1		1 1 1 0 0 . 1 1 1 1 0 0
	1 0 1 . 1 1 . . 0 0 0 1		0 1 1 0 0 1 1 1 1 1 0 0
	1 0 1 0 1 1 0 1 0 1 0 1	DEHIJK	0 . . 1 1 0 0 1 1 1 1 0
BCDFH	0 1 1 1 . 1 . 1 . . 0 0		1 0 0 1 1 0 0 1 1 1 1 0
	0 1 1 1 . 1 . 1 . 1 0 0	DFGIKL	0 . . 1 0 1 1 0 1 0 1 1
EFGHK	0 . . . 1 1 1 1 . . 0 1		1 . 0 1 0 1 1 0 1 0 1 1
	1 . 0 1 1 1 1 1 . . 0 1	ABDGIKL	1 1 0 1 0 0 1 0 1 0 1 1
	0 . . 0 1 1 1 1 . 1 1 0	ACFGHIJ	1 0 1 0 0 1 1 1 1 1 0 0
	1 . 0 0 1 1 1 1 . 1 1 0	BCDEHIJ	0 1 1 1 1 0 0 1 1 1 0 0
		BCDFGIL	0 1 1 1 0 1 1 0 1 0 0 1

Source: M. O. Locks and J. M. Wilson, "Note on Disjoint Products Algorithms," *IEEE Transactions on Reliability* (vol. 41, 1992, 81–84), with permission, © 1992 IEEE.

Figure 12.4. Sixty-one-term SDP formula for 12 components, 24 paths.

With continuous minimization, the terms *DFHK* and *BDHK* are both absorbed into *DH*, and the first and second steps are combined to yield the reordered and minimized result.

$$BJ + DH + GH + JK.$$

Rapid inversion of this expression at inner step 3 is performed one term at a time. At each step, the polynomial representing the combined inverse of the sum of all of the previous terms is multiplied by the inverse of the incumbent term. The combined inverse is a sum of monomial terms with all variables at their respective inverted values, whereas the inverse of the incumbent is a Boolean sum of inverses of single variables. This makes it possible to minimize continuously while inverting, as is illustrated by the following.

The combined inverse of the first two terms is

$$(\overline{B} + \overline{J})(\overline{D} + \overline{H}) = \overline{B}\,\overline{D} + \overline{B}\,\overline{H} + \overline{D}\,\overline{J} + \overline{H}\,\overline{J}.$$

This expression is multiplied by the inverse of GH to obtain

$$(\overline{B}\,\overline{D} + \overline{B}\,\overline{H} + \overline{D}\,\overline{J} + \overline{H}\,\overline{J})(\overline{G} + \overline{H}) = \overline{B}\,\overline{H} + \overline{H}\,\overline{J} + \overline{B}\,\overline{D}\,\overline{G} + \overline{D}\,\overline{G}\,\overline{J}.$$

This is multiplied by the inverse of *JK* to obtain the combined inverse

$$(\overline{B}\,\overline{H} + \overline{H}\,\overline{J} + \overline{B}\,\overline{D}\,\overline{G} + \overline{D}\,\overline{G}\,\overline{J})(\overline{J} + \overline{K})$$
$$= \overline{H}\,\overline{J} + \overline{B}\,\overline{H}\,\overline{K} + \overline{D}\,\overline{G}\,\overline{J} + \overline{B}\,\overline{D}\,\overline{G}\,\overline{K}.$$

At inner step 4, the disjoint form of the inverse obtained in inner step 3 is generated by a stepwise minimization and reinversion that is a miniaturized reversal of inner-loop steps 1, 2, and 3. At each step, the inverted variables in the incumbent term are compared to the inverted variables in all prior terms. A polynomial is formed with the terms being lists of those variables; this polynomial is minimized and then put into disjoint form.

The disjoint form has four terms. The first term is $\overline{H}\,\overline{J}$. By comparing $\overline{B}\,\overline{H}\,\overline{K}$ to $\overline{H}\,\overline{J}$, the variable to be reinverted is *J*, which results in $\overline{B}\,\overline{H}\,J\overline{K}$. Comparing $\overline{D}\,\overline{G}\,\overline{J}$ to $\overline{H}\,\overline{J}$ and $\overline{B}\,\overline{H}\,\overline{K}$, *H* must be reinverted and $\overline{D}\,\overline{G}\,H\overline{J}$ is obtained. At the last step both *H* and *J* are reinverted; this results in $\overline{B}\,\overline{D}\,\overline{G}\,HJ\overline{K}$.

At step 5, multiply every term of the disjoint inverted form obtained at step 4 by the variables of the incumbent path to form a subpolynomial representing the added contribution of the incumbent path *ACEIL* to the probability of success, and at step 6 augment the system polynomial with this subpolynomial. The terms of this subpolynomial are listed in Figure 12.4.

238 Chapter 12 ■ *Sum of Disjoint Products*

GHIJK: The minimized polynomial at step 2 resulting from comparing the variables *G, H, I, J,* and *K* to the 12 prior paths is

$$L + DF + EF + ABD + ACD + ACE.$$

The combined inverse of the first three terms, $L + DF + EF$, is

$$\overline{F}\,\overline{L} + \overline{D}\,\overline{E}\,\overline{L};$$

Multiply this by the inverse of *A B D* to obtain

$$(\overline{F}\,\overline{L} + \overline{D}\,\overline{E}\,\overline{L})(\overline{A} + \overline{B} + \overline{D}) = \overline{A}\,\overline{F}\,\overline{L} + \overline{B}\,\overline{F}\,\overline{L} + \overline{D}\,\overline{E}\,\overline{L} + \overline{D}\,\overline{F}\,\overline{L}.$$

Multiply again by the inverse of *A C D*; this yields

$$(\overline{A}\,\overline{F}\,\overline{L} + \overline{B}\,\overline{F}\,\overline{L} + \overline{D}\,\overline{E}\,\overline{L} + \overline{D}\,\overline{F}\,\overline{L})(\overline{A} + \overline{C} + \overline{D})$$
$$= \overline{A}\,\overline{F}\,\overline{L} + \overline{D}\,\overline{E}\,\overline{L} + \overline{D}\,\overline{F}\,\overline{L} + \overline{B}\,\overline{C}\,\overline{F}\,\overline{L}.$$

Multiply by the inverse of *ACE* to obtain the combined inverse

$$(\overline{A}\,\overline{F}\,\overline{L} + \overline{D}\,\overline{E}\,\overline{L} + \overline{D}\,\overline{F}\,\overline{L} + \overline{B}\,\overline{C}\,\overline{F}\,\overline{L})(\overline{A} + \overline{C} + \overline{E}) =$$
$$\overline{A}\,\overline{F}\,\overline{L} + \overline{D}\,\overline{E}\,\overline{L} + \overline{B}\,\overline{C}\,\overline{F}\,\overline{L} + \overline{C}\,\overline{D}\,\overline{F}\,\overline{L}.$$

The disjoint form of the combined inverse is derived stepwise at inner step 4. The first term is $\overline{A}\,\overline{F}\,\overline{L}$; since $\overline{D}\,\overline{E}\,\overline{L}$ does not contain either $\overline{A}$ or $\overline{F}$, the second and third terms of the disjoint inverse are, respectively, $A\overline{D}\,\overline{E}\,\overline{L}$ and $\overline{A}\,D\overline{E}\,\overline{F}\,\overline{L}$; since $\overline{B}\,\overline{C}\,\overline{F}\,\overline{L}$ does not contain $\overline{A}$ or $\overline{D}\,\overline{E}$, the fourth and fifth terms are, respectively, $A\,B\,\overline{C}\,D\,\overline{F}\,\overline{L}$ and $A\,B\,\overline{C}\,D\,E\,\overline{F}\,\overline{L}$; the sixth term is $A\,B\,\overline{C}\,D\,E\,F\,\overline{L}$. The results of multiplying the disjoint inverse by the letters $GHIJK$ can be seen in Figure 12.4.

B C G H I J: The minimized polynomial of terms at steps 1 and 2 is

$$K + L + AD + AE + DF + EF.$$

The minimized inverse obtained at step 3 is

$$\overline{A}\,\overline{F}\,\overline{K}\,\overline{L} + \overline{D}\,\overline{E}\,\overline{K}\,\overline{L}.$$

The disjoint form of the inverse is

$$\overline{A}\,\overline{F}\,\overline{K}\,\overline{L} + A\overline{D}\,\overline{E}\,\overline{K}\,\overline{L} + \overline{A}\,D\overline{E}\,F\overline{K}\,\overline{L}.$$

Multiplying every term of the disjoint inverse by the variables *B C G H I J*, the corresponding disjoint subpolynomial given in Figure 12.4 is obtained.

Reliability importance

It is useful for a system engineer to know how much each element contributes to the overall probability of success. By finding which component or components make the greatest contributions, the designer can substitute more reliable components for the weak ones that are strategically placed, or make some other redesign to increase the reliability. *Importance*, or *reliability importance*, deals with constructing a scale to measure the relative contribution of each component. Even though the computations can be tedious and require the use of computers, SDP offers a simple and convenient way to obtain an importance function and evaluate it for every element or component.

Like reliability, importance is a number between 0 and 1. The difference, however, is that reliability is a *cardinal* number with the meaning of probability of success; while importance is an *ordinal* number that means something only by comparison to the importance numbers obtained for other elements of the same system.

Importance: A partial derivative

Mathematically, the reliability importance of a component is the partial derivative of the system reliability with respect to the component reliability. Thus, it is a function of both the configuration of the system and the reliabilities of all components. Because of disjointness and the fact that all terms are monomials, the SDP system logic function and the system probability formula are 1:1, and numerical operations such as taking derivatives are performed upon a logic function as if it were a probability formula.

Every variable of an SDP monomial term of a system formula has the exponent 1. Since a term of a monomial function is a product of all of the variables in that term, in the importance formula of a specified variable x_j, the partial derivative of each term with respect to the reliability of x_j is the product in the reliabilities of all of the other variables in the term. This is because of the partial differentiation the exponent of x_j is 0; this means that 1 is substituted for x_j in the corresponding term of the importance formula.

The sign, + or −, of a term in an importance function derived from an SDP system formula is determined by whether the differentiated variable x_j is positive (1 or success) valued, or inverted. This follows from the rules of calculus because the probability of success for x_j is p_j and the probability of failure is $1-p_j$. Therefore, the sign of the derivative of the term in the importance formula for a variable is the same as the sign of the variable in that term.

EXAMPLE 12.3.

Measuring reliability importance: Derive a reliability importance formula for every one of the five components A, B, C, D, and E of the bridge circuit of Figure 11.1, and calculate the importance values if every component has a reliability of 0.9.

Let R denote the system reliability in disjoint form,

$$R = AD + \bar{A}BE + AB\bar{D}E + A\bar{B}C\bar{D}E + \bar{A}BCD\bar{E} = 0.97848,$$

and let I_A, I_B, I_C, I_D, and I_E denote the importances, respectively.
By taking the partial derivatives

$$I_A = \frac{\partial R}{\partial A} = D - BE + B\bar{D}E + \bar{B}C\bar{D}E - BCD\bar{E} = 0.1062$$

$$I_B = \frac{\partial R}{\partial B} = \bar{A}E + A\bar{D}E - AC\bar{D}E + \bar{A}CD\bar{E} = 0.1062$$

$$I_C = \frac{\partial R}{\partial C} = A\bar{B}\bar{D}E + \bar{A}BD\bar{E} = 0.0162$$

$$I_D = \frac{\partial R}{\partial D} = A - ABE - A\bar{B}CE + \bar{A}BC\bar{E} = 0.1062$$

$$I_E = \frac{\partial R}{\partial E} = \bar{A}B + AB\bar{D} + A\bar{B}C\bar{D} - \bar{A}BCD = 0.1062.$$

Because all terms are monomials, the notation is simplified by having the letters A, B, C, D, and E interchangeably mean either a component or its reliability. At a common component reliability of 0.9, A, B, D, and E all have the same reliability importance, and each of these has about 6.5 times as much importance as does C.

Because component C in this example has the smallest effect on the system reliability, there is little to gain from increasing the reliability of C. Suppose, for example, that the reliability of C is increased to 0.95, while the other four components remain at 0.9: then R increases slightly to 0.97929, I_C remains the same at 0.0162 and the other four importance numbers all decline slightly to 0.1026; there are virtually no changes either in R or in the importances. For a comparable change in one of the other four components: if the reliability of A is increased to 0.95 while the other four components all remain at 0.9, R increases significantly to 0.983790.

13

Topological Reliability (TR)

Introduction

This chapter describes topological reliability (TR), mathematical graph theory, and search methods applied to derivation of a system-reliability formula. Topological reliability, which was introduced in a seminal 1978 paper by Satyanarayana and Prabhaker (S & P), develops a compact nested formula such that, fully expanded, it would be the same as that obtained by IE. The procedural, terminological, and other differences between TR and IE are substantial, but what results from TR is a different version of the same formula.

The objective of this chapter is to clarify the concepts of TR, using terminology and notation that is adapted from the original S & P article, with some minor exceptions. For the reader who already has a working familiarity with chapter 11 on IE, it may help to start this description with a list of the principal differences between IE and TR.

The description of topological reliability and the S3T problem in this chapter are from M. O. Locks, "Recent Developments in the Computing of System Reliability," *IEEE Transactions on Reliability* (vol. 34, 1985, 425–36), with permission, © 1985 IEEE.

Principal differences between TR and IE

1. Inclusion-exclusion starts from a list of the minimal paths of an ST (source-to-terminal) system. TR starts from an ST system graph, where the edges of the graph stand for the components.
2. Each term of an IE formula represents either the components of a path or of an intersection of paths. The comparable concept in TR is a node of a search tree representing a subgraph that contains a subset of the edges of the system graph.
3. An IE system formula is a sum of products of the reliabilities of the components of terms, with the terms alternating + and − signs. A TR formula is a product of the reliabilities of all components multiplied by a nest consisting of the sum of products of inverses of the reliabilities of components, where the placement of elements in the nest follows the search rules; and the alternation of + and − signs is based upon the graph-theory concept of domination.
4. When a system graph or logic diagram has two-way edges or cycles, IE tends to generate many duplicated terms that do not contribute anything to the system-reliability formula and are ultimately cancelled out in processing. In TR, cycles are removed in any branch of the search tree before developing any subgraphs in that branch, and the formula is derived only from those ST subgraphs with no cycles.

Concepts of topological reliability

Denote the system-reliability ST graph, also known as the logic diagram, as G_0. A fundamental fact of TR is that there is a 1:1 relationship between certain subgraphs of G_0, called the p-acyclic subgraphs, and the noncancelling terms of the IE formula. Therefore, if the p-acyclic subgraphs are identified by an efficient search, processing time is saved in deriving the formula, because there are no duplications or cancellations.

TR decomposes G_0 by systematically stripping away edges or sequences of edges to find the p-acyclic subgraphs. With each p-acyclic subgraph, an incremental change is made in the system-reliability formula. Instead of being a sum of monomial terms, as in IE, the formula is in a nested and factored form that is equivalent factor by term to the

IE polynomial, and also follows the order of the search. Each factor also has the same sign, + or −, as the corresponding IE term.

Terminology

Acyclic and cyclic graphs: p-, p-*acyclic, and* p-*cyclic graphs*

An acyclic graph has no cycles; a cyclic graph has at least one cycle. A p-*graph* is a subgraph of G_0 with every edge on a path from source S to terminal T. A *p*-acyclic graph is a *p*-graph with no cycles; a *p*-cyclic graph is a *p*-graph with at least one cycle. Figure 13.1 has some examples of different graphs.

Tree search

Tree search identifies the subgraphs of G_0. Every *node* represents a subgraph, and every *internode* (line between nodes) is the removal of

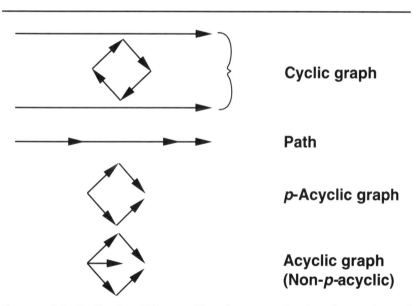

Source: M. O. Locks, "Recent Developments in the Computing of System Reliability," *IEEE Transactions on Reliability* (vol. 34, 1985, 425–36), with permission, © 1985 IEEE.

Figure 13.1. Examples of types of graphs.

either an edge or a consecutive sequence of edges to form the subgraph at the next node. If the subgraph from which a string is removed is p-acyclic, the string is called a *neutral sequence,* and the resulting subgraph is also p-acyclic.

A search tree may be rooted at any place on a page: left, right, top, or bottom. S & P make the starting point of tree search, called the *root,* representing G_0, at the top, and a tip of a branch, called a *leaf,* at the bottom. A leaf may be a path or else the end of a branch that is bypassed because of a backtracking rule.

Family relationships of nodes

In graph theory, the descriptive family names of nodes of a search tree have a genealogy like ordinary usage of the same terms. At any given node G_i a subgraph of G_0, the nodes obtained by branching outward and downward are the *children,* and G_i is the *father.* Nodes above G_i in the same branch are *ancestors* and nodes below G_i in the same branch are *descendants.* Two or more nodes with the same father are *brothers.* The order of priority in processing is determined by which brother is senior and which brother is junior. In this text, the order of processing is from right to left; an *elder brother* is on a right branch and a *younger brother* is on the left branch. Figure 13.2 depicts family relationships of the nodes of the search tree.

Neutral sequences

A *neutral sequence* is a consecutive string of vertices and edges in a p-acyclic subgraph with no internal vertices connecting to other parts of the subgraph; removing a single edge breaks the connection between S and T. In order that every child and every descendant of a p-acyclic subgraph shall also be p-acyclic, only neutral sequences are removed. A neutral sequence can also be a single component.

Formations and dominations: The sign of a factor in the formula

Two concepts from graph theory, formation and domination, help explain the signs of the factors in the system-reliability formula. Understanding these concepts provides insights into the connection between construction of a subgraph and its contribution to the formula, as well as the procedural shortcuts of TR.

A *formation* F of a p-graph G which is a subgraph of G_0 is an inclusive-or union of a set of paths from S to T, that includes all of the

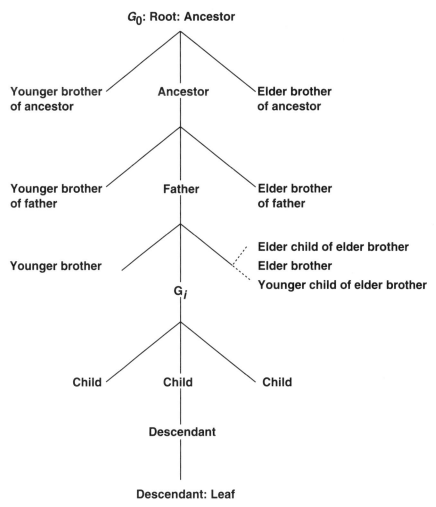

Source: M. O. Locks, "Recent Developments in the Computing of System Reliability," *IEEE Transactions on Reliability* (vol. 34, 1985, 425–36), with permission, © 1985 IEEE.

Figure 13.2. Family Tree of G_i.

vertices and edges of G. F is odd, with a sign of +1 if the number of paths in F is odd, and even, with a sign of −1, if the number of paths is even. The *domination D* is the sum of the signs of all of the formations of G.

It has been shown that for a p-acyclic graph, D is always either $+1$ or -1 (there is either one odd formation more than the number of even formations, or else one even formation more than the number of odd formations). For a p-cyclic graph, $D = 0$, meaning that there are exactly the same number of even formations as odd formations.

There is a relationship between the method of construction of the subgraph G_i and D that makes it possible to find the sign of the factor corresponding to G_i in the formula, without identifying or counting the formations. If $D = +1$, the factor has a + sign; if $D = -1$, the factor has a − sign; and if $D = 0$, the subgraph is p-cyclic. The simplification is that for a p-acyclic graph, it is only necessary to count the number of vertices and edges in order to find the value of D. If the difference between the number of vertices and the number of edges is odd, $D = +1$ and the sign of the factor is positive; if the difference is even, $D = -1$ and the sign is negative.

In order to find the sign, it is usually not necessary to count vertices and edges. Since the child of a p-acyclic father strips away a neutral sequence, the number of edges removed is one greater than the number of vertices removed. For example, if the neutral sequence is a single edge, no vertices are removed; if the neutral sequence removed is a string of two edges, the only vertex removed is the one connecting the two edges, and so on. As a result, the signs of the nodes at successive levels of the search tree alternate. Once a p-acyclic ancestor is attained, the signs of all of its descendants are obtained just by alternating plus and minus.

Formulating the system graph: Two-way edges

The system-reliability graph G_0 has standard success logic; a serial connection of components in a subpath means that all of the components in the subpath are needed for success, and a set of parallel connections means redundancy. A minor modification is needed when one or more of the edges is two-way. Since a two-way edge is a cycle, two one-way edges in opposite directions are substituted for each two-way edge, to facilitate decycling.

The rules of tree search

The weight restriction

The subgraphs and the corresponding factors of the TR formula are obtained in the search, without any duplications. The principal reason there are no duplications is a rule, called the *weight restriction* (WR). The set of edges stripped by G_i is the *weight* of G_i. The WR requires that G_i may not remove any edge that is in the weight of an elder brother, or the weight of an elder brother of the father or of the elder brother of any ancestor.

The four processing rules

The four processing rules of TR are applied in the following order: rule 1 first, if applicable; then rule 2, rule 3, or rule 4.

Rule 1: Cyclic subgraph

If G_i is cyclic, decycle it by removing the edges on the cycle one-by-one, except for those edges in the weight of an elder brother or of the elder brother of the father or of any ancestor. Explanation: Create children by stripping edges on the cycle, one edge per child.

The corollary to rule 1 is that if a child of a cyclic graph is also cyclic, continue applying rule 1 until an acyclic descendant is found.

Rule 2: Acyclic subgraph that is not a p-graph

Create a single child that strips all of the edges that are not on paths from S to T, except those that are in the weight of an elder brother or the elder brother of the father or of any ancestor; the edges removed need not be consecutive. Explanation: If all loose edges are removed, the subgraph is p-acyclic.

If an edge that is not on a path from S to T cannot be removed because of the WR, this branch of the tree is processed no further, and there is a backtrack to the next node in order of priority.

Rule 3: p-Acyclic subgraph with a non-p-acyclic father

If G_i is p-acyclic and the father is not p-acyclic, remove all neutral sequences except those containing edges that cannot be removed because of the WR. Each different neutral sequence removed results in a different child of G_i. In this text we follow a protocol that selects the edges for removal in alphabetical or numerical (that is, lexicographical) order. Rule 3 ensures that all of the descendants of a p-acyclic G_i will be p-acyclic.

Rule 4: Decomposition of a p-acyclic child of a p-acyclic father

If G_i and its father are both p-acyclic, the weights of the children of G_i are 1:1 identical with the younger brothers of G_i. The reason for this is that all of the neutral sequences were previously identified when the children of the father were generated. Since rule 4 is governed by the WR, it speeds up processing by eliminating the need to search for *p*-acyclic subgraphs.

Backtracking and stopping

The search has both forward and backward steps, and is known as *depth-first search*. In the forward steps, a child and the descendants of the child of an elder brother are visited before a younger brother. A backtrack takes place when a leaf is attained. A leaf may be either a path or an acyclic subgraph such that if any more edges were removed, either an open graph would result or a violation of the WR would occur.

In backtracking, the next node visited is the eldest younger brother, followed by forward search. If there are no younger brothers, move the pointer one level up the tree and to the left to the younger brother of the father, and continue forward. Backtracking stops when there are no ancestors with younger brothers.

Deriving the system formula

Node labels and processing order

Every subgraph descended from G_0 has an identifying label or sequence number assigned in the forward search. The order that nodes are processed, however, is different from the sequence number, because of depth-first search rules. All subgraphs are labeled, but a new factor is entered into the system-reliability formula only when the pointer is at a node representing a *p*-acyclic subgraph.

The system-reliability formula

Let Y be the probability that every single element of the system graph G_0 is successful, namely, the product of the reliabilities of all of the elements. Let G_i be a *p*-acyclic subgraph, D_i its domination, x_i the set of variables in its weight, and $\mathbf{x}_i$ the inverse of the product of the probabilities of success for all of the variables in the weight. The system formula is a nested sequence of products and sums of products of the form x_iD_i. The following example illustrates the principles involved.

$$R = Y(\mathbf{x}_1(D_1 + \mathbf{x}_2(D_2 + \mathbf{x}_3D_3 + \mathbf{x}_4D_4)) + \mathbf{x}_5(D_5 + \mathbf{x}_6(D_6 + \mathbf{x}_7D_7))). \quad (13.1)$$

A multiplication of the form $\mathbf{x}_i\mathbf{x}_j$ indicates a parent-child relationship.

For example, in Equation 13.1, G_2 is a child of G_1, and G_3 and G_4 are both children of G_2; likewise G_6 is a child of G_5, and G_7 is a child of G_6. Note that the signs of the dominations alternate from $+$ to $-$, or vice versa, from father to child. A plus (+) sign represents a backtrack: for example, the backtrack after G_3 is to its younger brother G_4; the backtrack after G_4 is to G_5. A subtraction ($-$) in the formula represents forward search, as is shown in the following example.

Relationship to the IE formula

The expansion of Equation 13.1 helps to explain the relationship of IE and TR; when all of the factors are multiplied out, we have

$$Y(x_1 D_1 + x_1 x_2 D_2 + x_1 x_2 x_3 D_3 + x_1 x_2 x_4 D_4 + x_5 D_5 + x_5 x_6 D_6 + x_5 x_6 x_7 D_7) \quad (13.2)$$

Equation 13.2 is the IE system-reliability formula. Each of the seven terms represents a p-acyclic subgraph. The domination D of a term is also its sign, $+$ or $-$; D is $+$ if there are an odd number of factors x_i in the term, and $-$ if there are an even number of factors. The x_i show (1) which components are in the term and subgraph and which ones are not; (2) ancestor-descendant relationships: G_1 is the ancestor of G_2, G_1 and G_2 are ancestors of G_3 and G_4, and so on.

Example: The bridge circuit

The bridge circuit example of Figure 11.1 is used again to illustrate the similarities and differences of IE and TR. This circuit has one two-way edge C which is now split into two one-way edges, C_1 downward and C_2 upward, in order to remove the cycle.

Figure 13.3 depicts the search tree of the bridge circuit. The diamond-shaped diagram at the top is the revised drawing, including the two-directional split-up of edge C. The bottom portion is the search tree, showing the sequence of processing, the node labels, and the names of the edges removed. The root at the top of the tree is the cyclic graph G_0; the p-acyclic subgraphs are the other numbered nodes of the tree, all shown as filled-in dots.

Figure 13.4 shows the p-acyclic subgraphs of the bridge circuit, in the order that they are searched under the depth-first protocol. The number at the left of each subgraph is the node label, the identification number assigned to it on the tree. For example, G_1 and G_2 are the children of the father G_0, and so on, even though the branches of G_1 are searched before G_2 and its branches. At the right of each entry in

250   Chapter 13 ■ *Topological Reliability (TR)*

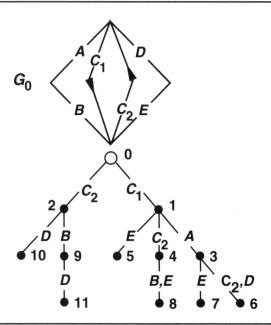

Figure 13.3. Search tree for the bridge circuit.

Figure 13.4 is (are) the name(s) of the edge(s) in the neutral sequence that was just removed.

Rule 1 requires that a cyclic graph be decycled. Since G_0 is cyclic, two children are formed, G_1 with C_1 removed and G_2 with C_2 removed. G_1 and G_2 are both p-acyclic and rule 2 is not invoked; likewise, all of the children of G_1 and G_2 and their other descendants are p-acyclic. The children of G_1 and G_2 are obtained by rule 3, in alphabetic order; the children of the children of G_1 and G_2 are obtained by rule 4. For example, G_3 with A removed, G_4 with C_2 removed, and G_5 with E removed are the children of G_1; in turn, the children of G_3 are G_6 with C_2 removed (as well as D, because together C_2 and D form a neutral sequence) and G_7 with E removed.

The dominations

The dominations, +1 or −1, for the p-acyclic subgraphs are obtained either from Figure 13.4 or from Figure 13.3. At the first tier, G_1 and G_2 have 5 edges and 4 vertices apiece and dominations of +1. At the second tier, G_3, G_4, G_5, G_9, and G_{10} have 4 vertices and 4 edges apiece

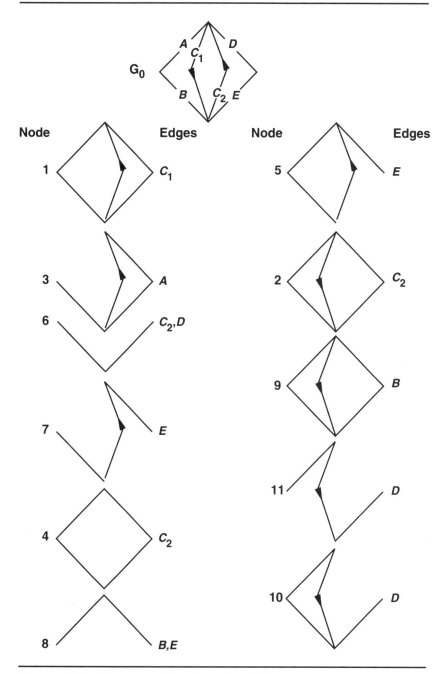

Figure 13.4. The p-acyclic subgraphs of G_0.

252 Chapter 13 ■ Topological Reliability (TR)

and dominations of -1. At the third level, G_6, G_7, G_8, and G_{11}, all derived by deleting neutral sequences from nodes with -1 dominations, are odd, with dominations of $+1$.

The system-reliability formula

Let A, B, C_1, C_2, D, and E now represent the respective reliabilities of the components of G_0 (C_1 and C_2 are of course the same component); the nested system formula is derived from the search tree, based on the components removed at each level and the dominations, using a format similar to Equation 13.1.

$$R = A \cdot B \cdot C_1 \cdot C_2 \cdot D \cdot E \left(\frac{1}{C_1} \left(1 - \frac{1}{A} \left(1 - \left(\frac{1}{C_2} \cdot \frac{1}{D} + \frac{1}{E} \right) \right) \right) \right.$$
$$\left. - \frac{1}{C_2} \left(1 - \frac{1}{B} \frac{1}{E} \right) - \frac{1}{E} \right) + \frac{1}{C_2} \left(1 - \frac{1}{B} \left(1 - \frac{1}{D} \right) - \frac{1}{D} \right) \right).$$

In the expansion of this nested form, C is substituted for either C_1 or C_2. The expansion results in

$$R = ABCDE - BCDE + BE + BCD - ABDE + AD - ABCD$$
$$+ ABCDE - ACDE + ACE - ABCE;$$

this is identical to the IE formula for the probability of success obtained for the bridge circuit in chapter 11, but with the terms in a different order. This expansion was done only for tutorial purposes. It is unnecessary to expand a nested form in practice because it is derived simultaneously with the growing of the search tree, by entering the appropriate arithmetic operation and sign and the reliability(ies) of the component(s) for each element.

Extensions of TR: Unified approach: *k*-Terminal network

Introduction

The previous portion of this chapter has dealt entirely with using TR for solving what is frequently called a *coherent ST system*. A coherent ST system has both a single source S and a single terminal T and every edge is on at least one minimal path. In 1982 Satyanarayana showed that non-ST graphs can be suitably modified and solved by TR. Two examples are

1. Systems such that every vertex is both a source and a terminal in

Chapter 13 ■ *Topological Reliability (TR)*

two-way communication with every other vertex (known as *overall reliability*)
2. Systems with mixtures of one- and two-way connections.

One special case that occurs in communication networks is the k-terminal problem. A subset of k of the vertices are compulsory recipients of every message, with two-way communication between every member of the k-subset, the other vertices primarily serving the function of relays.

The key idea that allows this type of structural solution of a generalized network problem is the unified approach—creating a supergraph with a dummy terminal T^* and a dummy source S^*, if necessary, and dummy one-way edges connecting members of the k-subset to S^* or T^* so that a system-reliability formula is generated by TR with the same technique as for coherent ST systems. The nodes of the search tree corresponding to removal of these dummy edges are elder brothers to every ancestor of every other node, and the dummy edges cannot be removed to form p-acyclic subgraphs, because of the weight restriction.

Example: Source-to-k-terminal problem

The source-to-k-terminal (SKT) problem example is a 4-vertex 5-edge diamond-shaped graph that differs from a bridge circuit in that there are two terminals rather than one, both terminals being in the middle rather than at the right-hand side, and one edge is directed from right to left rather than from left to right. We call this a k-graph, $S3T$, with $k = 3$, because there are 3 compulsory vertices. Since edge C goes both ways, it is replaced by two one-way edges, C_1 and C_2, in opposite directions. The original graph and its replacement are the first two figures in Figure 13.5.

Decomposing G_0

The system-reliability problem is to determine the probability that a message or signal from the source vertex S is received by both terminals T_1 and T_2. Form a supergraph which we shall call G_0, the third figure in Figure 13.5, with an artificial terminal T^* and the dummy edges F_1 and F_2 that connect T_1 and T_2 to T^*. Decompose G_0 to find the k-acyclic subgraphs that include both terminals T_1 and T_2, such that every edge is on a path from S to T^*. The search tree, including the dummy edges, is shown in Figure 13.6 and the subgraphs corresponding to the nodes of the tree, with the dummy edges deleted, and their dominations are displayed in Figure 13.7.

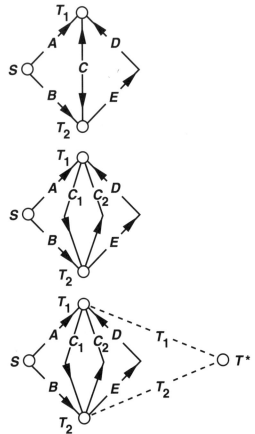

Source: M. O. Locks, "Recent Developments in the Computing of System Reliability," *IEEE Transactions on Reliability* (vol. 34, 1985, 425–36), with permission, © 1985 IEEE.

Figure 13.5. $S3T$ graph, modified graph, and supergraph.

The first two children of G_0, G_1, and G_2 are dummy nodes resulting from the removal of the dummy edges F_1 and F_2 and are elder brothers to an ancestor of every node on the tree. F_1 and F_2 are not removed to form any other subgraphs because of the weight restriction; this guarantees that both terminals T_1 and T_2 are included in every k-acyclic subgraph descended from G_0 (k-acyclic means a k-graph that is acyclic; however, to the computer, in this case, it would appear as a p-graph). Only the k-acyclic subgraphs of G_0 are shown on the tree as inked-in dots.

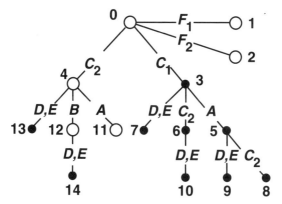

Source: M. O. Locks, "Recent Developments in the Computing of System Reliability," *IEEE Transactions on Reliability* (vol. 34, 1985, 425–36), with permission, © 1985 IEEE.

Figure 13.6. S3T search tree.

G_0 is cyclic; following rule 1, decycle G_0 by stripping edges C_1 and C_2 to form the children G_3 and G_4. By the depth-first protocol, the children of G_3 are searched before the children of G_4. G_3 is k-acyclic, and therefore all of its descendants are also k-acyclic, as is shown in Figure 13.7.

G_4 is cyclic; by the corollary to rule 1, the decycling continues. The children of G_4 are G_{11} with A removed, G_{12} with C_2 removed, and G_{13} with D and E removed. G_{11} is also cyclic, but cannot be stripped any further because (1) removing B would result in a graph that is not a k-graph, (2) removing C_1 would violate a weight restriction, and (3) removing D and E would result in another cyclical graph. G_{12} is also cyclic, but by removing D and E, we obtain a k-acyclic graph, G_{14}, with a domination of $+1$. G_{13} is k-acyclic.

The reliability formula

The reliability formula is

$$R = A \cdot B \cdot C_1 \cdot C_2 \cdot D \cdot E \left(\frac{1}{C_1} \left(1 + \frac{1}{A} \left(-1 + \frac{1}{C_2} + \frac{1}{DE} \right) \right. \right.$$
$$\left. \left. + \frac{1}{C_2} \left(-1 + \frac{1}{DE} \right) - \frac{1}{DE} \right) + \frac{1}{C_2} \left(\frac{1}{BDE} - \frac{1}{DE} \right) \right).$$

If all components have reliabilties of 0.9, the system reliability is 0.979284, to six significant figures.

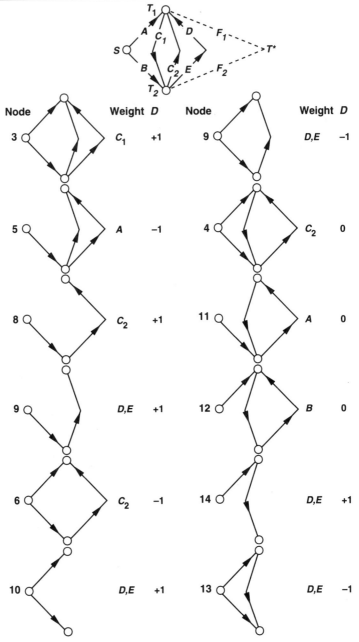

Source: M. O. Locks, "Recent Developments in the Computing of System Reliability," *IEEE Transactions on Reliability* (vol. 34, 1985, 425–36), with permission, © 1985 IEEE.

Figure 13.7. Decomposition of G_0: G_1, G_2, F_1, F_2, and T^* deleted.

Problems and References for Part V

Problems on system reliability

V.1. A system is composed of three components A, B, and C, with reliabilities of 0.5, 0.4, and 0.9, respectively. B is a standby that switches on only in case A fails, with a switching probability of 0.9, and C is a standby that switches on only in case both A and B fail, with a switching probability of 0.8. What is the reliability of the combined system?

V.2.
   a. A four-engine jet aircraft can land with two engines. If the probability that an engine can complete a trip is 0.95, what is the probability that the aircraft will complete the trip?
   b. Let a, b, c, and d, respectively, represent the probabilities of success for each of the four engines on a single trip, and let $(1-a)$, $(1-b)$, $(1-c)$ and $(1-d)$, respectively, be the probabilities of failure. Write a disjoint formula for the system probability of success as a function of these eight numbers in six terms.
   Hint: The numerical value is the same as in V.2a.

V.3. A coolant loop for a nuclear reactor has three subsystems—the pump, the power, and the piping. All three of these subsystems must work in order for the coolant loop to operate. The probability of a pump failure is 0.2; the probability of a power failure is 0.05; and the probability of a piping failure is 0.001.

a. What is the probability of a coolant-loop failure?
b. If the power fails, what is the probability of loop failure?
c. When there is standby operation, 60 percent of the pump failures cause loop failures and 40 percent of the pipe failures cause loop failures. If a loop fails and this was not caused by a power failure, what is the probability that the failure was caused by the pump? *Note:* This requires Bayes theorem.
d. If the reactor is designed with four coolant loops, but it can still continue to operate with three loops working, what is the probability that at least three coolant loops will work?

V.4. For the following circuit logic diagram,

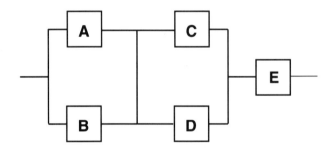

a. Let a, b, c, d, and e denote the reliabilities for components A, B, C, D, and E, respectively. Give the formula for the system reliability by the method of inclusion-exclusion.

$R =$

b. Give the formula for the system reliability by disjoint products.

$R =$

c. Give the formula by parallel, series, or series-parallel reductions.

$R =$

d. What is the reliability of the system if the component reliabilities a, b, c, d, and e are all 0.9?

V.5. For problem V.4,

a. Give the formula for the *importance* of each of the five components of the system, using the results of V.4b, by taking partial derivatives of the disjoint-products formula.

   A:
   B:
   C:
   D:
   E:

b. Give the numerical values of the importances of each of the five components.

V.6. A solar energy module has two blocks of solar collectors, in series. Each block consists of three parallel strings of two collectors each in series. The module is bridged by two series pairs of strings of two bypass diodes for each string.

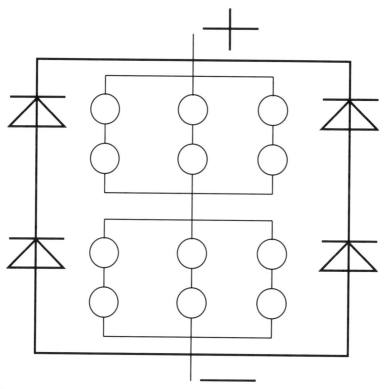

Let C denote collector reliability and D diode reliability. Assume that all component failures are independent.

a. Write a general reliability formula for a solar energy module.

$R =$

b. If $C = 0.6$ and $D = 0.5$, what is the reliability of the module?

The solution is obtained systematically by building up from the inside out. Start from the smallest unit, a string of two collectors in series, then a block of three strings in parallel, then the series string of two blocks, and then the two strings of diodes in parallel with the collectors. Determine

    i. The reliability of a string of collectors:
    ii. The reliability of a block:
    iii. The reliability of the string of two blocks:
    iv. The reliability of the diodes:
    v. The reliability of the solar module:

V.7. A ship's guidance system is controlled by an NDS computer with three major modules A, B, and C. All three modules must function in order for the guidance system to work. Modules A and B both have reliabilities of 0.970 and C has a reliability of 0.980.

    a. What is the reliability of the NDS (3 digits)?
    b. A backup NDS computer is to be installed, identical to the original one. If the backup goes on automatically in case the original fails, what is the reliability of the combined system?
    c. If the backup NDS computer is to be activated by a switch with a reliability of 0.8, what is the reliability of the combined system?

V.8. The following circuit diagram represents a ring circuit for a local area network with four nodes or terminals: 1, 2, 3, and 4, as well as five edges or transmission lines: A, B, C, D, E. Messages travel from node 1 at the top of the ring to the other three nodes. Success is defined by having the message from node 1 received at all of the other three nodes. Note that A from node 1 to node 2 and B from 1 to 3 are both one-way edges and that C, D, and E are all two-way edges, so that a message can be transmitted either from node 2 to node 3 or from node 3 to node 2.

Problems and References for Part V 261

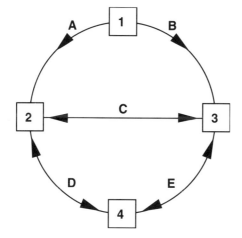

This network has four minimal paths: *ACE*, *ADE*, *BCD*, and *BDE*.

a. Assuming that A represents the reliability of edge A, B represents the reliability of edge B, and so on, show the buildup of the system-reliability formula in four steps by the method of inclusion-exclusion.

   i. $R(1) = ACE$
   ii. $R(2) = R(1) +$
   iii. $R(3) = R(2) +$
   iv. $R(4) = R(3) +$

b. If $A = B = C = D = E = 0.9$, what is the reliability of this network, using the formula derived in V.8a?

V.9 For the same network as problem V.8, and using the same assumptions and data,

a. Derive the system reliability formula in four steps by the method of disjoint products.

   i. $R(1) = ACE$
   ii. $R(2) = R(1) +$
   iii. $R(3) = R(2) +$
   iv. $R(4) = R(4) +$

b. What is the value of the system reliability, using the formula you derived in V.9a?

V.10.

a. Using the disjoint formula you obtained in V.9a, derive the formulas (the partial derivatives) for

   i. Importance of A:
   ii. Importance of B:
   iii. Importance of C:
   iv. Importance of D:
   v. Importance of E:

b. Give the relative numerical value of the importance of each of the five components, based on the partial derivatives in V.10a, assuming that $A = B = C = D = E = 0.9$.

V.11. A two-out-of-three system has three identical components, A, B, and C. The system works if at least two of three components operate. Let the symbols A, B, and C, respectively, denote the reliabilities of the three components.

a. What are the system minimal paths?
b. Give the reliability formula in disjoint form.
c. What is the numerical system reliability if $A = B = C = 0.9$?
d. Give the importance formula for each of the three components.
   A:
   B:
   C:

e. Give the relative importance value for each of the three components if $A = B = C = 0.9$.

References

Abraham, J. A. 1979. An improved method for network reliability. *IEEE Trans. Reliability* 28:58–61.

Aggarwal, K. K., K. B. Misra, and J. S. Gupta. 1975. A fast algorithm for reliability evaluation." *IEEE Trans. Reliability* 24:83–85.

Agrawal, A., and R. E. Barlow. 1984. A survey of network reliability and domination theory. *Operations Research* 32:478–92.

Ball, M. O. 1986. Computational complexity of network reliability analysis. *IEEE Trans. Reliability* 35:230–39.

Ball, M. O., and J. S. Provan. 1988. Disjoint products and efficient computation of reliability. *Operations Research.* 36:703–15.

Barlow, R. E., and F. Proschan. 1965. *Mathematical theory of reliability.* New York: Wiley.

———. 1975. *Statistical theory of reliability and life testing.* New York: Holt, Rinehart, and Winston.

Billinton, R., and R. N. Allan. 1983. *Reliability evaluation for engineering systems.* New York: Plenum Press.

Colbourn, C. J. 1987. *The combinatorics of network reliability.* New York: Oxford University Press.

Cooper, D. F., and G. B. Chapman. 1987. *Risk analysis for large projects.* New York: Wiley.

Dugan, J. B., S. J. Bavuso, and M. A. Boyd. 1992. Dynamic fault-tree models for fault-tolerant computer systems. *IEEE Trans. Reliability* 42:363–77.

Feller, W. 1968. *An introduction to probability theory and its applications.* Vol I. 3d ed. New York: Wiley. (In particular, see chapter 4.)

Frankel, E. G. 1988. *Systems reliability and risk analysis.* Boston: Kluwer.

Fratta, L., and U. G. Montanari. 1973. A recursive method based on case analysis for computing network terminal reliability. *IEEE Trans. Communications* COM-26:1166–77.

Fussell, J. B., E. F. Aber, and R. G. Rahl. 1976. On the quantitative analysis of priority and failure logic. *IEEE Trans. Reliability* 25:324–26.

Gnedenko, B., Y. Belyayev, and A. Solovyev. 1969. *Mathematical methods of reliability.* New York: Academic Press.

Grnarov A., L. Kleinrock, and M. Gerla. A new algorithm for network reliability computation. *Proceedings of the 1979 Computer Networking Symposium.* IEEE and U.S. Dept of Commerce. National Bureau of Standards 79CH1467OC.

Haasl, D. F. 1965. Advanced concepts in fault tree analysis. *System Safety Symposium.* Seattle, Wash.: The Boeing Company. Available from the University of Washington Library.

Heidtmann, K. D. 1989. Smaller sums of disjoint products by subproduct inversion. *IEEE Trans. Reliability* 38:305–11.

Henley, E. J., and H. Kumamoto. 1981. *Reliability engineering and risk assessment.* New York: Prentice Hall.

Ireson, W. G., and C. F. Coombs. 1988. *Handbook of reliability engineering and management.* New York: McGraw-Hill.

Lin, P. M., B. J. Leon, and T. C. Huang. 1976. A new algorithm for symbolic reliability analysis. *IEEE Trans. Reliability* R-25 (April): 2–15.

Locks, M. O. 1982. Recursive disjoint products: A review of three algorithms. *IEEE Trans. Reliability* 31: 36–39.

———. 1984. Comments on: Improved method of inclusion-exclusion applied to k-out-of-n systems. *IEEE Trans. Reliability* 33:321–22.

———. 1985. Recent developments in computing of system reliability. *IEEE Trans. Reliability* 34:425–36.

———. 1987. A minimizing algorithm for the sum of disjoint products. *IEEE Trans. Reliability* 36:445–53.

———. 1993. Anomalies in interpreting a fault tree. *IEEE Trans. Reliability* 42:314–17.

Locks, M. O., and J. M. Wilson. 1992. Note on disjoint products algorithms. *IEEE Trans. Reliability* 41:81–84.

Martz, H. F., and R. A. Waller. 1982. *Bayesian reliability analysis.* New York: Wiley.

Misra, K. B. 1992. *Reliability analysis and prediction.* Amsterdam and New York: Elsevier.

Pages, H., and M. Gondrain. 1986. *System reliability.* Translated by E. Griffin. Berlin: Springer-Verlag.

Parker, K. P., and E. J. McCluskey. 1975. Probabilistic treatment of general combinational networks. *IEEE Trans. Computers* C-24:668–70.

Provan, J. S., and M. O. Ball. 1984. Computing network reliability in time polynomial in the number of cuts. *Operations Research* 32:516–26.

Rai, S., and K. K. Aggarwal. 1978. An efficient method for reliability evaluation of a general network. *IEEE Trans. Reliability* R-27:206–11.

Satyanarayana, A. 1982. A unified formula for the analysis of some network reliability problems. *IEEE Trans. Reliability* 31:23–32.

Satyanarayana, A., and A. Prabhaker. 1978. New topological formula and rapid algorithm for reliability analysis of complex networks. *IEEE Trans. Reliability* 27:82–100.

Schneeweiss, W. 1989. *Boolean functions with engineering applications and computer programs.* Berlin: Springer-Verlag.

Shier, D. R. 1991. *Network reliability and algebraic structures.* Oxford, N.Y.: Clarendon Press.

Shooman, M. 1968. *Probabilistic reliability: An engineering approach.* New York: McGraw-Hill.

Soh, S., and S. Rai. 1990. CAREL: Computer aided reliability evaluator for distributed computing networks. *IEEE Trans. on Parallel and Distributed Systems* 2:199–213.

Srinivasan, S. K., and R. Subramanian. 1980. *Probabilistic analysis of redundant systems.* Berlin: Springer-Verlag.

Tsuboi, T., and K. Aihara. 1975. A new approach to computing terminal reliability in large complex networks. *Electronics and Communications in Japan* 58A:52–61.

Veeraraghavan, M., and K. Trivedi. 1991. An improved algorithm for symbolic reliability analysis. *IEEE Trans. Reliability* 40:347–58.

Villemeur, A. 1992. *Reliability, availability, maintainability, and safety assessment.* 2 vols. New York: Wiley.

Wilson, J. M. 1990. An improved minimizing algorithm for sum of disjoint products. *IEEE Trans. Reliability* 39:42–46.

Zelen, M. S. 1962. *Statistical theory of reliability.* Madison, Wis.: University of Wisconsin Press.

Solutions for Problems

Chapter 1

1.3 a. 0.0108
     b. 0.1296
     c. 0.1296
     d. 3·3·3·3 = 81
     e. 0.375

1.4. 0.8

1.5 a. 0.3
     b. 0.6
     c. 0.1

1.6. 0.1281

Chapter 2

2.3 a. a, b

  b. $\dfrac{b+a}{2}$

  c. $\dfrac{2}{b-a}$

2.6. $x = 0 \Rightarrow z = -1.225;\ x = 1 \Rightarrow z = 0.814$

2.11 a. 5.833
      b. 2.415

2.12 a. 0.21
      b. 1.05
      c. 2.10

2.13 a. $0.90
      b. $² 1.89
      c. $1.37

(*Note*: The variance has units of dollars-squared, while the standard deviation has units of dollars.)

2.14. $\dfrac{C(6,6) \cdot C(43,43)}{C(49,6)} = \dfrac{1}{13,983,816}$

Chapter 3

3.2. 0.9

3.3. The answer is slightly more than 200, with no failures; for 200 successes in 200 trials, there would be 90 percent confidence in 0.989 reliability.

3.4. 21

3.5. 37 trials, by interpolating between $n = 35$ and $n = 40$

3.6. $1 - P(X \leq 3 | p = 0.9, n = 5) = 1 - 0.0815 = 0.9185$

3.7. $1 - P(X \leq 5 | p = 0.7, n = 8) = 1 - 0.4482 = 0.5518$

3.8.
AQL%	P(accept)
0.5	0.9511
1.0	0.9044
2.0	0.8171
4.0	0.6648
6.0	0.5386

3.9. $\dfrac{2}{0.06} \approx 34$

Chapter 4

4.1. $D = 0.137 < D(0.10) = 0.239$

4.2. Mean = 501.4; s.d. = 11.8715. Lognormal is slightly better, because $D = 0.212$ vs. $D = 0.218$ for normal; also lognormal has a better straight line; the mean log is 2.70 and the s.d. log is 0.0102.

4.3 b. Lognormal is a little better; D is 0.124 vs. 0.152 for normal.
    c. $\mu = 178.417$; $\sigma = 39.410$
    d. $\mu = 2.242$; $\sigma = 0.09457$
    e. For the calculated normal distribution,
$$P(Z > -0.721) = 0.765$$
for the calculated lognormal distribution,
$$P(Z > -0.697) = 0.756$$

4.4 a. Est. $\mu = 0.2888$; est. $\sigma = 0.6167$
    b. Yes
    c. 0.13; good fit
    d. $10^\mu = 1.945$ days
    e. 8.044 days

4.5 a. 1617.88, 13.10
    c. $D = 0.238$
    d. Deviation is significant at the 5 percent level; $D(0.05) = 0.206$.

4.6. $F_b(290|0.95, 301) \cong F_N\left(\dfrac{290 - (0.95)(301) + 0.5}{((301)(0.95)(0.05))^{1/2}}\right) = 0.885$

4.7 b. 1.527
    c. 0.541
    f. 0.118
    g. 33.7 hours
    h. 9.68 hours

4.8 a. Mean = 1.885 µin; s.d.= 0.264 µin
    b. Yes
    c. 0.1761
    d. $D(0.10) = 0.174$; the fit is marginally good.

Chapter 5

5.4. 0.295

5.5 b. 2586 hours
c. 517.2 hours
d. 0.220
e. 0.406
f. Possibly

5.6 a. $D = 0.475$; $D(0.01) = 0.338$
b. Improved $D = 0.300$, but not good, since $D(0.05) = 0.287$

5.7 a. 2596 hours
b. 0.6803
c. $D = 0.153$, compared to $D(0.10) = 0.375$

5.8 a. 5410 hours
b. 0.000185 failures per hour
c. 0.603
e. $D = 0.142$, compared to $D(0.10) = 0.271$

5.9. $D = 0.244$, compared to $D(0.10) = 0.295$, indicating a marginal fit. The lognormal fit in problem 4.4 is better.

Chapter 6

6.1 a. 0.613
b. 0.582
c. $T = 148$; $\chi^2 = 15.338$; $\lambda = \dfrac{15.338}{296} = 0.05182$
d. $\text{LCL}(0.5) = 0.596$
e. $D = 0.267$; $D(0.10) = 0.329$
f. These are the same data as Example 6.5.

6.2 a. $\mu^* = 62.974$; $\lambda^* = 0.025$
b. 0.391
c. 0.376

6.3 a. $\mu^* = 414$; $\lambda^* = 0.0002$
c. $D = 0.102$. The two-parameter solution is better, with both a better graph and a smaller valuc of D.

6.4   b. $D = 0.16$, compared to $D(0.10) = 0.223$
     c. $\mu^* = 15.8$; $\lambda^* = 0.00898$
     d. $D = 0.145$
     e. The two-parameter solution is better than the one-parameter solution, because the nature of the dispersion indicates that there is a nonzero location parameter.

Chapter 7

7.1 a. MTTF = 5410 hours; λ = 000185

     b. $\exp\left\{-\dfrac{(4000)(33.196)}{129.846}\right\} = 0.3596$

     c. $\lambda = \dfrac{35.563}{137846} = 0.000258$; MTTF = 3876; $R = 0.3563$

7.2.   50 percent: $\exp\left\{-\dfrac{(37.335)(150)}{[2(2290+150)]}\right\} = 0.317$

      90 percent: $\exp\left\{-\dfrac{(49.513)(150)}{4880}\right\} = 0.218$

7.8 a. 39,413 hours
     b. 7 failures
     c. 16 failures
     d. $\lambda = \dfrac{7.962}{86826} = 0.000917$ to $\lambda = \dfrac{26.296}{86826} = 0.0003029$;

       $m = 3301$ to $m = 10905$; $R = 0.2977$ to $R = 0.6929$

Chapter 8

8.1 a. $D = 0.133$ vs. $D(0.10) = 0.174$; mean = 60.3; s.d. = 3.3576
     b. $\dfrac{70-50}{(2)(3.3576)} = 2.9783$
     c. $\gamma > 99\%$ for $R = 0.90$; $95\% < \gamma < 99\%$ for $R = 0.95$
     d. $D = 0.142 < 0.174$; mean = 1.7797; s.d. = 0.024261
     e. $\dfrac{\log 70 - \log 50}{(2)(0.024261)} = 3.012$
     f. Answers are the same as for 8.1c.

272 ■ Solutions for Problems

8.2 a. 0.126 inches
    b. 0.00359 inches
    c. 2.785
    d. 90 percent
    e. 2.537
    f. Yes

8.3 a. Graphical fit appears to be good.
    b. Mean = 19224; s.d. = 7790
    c. $D = 0.197$; accept at 10 percent level
    d. $k = 1.905$ indicates that there is not 90 percent coverage with 90 percent confidence, since the table tolerance limit factor is 2.219.

8.4 a. Mean = 0.50048 inches; s.d. = 0.000286 inches
    b. $D = 0.163$ vs. $D(10) = 0.239$
    c. Tolerance limit factor from Table A.5 is 3.379 s.d.'s vs. $k = 5.32$ s.d.'s; this shows 95 percent coverage with 95 percent confidence.
    d. There are no observations below the mean, but the lower tolerance limit 0.480 is 8.67 s.d.'s below the mean; hence we have 95-95 coverage on both sides.

8.5 a. 2.244
    b. The data show 95 percent reliability at 75 percent confidence.
    c. The data do not show 95 percent reliability at 90 percent confidence.
    d. The data do not show 95 percent reliability at 95 percent confidence.

Chapter 9

9.1 c. $\beta = 1.15$; $\delta = 170$
    d. Increasing failure rate
    e. Yes; Weibull has more flexibility than exponential.

9.2 b. Bathtub curve
    c. 10 or more years

9.3 a.

Years (t)	h(t)%	H(t)%
6		
(F) 6	7.14	7.14
8		
9		
(F) 10	9.09	16.24
(F) 12	10.00	26.24
12		
13		
(F) 16	14.29	40.52
(F) 17	16.67	57.19
(F) 19	20.00	77.19
20		
25		
(F) 28	50.00	127.19
31		

   c. The estimated Weibull shape parameter is 2, and the estimated characteristic lifetime is 23 years.

9.4 b.   i. $\hat{u} = 9.19138$
      ii. $\hat{b} = 0.81845$
      iii. $\hat{\delta} = 9812$ hrs
      iv. $\hat{\beta} = 1.222$
  c. V-statistic is 3.64.
  d.   i. 90 percent for $R = 0.90$
      ii. 60 percent for $R = 0.95$
      iii. 5 percent to 10 percent for $R = 0.99$

9.5 b. $\hat{u} = 5.017602$; $\hat{b} = 0.794787$; $\hat{\delta} = 151.050$; $\hat{\beta} = 1.258$
  c. $\hat{u} = 5.126838$; $\hat{b} = 0.882501$; $\hat{\delta} = 168.48$; $\hat{\beta} = 1.1133$
  d. $V_R = 2.263$;
     i. For $R = 0.90$, $25\% < \gamma < 50\%$
     ii. For $R = 0.95$, $10\% < \gamma < 25\%$
     iii. For $R = 0.99$, $\gamma < 2\%$
  e. $V_R = 2.162$
     i. For $R = 0.90$, $25\% < \gamma < 50\%$
     iii. For $R = 0.95$, $10\% = \gamma$
     iii. For $R = 0.99$, $\gamma < 2\%$

9.6 b. i. $\hat{u} = 3.054$
         ii. $\hat{b} = 0.7381$
         iii. $\hat{\delta} = 21.20$ months
         iv. $\hat{\beta} = 1.355$
     c. $V_R = 2.649$
         i. For $R = 0.90$, $50\% < \gamma < 60\%$
         ii. For $R = 0.95$, $25\% = \gamma$
         iii. For $R = 0.99$, $\gamma < 2\%$
     d. i. $\hat{u} = 3.478$
         ii. $\hat{b} = 1.044$
         iii. $\hat{\beta} = 0.958$
         iv. $\hat{\delta} = 32.39$ months
         $V_R = 2.279$
         v. For $R = 0.90$, $25\% < \gamma < 40\%$
         vi. For $R = 0.95$, $10\% = \gamma$
         vii. For $R = 0.99$, $\gamma < 2\%$

9.7 a. $\hat{u} = 8.585$; $\hat{b} = 1.028$
     b. 5765 hours = 5351 + 414
     c. $V_R = 4.018$;
         i. For $R = 0.90$, $\gamma = 95\%$;
         ii. For $R = 0.95$, $\gamma = 90\%$;
         iii. For $R = 0.99$, $10\% < \gamma < 25\%$

9.8 a. $\beta = 0.9$; $\delta = 3700$; $R = 0.317$
     b. $\beta = 1.7$; $\delta = 1500$; $R = 0.002$
     c. There is an apparent contradiction: The probability of a failure due to hard disk failure is greater than the probability of overall system failure. A possible explanation is that the disk failures are all bunched together; this results in increasing failure rate when the failure rate is actually down.

Solutions for Problems ■ 275

			Overall failures, including hard disk			Hard disk failures	
i	t(i)	h.d.?	Failed	h(i)%	H(i)%	h(i)%	H(i)%
1	96		96	3.448	3.448		
2	297						
3	336	Y	336	3.704	7.152	3.704	3.704
4	336	Y	336	3.846	10.998	3.846	7.550
5	480	Y	480	4.000	14.998	4.000	11.550
6	504						
7	504						
8	528						
9	648						
10	672		672	5.000	19.998		
11	744						
12	744						
-----	-----	-----	-----	-----	-----	-----	-----
28	1320						
29	1320						

All data are in hours at removal from service.

Part V

V.1. $1 - (1-0.5)(1-(0.4)(0.9))(1-(0.9)(0.8)) = 0.9104$

V.2 a. 0.9995
b. $ab + a(1-b)c + a(1-b)(1-c)d + (1-a)bc + (1-a)b(1-c)d + (1-a)(1-b)cd$

V.3 a. 0.24076
b. 1.0
c. 0.9967
d. 0.7537

V.4 a. $R = ace + ade - acde + bce - abce + bde - abde - bcde + abcde$
b. $R = ace + ac'de + a'bce + a'bc'de$
c. $R = e[1-(1-a)(1-b)][1-(1-c)(1-d)]$
d. 0.882090

V.5 a. A: $ce + c'de - bce - bc'de$
B: $a'ce + a'c'de$
C: $ae - ade + a'be - a'bde$
D: $ac'e + a'bc'e$
E: $ac + ac'd + a'bc + a'bc'd$
b. $\text{imp}(a) = \text{imp}(b) = \text{imp}(c) = \text{imp}(d) = 0.0891$;
$\text{imp}(e) = 0.98$

V.6 a. $R = 1 - \{[1-(1-C^2)^3]^2 \, [1-(1-D^2)^2]\}$
b. i. The reliability of a string of collectors is $C^2 = 0.36$.
ii. The reliability of a block is $1 - (1-C^2)^3 = 0.7378$.
iii. The reliability of the string of two blocks is 0.5444.
iv. The reliability of the diodes is 0.4375.
v. The reliability of the solar module is 0.7618.

V.7 a. 0.922
b. 0.994
c. 0.980

V.8 a. i. $R(1) = ACE$
ii. $R(2) = R(1) + ADE - ACDE$
iii. $R(3) = R(2) + BCD - ABCDE$
iv. $R(4) = R(3) + BDE - ABDE - BCDE + ABCDE$
b. 0.9477

V.9 a. i. $R(1) = ACE$
ii. $R(2) = R(1) + AC'DE$
iii. $R(3) = R(2) + A'BCD + ABCDE'$
iv. $R(4) = R(4) + A'BC'DE$
b. 0.9477

V.10 a. i. $\text{imp}(A) = CE + C'DE - BCD + BCDE' - BC'DE$
ii. $\text{imp}(B) = A'CD + ACDE' + A'C'DE$
iii. $\text{imp}(C) = AE - ADE + A'BD + ABDE' - A'BDE$
iv. $\text{imp}(D) = AC'E + A'BC + ABCE' + A'BC'E$
v. $\text{imp}(E) = AC + AC'D - ABCD + A'BC'D$
b. $\text{imp}(A) = \text{imp}(B) = \text{imp}(C) = 0.162$
$\text{imp}(D) = \text{imp}(E) = 0.243$.

V.11 a. AB, AC, BC
   b. $AB + AB'C + A'BC$
   c. 0.972
   d. $\text{imp}(A) = B + B'C - BC$
      $\text{imp}(B) = A - AC + A'C$
      $\text{imp}(C) = AB' + A'B$
   e. $\text{imp}(A) = \text{imp}(B) = \text{imp}(C) = 0.18$

Additional References

AGREE (Advisory Group on Reliability of Electronic Equipment, Office of Assistant Secretary of Defense). 1957. *Reliability of military electronic equipment.* Washington, D.C.: AGREE.

ARINC Research Corporation. 1963. *Reliability engineering.* Englewood Cliffs, N.J.: Prentice Hall.

Bily, Matej. 1989. *Dependability of mechanical systems.* Amsterdam: Elsevier.

Caterneau, V., and A. N. Mihalaski. 1989. *Reliability fundamentals.* Amsterdam: Elsevier.

Chapouille, P. 1968. *Fiabilité des systèmes.* Paris: Masson et cie.

Charette, R. N. 1990. *Application strategies for risk analysis.* New York: McGraw-Hill Intertext.

Chorofas, D. N. 1960. *Statistical processes and reliability engineering.* New York: Van Nostrand.

Christian, N. L. 1989. *Fiber optic design, operation, testing, reliability, and maintainability.* Park Ridge, N.J.: Noyes Data Corp.

Dhillon, B. S., and C. Singh. 1981. *Engineering reliability.* New York: Wiley.

Dovich, R. A. 1990. *Reliability statistics.* Milwaukee, Wis.: ASQC Quality Press.

Dummer, G. W., and N. R. Griffin. 1966. *Electronic reliability: Calculation and design.* Oxford: Pergamon Press.

Dummer, G. W., and R. C. Winton. 1990. *An elementary guide to reliability.* 4th ed. Oxford: Pergamon Press.

Enrick, N. L. 1966. *Quality control and reliability.* New York: Industrial Press.

———. 1985. *Reliability and process improvement.* 8th ed. New York: Industrial Press.

Fuqua, N. B. 1987. *Reliability engineering for electronic design.* New York: Marcel Dekker.

Gedye, G. R. 1968. *A manager's guide to quality and reliability.* New York: Wiley.

Gilmore, H. L. 1986. *Integrated product testing and evaluation.* Milwaukee, Wis.: ASQC Quality Press.

Gnedenko, B., J. Belyayev, and A. D. Solovyev. 1969. *Mathematical methods of reliability.* Edited by R. E. Barlow. New York: Wiley.

Green, A. E. 1983. *Safety system reliability.* New York: Wiley.

Green, A. E., and A. J. Bourne. 1972. *Reliability technology.* New York: Wiley Interscience.

Grosh, D. L. 1989. *A primer of reliability theory.* New York: Wiley.

Häckler, J. 1965. *Methoden für die untersuchung der zuverlässigkeit von bauelementen.* Dresden: Institut für Datenverarbeitung.

Halpern, S. 1978. *The assurance sciences: An introduction to quality and reliability.* Englewood Cliffs, N.J.: Prentice Hall.

Haugen, E. B. 1968. *Probabilistic approaches to design.* New York: Wiley.

Henley, E. J., and H. Kumamoto. 1981. *Reliability engineering and risk assessment.* New York: Prentice Hall.

Hnatek, E. R. 1984. *Integrated circuit quality and reliability.* New York: Marcel Dekker.

International Electrotechnical Commission. 1969. *Managerial aspects of reliability.* Geneva: International Electrotechnical Commission.

Kapur, K. C., and L. R. Lamberson. 1977. *Reliability in engineering design.* New York: Wiley.

Kececioglu, D. 1991. *Reliability engineering handbook.* Englewood Cliffs, N.J.: Prentice Hall.

Klion, J. 1992. *Practical electronic reliability engineering.* New York: Van Nostrand Reinhold.

Kohlas, J. 1987. *Zuverlässigkeit und verfügbarkeit.* Stuttgart: Teubner.

Krishnamoorthi, K. S. 1992. *Reliability methods for engineers.* Milwaukee, Wis.: ASQC Quality Press.

Landers, R. R. 1963. *Reliability and product assurance.* Englewood Cliffs, N.J.: Prentice Hall.

Leitch, R. D. 1988. *Basic reliability engineering analysis.* London and Boston: Butterworth.

Lewis, E. E. 1987. *Introduction to reliability engineering.* New York: Wiley.

Lloyd, D. K., and M. Lipow. 1984. *Reliability: Management, methods, and mathematics.* 2d ed. Milwaukee, Wis.: ASQC Quality Press.

Longbottom, R. 1980. *Computer system reliability.* New York: Wiley.

McCormick, N. J. 1981. *Reliability and risk analysis.* New York: Academic Press.

Misra, K. B. 1992. *Reliability analysis and prediction.* Amsterdam and New York: Elsevier.

Moura, E. C. 1991. *Volume 15: How to determine sample size and estimate failure rate in life testing.* Milwaukee, Wis.: ASQC Statistics Division and ASQC Quality Press.

Musa, J. D., A. Iannino, and K. Okumoto. 1987. *Software reliability: Measurement, prediction, application.* New York: McGraw-Hill.

Myers, R. H., ed. 1964. *Reliability engineering for electronic systems.* New York: Wiley.

National Semiconductor Corp. 1979. *The reliability handbook.* Santa Clara, Calif.: National Semiconductor.

Nelson, W. 1983. *Volume 6: How to analyze reliability data.* Milwaukee, Wis.: ASQC Statistics Division and ASQC Quality Press.

Pieruschka, Erich. 1963. *Principles of reliability.* Englewood Cliffs, N.J.: Prentice Hall.

Polovko, A. M. 1968. *Fundamentals of reliability theory.* Translation edited by W. H. Pierce. New York: Academic Press.

Priest, J. W. 1988. *Engineering design for producibility and reliability.* New York: Marcel Dekker.

Roberts, N. H. 1964. *Mathematical methods in reliability engineering.* New York: McGraw-Hill.

Rook, Paul. 1990. *Software reliability handbook.* London and New York: Elsevier.

Sandler, G. H. 1963. *System reliability engineering.* Englewood Cliffs, N.J.: Prentice Hall.

Smith, C. S. 1969. *Quality and reliability: An integrated approach.* New York: Pitman.

Srinivasan, S. K., and R. Subramanian. 1980. *Probabilistic analysis of redundant systems.* Berlin: Springer-Verlag.

Stewart, D. A. 1966. *Probability, statistics, and reliability.* Richmond, Surrey, England: Draughtsmen's and Allied Technicians Association.

Störmer, Hans. 1970. *Mathematisch theorie der zuverlässigkeit.* Munich: R. Oldenburgh.

Thomason, R. 1969. *An introduction to quality and reliability.* Brighton, England: Machinery Publishing Company.

Zelen, M. S. 1962. *Statistical theory of reliability.* Madison, Wis.: University of Wisconsin Press.

Appendix

Table A.1. Fixed sample size (n) and confidence level (γ): Reliability (p) as a function of the number of successes (r) or the number of failures ($n-r$).

n	r	$n-r$	\multicolumn{6}{c}{γ}	n	r	$n-r$	\multicolumn{6}{c}{γ}
			0.50	0.75	0.90	0.95	0.99	0.995				0.50	0.75	0.90	0.95	0.99	0.995
1	0	1	0.293	0.134	0.051	0.025	0.005	0.002	8	6	2	0.714	0.609	0.510	0.450	0.344	0.307
	1	0	0.707	0.500	0.316	0.224	0.100	0.071		7	1	0.820	0.728	0.632	0.571	0.456	0.415
2	0	2	0.206	0.092	0.034	0.017	0.003	0.002		8	0	0.926	0.857	0.774	0.717	0.600	0.555
	1	1	0.500	0.326	0.196	0.135	0.059	0.041	9	7	2	0.742	0.645	0.550	0.493	0.388	0.352
	2	0	0.794	0.630	0.464	0.368	0.216	0.171		8	1	0.838	0.753	0.663	0.606	0.496	0.456
3	1	2	0.386	0.243	0.142	0.098	0.042	0.029		9	0	0.933	0.871	0.794	0.741	0.631	0.589
	2	1	0.614	0.456	0.320	0.249	0.141	0.111	10	7	3	0.676	0.580	0.489	0.436	0.340	0.307
	3	0	0.841	0.707	0.562	0.473	0.316	0.266		8	2	0.764	0.674	0.585	0.530	0.428	0.392
4	2	2	0.500	0.359	0.247	0.189	0.106	0.083		9	1	0.852	0.773	0.690	0.636	0.530	0.492
	3	1	0.686	0.546	0.416	0.343	0.222	0.185		10	0	0.939	0.882	0.811	0.762	0.658	0.618
	4	0	0.871	0.758	0.631	0.549	0.398	0.346	12	9	3	0.725	0.638	0.556	0.505	0.412	0.379
5	3	2	0.579	0.447	0.333	0.271	0.173	0.144		10	2	0.800	0.720	0.640	0.590	0.494	0.459
	4	1	0.736	0.610	0.490	0.418	0.294	0.254		11	1	0.874	0.806	0.732	0.684	0.587	0.551
	5	0.	0.891	0.794	0.681	0.607	0.464	0.414		12	0	0.948	0.899	0.838	0.794	0.702	0.665
6	4	2	0.636	0.514	0.404	0.341	0.236	0.203	15	12	3	0.775	0.702	0.629	0.583	0.497	0.466
	5	1	0.772	0.659	0.547	0.479	0.357	0.315		13	2	0.836	0.769	0.700	0.656	0.570	0.537
	6	0	0.906	0.820	0.720	0.652	0.518	0.469		14	1	0.897	0.840	0.778	0.736	0.651	0.619
7	5	2	0.679	0.567	0.462	0.400	0.293	0.258		15	0	0.958	0.917	0.866	0.829	0.750	0.718
	6	1	0.799	0.697	0.594	0.529	0.410	0.368	20	17	3	0.828	0.769	0.709	0.671	0.596	0.568
	7	0	0.917	0.841	0.750	0.688	0.562	0.516		18	2	0.875	0.822	0.766	0.729	0.656	0.628

281

Table A.1. *(continued)*

n	r	n−r	γ 0.50	0.75	0.90	0.95	0.99	0.995	n	r	n−r	γ 0.50	0.75	0.90	0.95	0.99	0.995
	19	1	0.921	0.877	0.827	0.793	0.723	0.696	60	56	4	0.924	0.899	0.873	0.856	0.821	0.808
	20	0	0.968	0.936	0.896	0.867	0.803	0.777		57	3	0.940	0.918	0.894	0.878	0.845	0.831
25	22	3	0.861	0.811	0.761	0.728	0.663	0.638		58	2	0.956	0.937	0.915	0.900	0.869	0.857
	23	2	0.898	0.855	0.808	0.777	0.714	0.069		59	1	0.973	0.956	0.938	0.925	0.896	0.884
	24	1	0.936	0.900	0.858	0.830	0.771	0.747		60	0	0.989	0.978	0.963	0.952	0.927	0.917
	25	0	0.974	0.948	0.915	0.891	0.838	0.816	75	71	4	0.939	0.919	0.898	0.884	0.855	0.843
30	27	3	0.883	0.841	0.797	0.768	0.711	0.689		72	3	0.952	0.934	0.914	0.901	0.874	0.863
	28	2	0.915	0.877	0.837	0.810	0.755	0.734		73	2	0.965	0.949	0.932	0.920	0.894	0.884
	29	1	0.946	0.916	0.880	0.856	0.804	0.784		74	1	0.978	0.965	0.950	0.939	0.916	0.906
	30	0	0.978	0.956	0.928	0.908	0.862	0.843		75	0	0.991	0.982	0.970	0.961	0.941	0.933
35	31	4	0.942	0.917	0.888	0.869	0.830	0.814	100	95	5	0.944	0.928	0.910	0.899	0.875	0.866
	32	3	0.899	0.862	0.824	0.798	07.47	0.727		96	4	0.954	0.939	0.922	0.912	0.889	0.880
	33	2	0.926	0.894	0.859	0.835	0.786	0.767		97	3	0.964	0.950	0.935	0.926	0.904	0.896
	34	1	0.954	0.927	0.896	0.875	0.829	0.811		98	2	0.974	0.962	0.948	0.939	0.919	0.911
	35	0	0.981	0.962	0.938	0.920	0.880	0.863		99	1	0.983	0.974	0.962	0.954	0.936	0.929
40	36	4	0.887	0.851	0.814	0.790	0.742	0.724		100	0	0.993	0.986	0.978	0.971	0.955	0.949
	37	3	0.911	0.879	0.844	0.822	0.775	0.757	150	145	5	0.962	0.951	0.939	0.932	0.916	0.909
	38	2	0.935	0.907	0.875	0.854	0.810	0.793		146	4	0.969	0.959	0.948	0.940	0.925	0.919
	39	1	0.959	0.936	0.908	0.889	0.849	0.832		147	3	0.976	0.966	0.956	0.950	0.935	0.929
	40	0	0.983	0.967	0.945	0.930	0.894	0.879		148	2	0.982	0.974	0.965	0.959	0.946	0.940
45	41	4	0.899	0.867	0.834	0.812	0.768	0.751		149	1	0.989	0.982	0.974	0.969	0.957	0.952
	42	3	0.921	0.892	0.860	0.840	0.798	0.781	200	195	5	0.972	0.963	0.954	0.948	0.936	0.931
	43	2	0.942	0.917	0.888	0.869	0.830	0.814		196	4	0.977	0.969	0.961	0.955	0.943	0.939
	44	1	0.964	0.942	0.918	0.901	0.864	0.814		197	3	0.982	0.975	0.967	0.962	0.951	0.946
	45	0	0.985	0.970	0.951	0.937	0.905	0.891		198	2	0.987	0.981	0.974	0.969	0.959	0.955
50	46	4	0.909	0.880	0.849	0.829	0.789	0.773		199	1	0.992	0.987	0.981	0.977	0.967	0.964
	47	3	0.928	0.902	0.874	0.855	0.816	0.801		200	0	0.996	0.993	0.989	0.985	0.977	0.974
	48	2	0.948	0.925	0.899	0.882	0.845	0.831									
	49	1	0.967	0.948	0.926	0.910	0.877	0.863									
	50	0	0.985	0.973	0.956	0.943	0.914	0.901									

Table A.2. Standard normal distribution function $F_N(z) = P(Z \leq z)$.

z	0	1	2	3	4	5	6	7	8	9
−3.0	0.0013	0.0013	0.0013	0.0012	0.0012	0.0011	0.0011	0.0011	0.0010	0.0010
−2.9	0.0019	0.0018	0.0018	0.0017	0.0016	0.0016	0.0015	0.0015	0.0014	0.0014
−2.8	0.0026	0.0025	0.0024	0.0023	0.0023	0.0022	0.0021	0.0021	0.0020	0.0019
−2.7	0.0035	0.0034	0.0033	0.0032	0.0031	0.0030	0.0029	0.0028	0.0027	0.0026
−2.6	0.0047	0.0045	0.0044	0.0043	0.0041	0.0040	0.0039	0.0038	0.0037	0.0036
−2.5	0.0062	0.0060	0.0059	0.0057	0.0055	0.0054	0.0052	0.0051	0.0049	0.0048
−2.4	0.0082	0.0080	0.0078	0.0075	0.0073	0.0071	0.0069	0.0068	0.0066	0.0064
−2.3	0.0107	0.0104	0.0102	0.0099	0.0096	0.0094	0.0091	0.0089	0.0087	0.0084
−2.2	0.0139	0.0136	0.0132	0.0129	0.0125	0.0122	0.0119	0.0116	0.0113	0.0110
−2.1	0.0179	0.0174	0.0170	0.0166	0.0162	0.0158	0.0154	0.0150	0.0146	0.0143
−2.0	0.0228	0.0222	0.0217	0.0212	0.0207	0.0202	0.0197	0.0192	0.0188	0.0183
−1.9	0.0287	0.0281	0.0274	0.0268	0.0262	0.0256	0.0250	0.0244	0.0239	0.0233
−1.8	0.0359	0.0351	0.0344	0.0336	0.0329	0.0322	0.0314	0.0307	0.0301	0.0294
−1.7	0.0446	0.0436	0.0427	0.0418	0.0409	0.0401	0.0392	0.0384	0.0375	0.0367
−1.6	0.0548	0.0537	0.0526	0.0516	0.0505	0.0495	0.0485	0.0475	0.0465	0.0455
−1.5	0.0668	0.0655	0.0643	0.0630	0.0618	0.0606	0.0594	0.0582	0.0571	0.0559
−1.4	0.0808	0.0793	0.0778	0.0764	0.0749	0.0735	0.0721	0.0708	0.0694	0.0681
−1.3	0.0968	0.0951	0.0934	0.0918	0.0901	0.0885	0.0869	0.0853	0.0838	0.0823
−1.2	0.1151	0.1131	0.1112	0.1093	0.1075	0.1056	0.1038	0.1020	0.1003	0.0985
−1.1	0.1357	0.1335	0.1314	0.1292	0.1271	0.1251	0.1230	0.1210	0.1190	0.1170
−1.0	0.1587	0.1562	0.1539	0.1515	0.1492	0.1469	0.1446	0.1423	0.1401	0.1379
−0.9	0.1841	0.1814	0.1788	0.1762	0.1736	0.1711	0.1685	0.1660	0.1635	0.1611
−0.8	0.2119	0.2090	0.2061	0.2033	0.2005	0.1977	0.1949	0.1922	0.1894	0.1867
−0.7	0.2420	0.2389	0.2358	0.2327	0.2296	0.2266	0.2236	0.2206	0.2177	0.2148
−0.6	0.2743	0.2709	0.2676	0.2643	0.2611	0.2578	0.2546	0.2514	0.2483	0.2451
−0.5	0.3085	0.3050	0.3015	0.2981	0.2946	0.2912	0.2877	0.2843	0.2810	0.2776
−0.4	0.3446	0.3409	0.3372	0.3336	0.3300	0.3264	0.3228	0.3192	0.3156	0.3121
−0.3	0.3821	0.3783	0.3745	0.3707	0.3669	0.3632	0.3594	0.3557	0.3520	0.3483
−0.2	0.4207	0.4168	0.4129	0.4090	0.4052	0.4013	0.3974	0.3936	0.3897	0.3859
−0.1	0.4602	0.4562	0.4522	0.4483	0.4443	0.4404	0.4364	0.4325	0.4286	0.4247
−0.0	0.5000	0.4960	0.4920	0.4880	0.4840	0.4801	0.4761	0.4721	0.4681	0.4641

Adapted from M. Abramowitz and I. A. Stegun, eds. 1964. *Handbook of mathematical functions.* National Bureau of Standards Publication AMS 55. Washington, D.C.: U.S. Government Printing Office, public domain.

Table A.2. *(continued)*

Z	0	1	2	3	4	5	6	7	8	9
0.0	0.5000	0.5040	0.5080	0.5120	0.5160	0.5199	0.5239	0.5279	0.5319	0.5359
0.1	0.5398	0.5438	0.5478	0.5517	0.5557	0.5596	0.5636	0.5675	0.5714	0.5753
0.2	0.5793	0.5832	0.5871	0.5910	0.5948	0.5987	0.6026	0.6064	0.6103	0.6141
0.3	0.6179	0.6217	0.6255	0.6293	0.6331	0.6368	0.6406	0.6443	0.6480	0.6517
0.4	0.6554	0.6591	0.6628	0.6664	0.6700	0.6736	0.6772	0.6808	0.6844	0.6879
0.5	0.6915	0.6950	0.6985	0.7019	0.7054	0.7088	0.7123	0.7157	0.7190	0.7224
0.6	0.7257	0.7291	0.7324	0.7357	0.7389	0.7422	0.7454	0.7486	0.7517	0.7549
0.7	0.7580	0.7611	0.7642	0.7673	0.7704	0.7734	0.7764	0.7794	0.7823	0.7852
0.8	0.7881	0.7910	0.7939	0.7967	0.7995	0.8023	0.8051	0.8078	0.8106	0.8133
0.9	0.8159	0.8186	0.8212	0.8238	0.8264	0.8289	0.8315	0.8340	0.8365	0.8389
1.0	0.8413	0.8438	0.8461	0.8485	0.8508	0.8531	0.8554	0.8577	0.8599	0.8621
1.1	0.8643	0.8665	0.8686	0.8708	0.8729	0.8749	0.8770	0.8790	0.8810	0.8830
1.2	0.8849	0.8869	0.8888	0.8907	0.8925	0.8944	0.8962	0.8980	0.8997	0.9015
1.3	0.9032	0.9049	0.9066	0.9082	0.9099	0.9115	0.9131	0.9147	0.9162	0.9177
1.4	0.9192	0.9207	0.9222	0.9236	0.9251	0.9265	0.9279	0.9292	0.9306	0.9319
1.5	0.9332	0.9345	0.9357	0.9370	0.9382	0.9394	0.9406	0.9418	0.9429	0.9441
1.6	0.9452	0.9463	0.9474	0.9484	0.9495	0.9505	0.9515	0.9525	0.9535	0.9545
1.7	0.9554	0.9564	0.9573	0.9582	0.9591	0.9599	0.9608	0.9616	0.9625	0.9633
1.8	0.9641	0.9649	0.9656	0.9664	0.9671	0.9678	0.9686	0.9693	0.9699	0.9706
1.9	0.9713	0.9719	0.9726	0.9732	0.9738	0.9744	0.9750	0.9756	0.9761	0.9767
2.0	0.9772	0.9778	0.9783	0.9788	0.9793	0.9798	0.9803	0.9808	0.9812	0.9817
2.1	0.9821	0.9826	0.9830	0.9834	0.9838	0.9842	0.9846	0.9850	0.9854	0.9857
2.2	0.9861	0.9864	0.9868	0.9871	0.9875	0.9878	0.9881	0.9884	0.9887	0.9890
2.3	0.9893	0.9896	0.9898	0.9901	0.9904	0.9906	0.9909	0.9911	0.9913	0.9916
2.4	0.9918	0.9920	0.9922	0.9925	0.9927	0.9929	0.9931	0.9932	0.9934	0.9936
2.5	0.9938	0.9940	0.9941	0.9943	0.9945	0.9946	0.9948	0.9949	0.9951	0.9952
2.6	0.9953	0.9955	0.9956	0.9957	0.9959	0.9960	0.9961	0.9962	0.9963	0.9964
2.7	0.9965	0.9966	0.9967	0.9968	0.9969	0.9970	0.9971	0.9972	0.9973	0.9974
2.8	0.9974	0.9975	0.9976	0.9977	0.9977	0.9978	0.9979	0.9979	0.9980	0.9981
2.9	0.9981	0.9982	0.9982	0.9983	0.9984	0.9984	0.9985	0.9985	0.9986	0.9986
3.0	0.9987	0.9987	0.9987	0.9988	0.9988	0.9989	0.9989	0.9989	0.9990	0.9990

Table A.3. Plotting points $\hat{F}_i$ for probability plotting.

i\n	1	2	3	4	5	6	7	8	9	10	11	12	13	14	15	16
1	0.500															
2	0.333	0.667														
3	0.250	0.500	0.750													
4	0.200	0.400	0.600	0.800												
5	0.167	0.333	0.500	0.667	0.833											
6	0.143	0.286	0.429	0.571	0.714	0.857										
7	0.125	0.250	0.375	0.500	0.625	0.750	0.875									
8	0.111	0.222	0.333	0.444	0.556	0.667	0.778	0.889								
9	0.100	0.200	0.300	0.400	0.500	0.600	0.700	0.800	0.900							
10	0.091	0.182	0.273	0.364	0.455	0.545	0.636	0.727	0.818	0.909						
11	0.0.83	0.167	0.250	0.333	0.417	0.500	0.583	0.667	0.750	0.833	0.917					
12	0.077	0.154	0.231	0.308	0.385	0.462	0.538	0.615	0.692	0.769	0.846	0.923				
13	0.071	0.143	0.214	0.286	0.357	0.429	0.500	0.571	0.643	0.714	0.786	0.857	0.929			
14	0.067	0.133	0.200	0.267	0.333	0.400	0.467	0.533	0.600	0.667	0.733	0.800	0.867	0.933		
15	0.063	0.125	0.188	0.250	0.312	0.375	0.438	0.500	0.563	0.625	0.688	0.750	0.812	0.875	0.937	
16	0.059	0.118	0.176	0.235	0.294	0.353	0.412	0.471	0.529	0.588	0.647	0.706	0.765	0.824	0.882	0.941
17	0.056	0.111	0.167	0.222	0.278	0.333	0.389	0.444	0.500	0.556	0.611	0.667	0.722	0.778	0.833	0.889
18	0.053	0.105	0.158	0.211	0.263	0.316	0.368	0.421	0.474	0.526	0.579	0.632	0.684	0.737	0.789	0.842
19	0.050	0.100	0.150	0.200	0.250	0.300	0.350	0.400	0.450	0.500	0.550	0.600	0.650	0.700	0.750	0.800
20	0.048	0.095	0.143	0.190	0.238	0.286	0.333	0.381	0.429	0.476	0.524	0.571	0.619	0.667	0.714	0.762
21	0.045	0.091	0.136	0.182	0.227	0.273	0.318	0.364	0.409	0.455	0.500	0.545	0.591	0.636	0.682	0.727
22	0.043	0.087	0.130	0.174	0.217	0.261	0.304	0.348	0.391	0.435	0.478	0.522	0.565	0.609	0.652	0.696
23	0.042	0.083	0.125	0.167	0.208	0.250	0.292	0.333	0.375	0.417	0.458	0.500	0.542	0.583	0.625	0.667
24	0.040	0.080	0.120	0.160	0.200	0.240	0.280	0.320	0.360	0.400	0.440	0.480	0.520	0.560	0.600	0.640
25	0.038	0.077	0.115	0.154	0.192	0.231	0.269	0.308	0.346	0.385	0.423	0.462	0.500	0.538	0.577	0.615
26	0.037	0.074	0.111	0.148	0.185	0.222	0.259	0.296	0.333	0.370	0.407	0.444	0.481	0.519	0.556	0.593
27	0.036	0.071	0.107	0.143	0.179	0.214	0.250	0.286	0.321	0.357	0.393	0.429	0.464	0.500	0.536	0.571
28	0.034	0.069	0.103	0.138	0.172	0.207	0.241	0.276	0.310	0.345	0.379	0.414	0.448	0.483	0.517	0.552
29	0.033	0.067	0.100	0.133	0.167	0.200	0.233	0.267	0.300	0.333	0.367	0.400	0.433	0.467	0.500	0.533
30	0.032	0.065	0.097	0.129	0.161	0.194	0.226	0.258	0.290	0.323	0.355	0.387	0.419	0.452	0.484	0.516

Table A.4. Percentage points of the χ^2 distribution.

v	0.01	0.025	0.05	0.10	0.20	0.80	0.90	0.95	0.975	0.99
2	0.020	0.051	0.103	0.211	0.446	3.219	4.605	5.991	7.378	9.210
4	0.297	0.484	0.711	1.064	1.649	5.989	7.779	9.488	11.143	13.277
6	0.872	1.237	1.635	2.204	3.070	8.558	10.645	12.592	14.449	16.812
8	1.646	2.180	2.733	3.490	4.594	11.030	13.362	15.507	17.534	20.090
10	2.558	3.247	3.940	4.865	6.179	13.442	15.987	18.307	20.483	23.209
12	3.571	4.404	5.226	6.304	7.807	15.812	18.549	21.026	23.337	26.217
14	4.660	5.629	6.571	7.790	9.467	18.151	21.064	23.685	26.119	29.141
16	5.812	6.908	7.962	9.312	11.152	20.465	23.542	26.296	28.845	32.000
18	7.015	8.231	9.390	10.865	12.857	22.760	25.989	28.869	31.526	34.805
20	8.260	9.591	10.851	12.443	14.578	25.038	28.412	31.410	34.170	37.566
22	9.543	10.982	12.338	14.042	16.314	27.302	30.813	33.924	36.781	40.289
24	10.856	12.401	13.848	15.659	18.062	29.553	33.196	36.415	39.364	42.980
26	12.198	13.844	15.379	17.292	19.820	31.795	35.563	38.885	41.923	45.642
28	13.565	15.308	16.928	18.939	21.588	34.027	37.916	41.337	44.461	48.278
30	14.954	16.791	18.493	20.599	23.364	36.250	40.256	43.773	46.979	50.892
32	16.362	18.291	20.072	22.271	25.148	38.466	42.585	46.194	49.480	53.486
34	17.789	19.806	21.664	23.952	26.938	40.676	44.903	48.602	51.966	56.061
36	19.233	21.336	23.269	25.643	28.735	42.879	47.212	50.999	54.437	58.619
38	20.691	22.878	24.884	27.343	30.537	45.076	49.513	53.384	56.896	61.162
40	22.164	24.433	26.509	29.050	32.345	47.268	51.805	55.758	59.342	63.691
42	23.650	25.999	28.144	30.765	34.157	49.456	54.090	58.124	61.777	66.206
44	25.148	27.575	29.788	32.487	35.974	51.639	56.368	60.481	64.201	68.710
46	26.657	29.160	31.439	34.215	37.796	53.818	58.640	62.830	66.616	71.201
48	28.177	30.754	33.098	35.949	39.620	55.993	60.907	65.171	69.023	73.683
50	29.707	32.357	34.764	37.689	41.449	58.164	63.167	67.505	71.420	76.154
52	31.246	33.968	36.437	39.433	43.281	60.332	65.422	69.832	73.810	78.616
54	32.793	35.586	38.116	41.183	45.117	62.496	67.673	72.153	76.192	81.069
56	34.350	37.212	39.801	42.937	46.955	64.658	69.918	74.468	78.567	83.513
58	35.914	34.844	41.492	44.696	48.796	66.816	72.160	76.778	80.936	85.950
60	37.485	40.482	43.188	46.459	50.641	68.972	74.397	79.082	83.298	88.379
62	39.063	42.126	44.889	48.226	52.487	71.125	76.630	81.381	85.654	90.801
64	40.649	43.776	46.595	49.996	54.336	73.276	78.860	83.675	88.004	93.217
66	42.240	45.431	48.305	51.770	56.188	75.424	81.086	85.965	90.349	95.626
68	43.838	47.092	50.020	53.548	58.042	77.571	83.308	88.250	92.688	98.028
70	45.442	48.758	51.739	55.329	59.898	79.715	85.527	90.531	95.023	100.425
72	47.051	50.428	53.462	57.113	61.756	81.857	87.743	92.808	97.353	102.816
74	48.666	52.103	55.189	58.900	63.616	83.996	89.956	95.082	99.678	105.202
76	50.286	53.782	56.920	60.690	65.478	86.135	92.166	97.351	101.999	107.582
78	51.910	55.466	58.654	62.482	67.342	88.271	94.374	99.617	104.316	109.958
80	53.540	57.153	60.392	64.278	69.207	90.405	96.578	101.879	106.629	112.329

Adapted from H. Leon Harter. 1964. *New tables of the incomplete gamma-function ratio and of percentage points of the chi-square and beta distributions.* Aerospace Research Laboratories, Office of Aerospace Research, U.S. Air Force, Superintendent of Documents. Washington, D.C.: U.S. Government Printing Office, public domain.

Table A.5. One-sided tolerance limit factors for a normal distribution.

	Values of k for $\gamma = 0.90$ and $n = f + 1$					Values of k for $\gamma = 0.95$ and $n = f + 1$				
n	0.900	0.950	0.975	0.990	0.999	n	0.900	0.950	0.975	0.990	0.999
2	10.253	13.090	15.586	18.500	24.582	2	20.581	26.260	31.257	37.094	49.276
3	4.258	5.311	6.244	7.340	9.651	3	6.155	7.656	8.986	10.553	13.857
4	3.188	3.957	4.637	5.438	7.129	4	4.162	5.144	6.015	7.042	9.214
5	2.744	3.401	3.983	4.668	6.113	5	3.413	4.210	4.916	5.749	7.509
6	2.494	3.093	3.621	4.243	5.556	6	3.008	3.711	4.332	5.065	6.614
7	2.333	2.893	3.389	3.972	5.201	7	2.756	3.401	3.971	4.643	6.064
8	2.219	2.754	3.227	3.783	4.955	8	2.582	3.188	3.724	4.355	5.689
9	2.133	2.650	3.106	3.641	4.771	9	2.454	3.032	3.543	4.144	5.414
10	2.066	2.568	3.011	3.532	4.628	10	2.355	2.911	3.403	3.981	5.204
11	2.012	2.503	2.936	3.444	4.515	11	2.275	2.815	3.291	3.852	5.036
12	1.966	2.448	2.872	3.371	4.420	12	2.210	2.736	3.201	3.747	4.900
13	1.928	2.403	2.820	3.310	4.341	13	2.155	2.670	3.125	3.659	4.787
14	1.895	2.363	2.774	3.257	4.274	14	2.108	2.614	3.060	3.585	4.690
15	1.866	2.329	2.735	3.212	4.215	15	2.068	2.566	3.005	3.520	4.607
16	1.842	2.299	2.700	3.172	4.164	16	2.032	2.523	2.956	3.463	4.534
17	1.819	2.272	2.670	3.137	4.118	17	2.002	2.486	2.913	3.414	4.471
18	1.800	2.249	2.643	3.106	4.078	18	1.974	2.453	2.875	3.370	4.415
19	1.781	2.228	2.618	3.078	4.041	19	1.949	2.423	2.840	3.331	4.364
20	1.765	2.208	2.597	3.052	4.009	20	1.926	2.396	2.809	3.295	4.319
21	1.750	2.190	2.575	3.028	3.979	21	1.905	2.371	2.781	3.262	4.276
22	1.736	2.174	2.557	3.007	3.952	22	1.887	2.350	2.756	3.233	4.238
23	1.724	2.159	2.540	2.987	3.927	23	1.869	2.329	2.732	3.206	4.204
24	1.712	2.145	2.525	2.969	3.904	24	1.853	2.309	2.711	3.181	4.171
25	1.702	2.132	2.510	2.952	3.882	25	1.838	2.292	2.691	3.158	4.143
30	1.657	2.080	2.450	2.884	3.794	30	1.778	2.220	2.608	3.064	4.022
35	1.623	2.041	2.406	2.833	3.730	35	1.732	2.166	2.548	2.994	3.934
40	1.598	2.010	2.371	2.793	3.679	40	1.697	2.126	2.501	2.941	3.866
45	1.577	1.986	2.344	2.762	3.638	45	1.669	2.092	2.463	2.897	3.811
50	1.560	1.965	2.320	2.735	3.604	50	1.646	2.065	2.432	2.863	3.766
60	1.532	1.933	2.284	2.694	3.552	60	1.609	2.022	2.384	2.807	3.695
70	1.511	1.909	2.257	2.663	3.513	70	1.581	1.990	2.348	2.766	3.643
80	1.495	1.890	2.235	2.638	3.482	80	1.560	1.965	2.319	2.733	3.601
90	1.481	1.874	2.217	2.618	3.456	90	1.542	1.944	2.295	2.706	3.567
100	1.470	1.861	2.203	2.601	3.435	100	1.527	1.927	2.276	2.684	3.539
120	1.452	1.841	2.179	2.574	3.402	120	1.503	1.899	2.245	2.649	3.495
145	1.436	1.821	2.158	2.550	3.371	145	1.481	1.874	2.217	2.617	3.455
300	1.386	1.765	2.094	2.477	3.280	300	1.417	1.800	2.133	2.522	3.335
500	1.362	1.736	2.062	2.442	3.235	500	1.385	1.763	2.092	2.475	3.277
∞	1.282	1.645	1.960	2.326	3.090	∞	1.282	1.645	1.960	2.326	3.090

From Donald B. Owen. 1962. *Handbook of statistical tables*, Reprinted with permission of Addison-Wesley Publishing Company, Inc., 117–26.

Table A.6. Two-sided tolerance factors for a normal distribution.

	γ = 0.75					γ = 0.90					γ = 0.95					γ = 0.99				
r	0.75	0.90	0.95	0.99	0.999	0.75	0.90	0.95	0.99	0.999	0.75	0.90	0.95	0.99	0.999	0.75	0.90	0.95	0.99	0.999
2	4.498	6.301	7.414	9.531	11.920	11.407	15.978	18.800	24.167	30.227	22.858	32.019	37.674	48.430	60.573	114.363	160.193	188.491	242.300	303.054
3	2.501	3.538	4.187	5.431	6.844	4.132	5.847	6.919	8.974	11.309	5.922	8.380	9.916	12.861	16.208	13.378	18.930	22.401	29.055	36.616
4	2.035	2.892	3.431	4.471	5.657	2.932	4.166	4.943	6.440	8.149	3.779	5.369	6.370	8.299	10.502	6.614	9.398	11.150	14.527	18.383
5	1.825	2.599	3.088	4.033	5.117	2.454	3.494	4.152	5.423	6.879	3.002	4.275	5.079	6.634	8.415	4.643	6.612	7.855	10.260	13.015
6	1.704	2.429	2.889	3.779	4.802	2.196	3.131	3.723	4.870	6.188	2.604	3.712	4.414	5.775	7.337	3.743	5.337	6.345	8.301	10.548
7	1.624	2.318	2.757	3.611	4.593	2.034	2.902	3.452	4.521	5.750	2.361	3.369	4.007	5.248	6.676	3.233	4.613	5.488	7.187	9.142
8	1.568	2.238	2.663	3.491	4.444	1.921	2.743	3.264	4.278	5.446	2.197	3.136	3.732	4.891	6.226	2.905	4.147	4.936	6.468	8.234
9	1.525	2.178	2.593	3.400	4.330	1.839	2.626	3.125	4.098	5.220	2.078	2.967	3.532	4.631	5.899	2.677	3.822	4.550	5.966	7.600
10	1.492	2.131	2.537	3.328	4.241	1.775	2.535	3.018	3.959	5.046	1.987	2.839	3.379	4.433	5.649	2.508	3.582	4.265	5.594	7.129
11	1.465	2.093	2.493	3.271	4.169	1.724	2.463	2.933	3.849	4.906	1.916	2.737	3.259	4.277	5.452	2.378	3.397	4.045	5.308	6.766
12	1.443	2.062	2.456	3.223	4.110	1.683	2.404	2.863	3.758	4.792	1.858	2.655	3.162	4.150	5.291	2.274	3.250	3.870	5.079	6.477
13	1.425	2.036	2.424	3.183	4.059	1.648	2.355	2.805	3.682	4.697	1.810	2.587	3.081	4.044	5.158	2.190	3.130	3.727	4.893	6.240
14	1.409	2.013	2.398	3.148	4.016	1.619	2.314	2.756	3.618	4.615	1.770	2.529	3.012	3.955	5.045	2.120	3.029	3.608	4.737	6.043
15	1.395	1.994	2.375	3.118	3.979	1.594	2.278	2.713	3.562	4.545	1.735	2.480	2.954	3.878	4.949	2.060	2.945	3.507	4.605	5.876
16	1.383	1.977	2.355	3.092	3.946	1.572	2.246	2.676	3.514	4.484	1.705	2.437	2.903	3.812	4.865	2.009	2.872	3.421	4.492	5.732
17	1.372	1.962	2.337	3.069	3.917	1.552	2.219	2.643	3.471	4.430	1.679	2.400	2.858	3.754	4.791	1.965	2.808	3.345	4.393	5.607
18	1.363	1.948	2.321	3.048	3.891	1.535	2.194	2.614	3.433	4.382	1.655	2.366	2.819	3.702	4.725	1.926	2.753	3.279	4.307	5.497
19	1.355	1.936	2.307	3.030	3.867	1.520	2.172	2.588	3.399	4.339	1.635	2.337	2.784	3.656	4.667	1.891	2.703	3.221	4.230	5.399
20	1.347	1.925	2.294	3.013	3.846	1.506	2.152	2.564	3.368	4.300	1.616	2.310	2.752	3.615	4.614	1.860	2.659	3.168	4.161	5.312
21	1.340	1.915	2.282	2.998	3.827	1.493	2.135	2.543	3.340	4.264	1.599	2.286	2.723	3.577	4.567	1.833	2.620	3.121	4.100	5.234
22	1.334	1.906	2.271	2.984	3.809	1.482	2.118	2.524	3.315	4.232	1.584	2.264	2.697	3.543	4.523	1.808	2.584	3.078	4.044	5.163
23	1.328	1.898	2.261	2.971	3.793	1.471	2.103	2.506	3.292	4.203	1.570	2.244	2.673	3.512	4.484	1.785	2.551	3.040	3.993	5.098
24	1.322	1.891	2.252	2.959	3.778	1.462	2.089	2.489	3.270	4.176	1.557	2.225	2.651	3.483	4.447	1.764	2.522	3.004	3.947	5.039
25	1.317	1.883	2.244	2.948	3.764	1.453	2.077	2.474	3.251	4.151	1.545	2.208	2.631	3.457	4.413	1.745	2.494	2.972	3.904	4.985
26	1.313	1.877	2.236	2.938	3.751	1.444	2.065	2.460	3.232	4.127	1.534	2.193	2.612	3.432	4.382	1.727	2.469	2.941	3.865	4.935
27	1.309	1.871	2.229	2.929	3.740	1.437	2.054	2.447	3.215	4.106	1.523	2.178	2.595	3.409	4.353	1.711	2.446	2.914	3.828	4.888
30	1.297	1.855	2.210	2.904	3.708	1.417	2.025	2.413	3.170	4.049	1.497	2.140	2.549	3.350	4.278	1.668	2.385	2.841	3.733	4.768
35	1.283	1.834	2.185	2.871	3.667	1.390	1.988	2.368	3.112	3.974	1.462	2.090	2.490	3.272	4.179	1.613	2.306	2.748	3.611	4.611

Reproduced, with permission, from C. Eisenhart, M. W. Hastay, and W. A. Wallis. 1947. *Techniques of statistical analysis.* New York: McGraw-Hill, 102–3.

Appendix ■ 289

r	γ = 0.75				γ = 0.90					γ = 0.95					γ = 0.99				
	0.75	0.90	0.95	0.99	0.999	0.75	0.90	0.95	0.99	0.999	0.75	0.90	0.95	0.99	0.999	0.75	0.90	0.95	0.99	0.999
40	1.271	1.818	2.166	2.846	3.635	1.370	1.959	2.334	3.066	3.917	1.435	2.052	2.445	3.213	4.104	1.571	2.247	2.677	3.518	4.493
45	1.262	1.805	2.150	2.826	3.609	1.354	1.935	2.306	3.030	3.871	1.414	2.021	2.408	3.165	4.042	1.539	2.200	2.621	3.444	4.399
50	1.255	1.794	2.138	2.809	3.588	1.340	1.916	2.284	3.001	3.833	1.396	1.996	2.379	3.126	3.993	1.512	2.162	2.576	3.385	4.323
55	1.249	1.785	2.127	2.795	3.571	1.329	1.901	2.265	2.976	3.801	1.382	1.976	2.354	3.094	3.951	1.490	2.130	2.538	3.335	4.260
60	1.243	1.778	2.118	2.784	3.556	1.320	1.887	2.248	2.955	3.774	1.369	1.958	2.333	3.066	3.916	1.471	2.103	2.506	3.293	4.206
65	1.239	1.771	2.110	2.773	3.543	1.312	1.875	2.235	2.937	3.751	1.359	1.943	2.315	3.042	3.886	1.455	2.080	2.478	3.257	4.160
70	1.235	1.765	2.104	2.764	3.531	1.304	1.865	2.222	2.920	3.730	1.349	1.929	2.299	3.021	3.859	1.440	2.060	2.454	3.225	4.120
75	1.231	1.760	2.098	2.757	3.521	1.298	1.856	2.211	2.906	3.712	1.341	1.917	2.285	3.002	3.835	1.428	2.042	2.433	3.197	4.084
80	1.228	1.756	2.092	2.749	3.512	1.292	1.848	2.202	2.894	3.696	1.334	1.907	2.272	2.986	3.814	1.417	2.026	2.414	3.173	4.053
85	1.225	1.752	2.087	2.743	3.504	1.287	1.841	2.193	2.882	3.682	1.327	1.897	2.261	2.971	3.795	1.407	2.012	2.397	3.150	4.024
90	1.223	1.748	2.083	2.737	3.497	1.283	1.834	2.185	2.872	3.669	1.321	1.889	2.251	2.958	3.778	1.398	1.999	2.382	3.130	3.999
95	1.220	1.745	2.079	2.732	3.490	1.278	1.828	2.178	2.863	3.657	1.315	1.881	2.241	2.945	3.763	1.390	1.987	2.368	3.112	3.976
100	1.218	1.742	2.075	2.727	3.484	1.275	1.822	2.172	2.854	3.646	1.311	1.874	2.233	2.934	3.748	1.383	1.977	2.355	3.096	3.954
110	1.214	1.736	2.069	2.719	3.473	1.268	1.813	2.160	2.839	3.626	1.302	1.861	2.218	2.915	3.723	1.369	1.958	2.333	3.066	3.917
120	1.211	1.732	2.063	2.712	3.464	1.262	1.804	2.150	2.826	3.610	1.294	1.850	2.205	2.898	3.702	1.358	1.942	2.314	3.041	3.885
130	1.208	1.728	2.059	2.705	3.456	1.257	1.797	2.141	2.814	3.595	1.288	1.841	2.194	2.883	3.683	1.349	1.928	2.298	3.019	3.857
140	1.206	1.724	2.054	2.700	3.449	1.252	1.791	2.134	2.804	3.582	1.282	1.833	2.184	2.870	3.666	1.340	1.916	2.283	3.000	3.833
150	1.204	1.721	2.051	2.695	3.443	1.248	1.785	2.127	2.795	3.571	1.277	1.825	2.175	2.859	3.652	1.332	1.905	2.270	2.983	3.811
160	1.202	1.718	2.047	2.691	3.437	1.245	1.780	2.121	2.787	3.561	1.272	1.819	2.167	2.848	3.638	1.326	1.896	2.259	2.968	3.792
170	1.200	1.716	2.044	2.687	3.432	1.242	1.775	2.116	2.780	3.552	1.268	1.813	2.160	2.839	3.627	1.320	1.887	2.248	2.955	3.774
180	1.198	1.713	2.042	2.683	3.427	1.239	1.771	2.111	2.774	3.543	1.264	1.808	2.154	2.831	3.616	1.314	1.879	2.239	2.942	3.759
190	1.197	1.711	2.039	2.680	3.423	1.236	1.767	2.106	2.768	3.536	1.261	1.803	2.148	2.823	3.606	1.309	1.872	2.230	2.931	3.744
200	1.195	1.709	2.037	2.677	3.419	1.234	1.764	2.102	2.762	3.529	1.258	1.798	2.143	2.816	3.597	1.304	1.865	2.222	2.921	3.731
250	1.190	1.702	2.028	2.665	3.404	1.224	1.750	2.085	2.740	3.501	1.245	1.780	2.121	2.788	3.561	1.286	1.839	2.191	2.880	3.678
300	1.186	1.696	2.021	2.656	3.393	1.217	1.740	2.073	2.725	3.481	1.236	1.767	2.106	2.767	3.535	1.273	1.820	2.169	2.850	3.641
400	1.181	1.688	2.012	2.644	3.378	1.207	1.726	2.057	2.703	3.453	1.223	1.749	2.084	2.739	3.499	1.255	1.794	2.138	2.809	3.589
500	1.177	1.683	2.006	2.636	3.368	1.201	1.717	2.046	2.689	3.434	1.215	1.737	2.070	2.721	3.475	1.243	1.777	2.117	2.783	3.555
600	1.175	1.680	2.002	2.631	3.360	1.196	1.710	2.038	2.678	3.421	1.209	1.729	2.060	2.707	3.458	1.234	1.764	2.102	2.763	3.530
700	1.173	1.677	1.998	2.626	3.355	1.192	1.705	2.032	2.670	3.411	1.204	1.722	2.052	2.697	3.445	1.227	1.755	2.091	2.748	3.511
800	1.171	1.675	1.996	2.623	3.350	1.189	1.701	2.027	2.663	3.402	1.201	1.717	2.046	2.688	3.434	1.222	1.747	2.082	2.736	3.495
900	1.170	1.673	1.993	2.620	3.347	1.187	1.697	2.023	2.658	3.396	1.198	1.712	2.040	2.682	3.426	1.218	1.741	2.075	2.726	3.483
1000	1.169	1.671	1.992	2.617	3.344	1.185	1.695	2.019	2.654	3.390	1.195	1.709	2.036	2.676	3.418	1.214	1.736	2.068	2.718	3.472
∞	1.150	1.645	1.960	2.576	3.291	1.150	1.645	1.960	2.576	3.291	1.150	1.645	1.960	2.576	3.291	1.150	1.645	1.960	2.576	3.291

Table A.7. Tables of weights for best-linear Weibull estimates.
$A(N,M,I)$ = weight for estimating u
$C(N,M,I)$ = weight for estimating b

N	M	I	A(N,M,I)	C(N,M,I)	N	M	I	A(N,M,I)	C(N,M,I)
2	2	1	0.110731	−0.421383			4	0.163591	−0.064666
		2	0.889269	0.421383			5	0.226486	0.031796
3	2	1	−0.166001	−0.452110			6	0.368179	0.405733
		2	1.166001	0.452110	7	2	1	−0.676894	−0.481140
3	3	1	0.081063	−0.278666			2	1.676894	0.481140
		2	0.251001	−0.190239	7	3	1	−0.272195	−0.315369
		3	0.667936	0.468904			2	−0.184061	−0.281139
4	2	1	−0.346974	−0.465455			3	1.456255	0.596507
		2	1.346974	0.465455	7	4	1	−0.110274	−0.229691
4	3	1	−0.044975	−0.297651			2	−0.060226	−0.215613
		2	0.088057	−0.234054			3	0.018671	−0.164168
		3	0.956918	0.531705			4	1.151829	0.609472
4	4	1	0.064336	−0.203052	7	5	1	−0.030368	−0.176203
		2	0.147340	−0.182749			2	0.004333	−0.172399
		3	0.261510	−0.070109			3	0.052957	−0.141218
		4	0.526813	0.455910			4	0.117599	−0.082820
5	2	1	−0.481434	−0.472962			5	0.855480	0.572640
		2	1.481434	0.472962	7	6	1	0.013524	−0.138436
5	3	1	−0.137958	−0.306562			2	0.041588	−0.140342
		2	−0.025510	−0.257087			3	0.075499	−0.121821
		3	1.163468	0.563650			4	0.117461	−0.082938
5	4	1	−0.006983	−0.217766			5	0.172092	−0.015394
		2	0.059652	−0.199351			6	0.579835	0.498931
		3	0.156664	−0.118927	7	7	1	0.038743	−0.108323
		4	0.790668	0.536044			2	0.064086	−0.113479
5	5	1	0.052975	−0.158131			3	0.090785	−0.103569
		2	0.103531	−0.155707			4	0.120971	−0.078748
		3	0.163808	−0.111820			5	0.157657	−0.032632
		4	0.246092	−0.005600			6	0.207825	0.054727
		5	0.433593	0.431259			7	0.319934	0.382022
6	2	1	−0.588298	−0.477782	8	2	1	−0.752513	−0.483616
		2	1.588298	0.477782			2	0.752513	0.483616
6	3	1	−0.211474	−0.311847	8	3	1	−0.323875	−0.317890
		2	−0.112994	−0.271381			2	−0.243808	−0.288231
		3	1.324468	0.583229			3	1.567683	0.606120
6	4	1	−0.063569	−0.225141	8	4	1	−0.149973	−0.232805
		2	−0.006726	−0.209083			2	−0.105015	−0.220324
		3	0.079882	−0.146386			3	−0.032257	−0.176675
		4	0.990412	0.580610			4	1.287245	0.629805
6	5	1	0.007521	−0.169920	8	5	1	−0.062656	−0.180231
		2	0.048328	−0.166319			2	−0.032248	−0.176510
		3	0.101608	−0.129510			3	0.012767	−0.149566
		4	0.172859	−0.054453			4	0.072446	−0.101642
		5	0.669685	0.520201			5	1.009691	0.607948
6	6	1	0.044826	−0.128810	8	6	1	−0.013509	−0.143834
		2	0.079377	−0.132102			2	0.010292	−0.145006
		3	0.117541	−0.111951			3	0.041357	−0.128393
							4	0.080475	−0.095696

Reproduced, with permission, from Nancy R. Mann. 1966. *Tables for attaining the best linear estimates of the parameters of the Weibull distribution*. Research Report No. 66-9. Canoga Park, Calif.: Rocketdyne Division of North American Aviation, public domain.

Table A.7. (continued)

N	M	I	A(N,M,I)	C(N,M,I)
		5	0.130327	−0.043280
		6	0.751058	0.556209
8	7	1	0.015973	−0.116317
		2	0.036729	−0.120331
		3	0.060439	−0.110582
		4	0.088239	−0.088450
		5	0.122062	−0.050995
		6	0.165529	0.009700
		7	0.511030	0.476975
8	8	1	0.034052	−0.093270
		2	0.053552	−0.098886
		3	0.073452	−0.093994
		4	0.095062	−0.079752
		5	0.119768	−0.053918
		6	0.149934	−0.010179
		7	0.191236	0.069325
		8	0.282943	0.360675
9	2	1	−0.818444	−0.485517
		2	1.818444	0.485517
9	3	1	−0.368833	−0.319786
		2	−0.295280	−0.293621
		3	1.664113	0.613407
9	4	1	−0.184461	−0.235080
		2	−0.143505	−0.223891
		3	−0.075815	−0.185970
		4	1.403781	0.644941
9	5	1	−0.090726	−0.183061
		2	−0.063541	−0.179515
		3	−0.021495	−0.155825
		4	0.034159	−0.115133
		5	1.141604	0.633534
9	6	1	−0.037118	−0.147411
		2	−0.016377	−0.148150
		3	0.012499	−0.133219
		4	0.049305	−105060
		5	0.095614	−0.062073
		6	0.896078	0.595913
9	7	1	−0.004220	−0.120988
		2	0.013386	−0.124245
		3	0.035068	−0.115091
		4	0.061198	−0.095508
		5	0.093013	−0.064162
		6	0.132740	−0.017187
		7	0.668815	0.537180
9	8	1	0.016797	−0.100011
		2	0.032919	−0.104750
		3	0.050582	−0.099608
		4	0.070497	−0.086226
		5	0.093635	−0.063541
		6	0.121560	−0.028346
		7	0.157175	0.026525
		8	0.456836	0.455956
9	9	1	0.030338	−0.081777
		2	0.045872	−0.087308
		3	0.061368	−0.085084
		4	0.077742	−0.076470
		5	0.095769	−0.060667
		6	0.116517	−0.035136
		7	0.141932	0.006001
		8	0.176764	0.078828
		9	0.253697	0.341614
10	2	1	−0.876869	−0.487022
		2	1.876869	0.487022
10	3	1	−0.408602	−0.321265
		2	−0.340443	−0.297858
		3	1.749045	0.619124
10	4	1	−0.214930	−0.236817
		2	−0.177223	−0.226688
		3	−0.113820	−0.193159
		4	1.505973	0.656663
10	5	1	−0.115524	−0.185169
		2	−0.090868	−0.181821
		3	−0.051341	−0.160697
		4	0.000925	−0.125311
		5	1.256809	0.652997
10	6	1	−0.058017	−0.149985
		2	−0.039595	−0.150451
		3	−0.012513	−0.136941
		4	0.022314	−0.112224
		5	0.065750	−0.075721
		6	1.022062	0.625321
10	7	1	−0.022198	−0.124170
		2	−0.006909	−0.126894
		3	0.013224	−0.118392
		4	0.037994	−0.100924
		5	0.068153	−0.073988
		6	0.105164	−0.035501
		7	0.804572	0.579868
10	8	1	0.001179	−0.104082
		2	0.014889	−0.108163
		3	0.030998	−0.103119
		4	0.049734	−0.090835
		5	0.071745	−0.070902
		6	0.098114	−0.041560
		7	0.130649	0.000799
		8	0.602692	0.517864
10	9	1	0.016841	−0.087538
		2	0.029807	−0.092405
		3	0.043570	−0.089839
		4	0.058640	−0.081428
		5	0.075576	−0.066855
		6	0.095169	−0.044670
		7	0.118570	−0.011816
		8	0.148575	0.038159
		9	0.413116	0.436394
10	10	1	0.027331	−0.072734
		2	0.040034	−0.077971

Table A.7. *(continued)*

N	M	I	A(N,M,I)	C(N,M,I)	N	M	I	A(N,M,I)	C(N,M,I)
		3	0.052496	−0.077242			3	0.038291	−0.081388
		4	0.065408	−0.071876			4	0.050160	−0.075977
		5	0.079263	−0.061652			5	0.063170	−0.066222
		6	0.094638	−0.045420			6	0.077772	−0.051429
		7	0.112414	−0.020698			7	0.094625	−0.030120
		8	0.134239	0.017927			8	0.114811	0.000537
		9	0.164178	0.085070			9	0.140333	0.046381
		10	0.230001	0.324597			10	0.377130	0.418384
11	2	1	−0.929310	−0.488243	11	11	1	0.024850	−0.065444
		2	1.929310	0.488243			2	0.035456	−0.070318
							3	0.045727	−0.070456
11	3	1	−0.444245	−0.322452			4	0.056215	−0.067076
		2	−0.380642	−0.301277			5	0.067261	−0.060207
		3	1.824887	0.623729			6	0.079220	−0.049300
							7	0.092560	−0.033156
11	4	1	−0.242206	−0.238188			8	0.108034	−0.009427
		2	−0.207204	−0.228941			9	0.127068	0.026879
		3	−0.147490	−0.198888			10	0.153197	0.089148
		4	1.596900	0.666017			11	0.210412	0.309357
11	5	1	−0.137718	−0.186803	12	2	1	−0.976872	−0.489254
		2	−0.115110	−0.183651			2	1.976872	0.489254
		3	−0.077762	−0.164597					
		4	−0.028411	−0.133278	12	3	1	−0.476530	−0.323426
		5	1.359000	0.668329			2	−0.416836	−0.304093
							3	1.893367	0.627519
11	6	1	−0.076739	−0.151936					
		2	−0.060142	−0.152221	12	4	1	−0.266888	−0.239300
		3	−0.034581	−0.139907			2	−0.234180	−0.230796
		4	−0.001490	−0.117886			3	−0.177681	−0.203562
		5	0.039518	−0.086131			4	1.678749	0.673657
		6	1.133434	0.648081					
					12	5	1	−0.157792	−0.188109
11	7	1	−0.038349	−0.126507			2	−0.136884	−0.185142
		2	−0.024842	−0.128838			3	−0.101445	−0.167790
		3	−0.005964	−0.120951			4	−0.054640	−0.139693
		4	0.017632	−0.105219			5	1.450761	0.680734
		5	0.046354	−0.081602					
		6	0.081182	−0.048929	12	6	1	−0.093679	−0.153471
		7	0.923987	0.612047			2	−0.078561	−0.153632
							3	−0.054320	−0.142329
11	8	1	−0.012943	−0.106922			4	−0.022769	−0.122474
		2	−0.001050	−0.110498			5	0.016136	−0.094355
		3	0.013869	−0.105662			6	1.233193	0.666261
		4	0.031661	−0.094405					
		5	0.052723	−0.076693	12	7	1	−0.052987	−0.128308
		6	0.077815	−0.051525			2	−0.040893	−0.130339
		7	0.108161	−0.016860			3	−0.023072	−0.123007
		8	0.729765	0.562564			4	−0.000515	−0.108712
							5	0.026930	−0.087681
11	9	1	0.004425	−0.091115			6	0.059918	−0.059256
		2	0.015498	−0.095437			7	1.030620	0.637304
		3	0.028023	−0.092780					
		4	0.042178	−0.084833	12	8	1	−0.025785	−0.109045
		5	0.058340	−0.071581			2	−0.015312	−0.112224
		6	0.077093	−0.052182			3	−0.001353	−0.107627
		7	0.099349	−0.024880			4	0.015634	−0.097276
		8	0.126592	0.013606			5	0.035853	−0.081361
		9	0.548502	0.499201			6	0.059835	−0.059315
							7	0.088444	−0.029900
11	10	1	0.016502	−0.077717			8	0.842684	0.596748
		2	0.027205	−0.082449					

Table A.7. (continued)

N	M	I	A(N,M,I)	C(N,M,I)	N	M	I	A(N,M,I)	C(N,M,I)
12	9	1	−0.006944	−0.093658	13	6	1	−0.109140	−0.154711
		2	0.002669	−0.097540			2	−0.095246	−0.154785
		3	0.014239	−0.094893			3	−0.072165	−0.144347
		4	0.027669	−0.087448			4	−0.041997	−0.126268
		5	0.043189	−0.075371			5	−0.004940	−0.101028
		6	0.061225	−0.058180			6	1.323488	0.681140
		7	0.082441	−0.034802					
		8	0.107856	−0.003342	13	7	1	−0.066358	−0.129743
		9	0.667655	0.545234			2	−0.055414	−0.131538
12	10	1	0.006411	−0.080881			3	−0.038503	−0.124701
		2	0.015598	−0.085171			4	−0.016879	−0.111609
		3	0.025675	−0.083952			5	0.009416	−0.092649
		4	0.036799	−0.078714			6	0.040810	−0.067475
		5	0.049211	−0.069610			7	1.126930	0.657714
		6	0.063256	−0.056237	13	8	1	−0.037540	−0.110704
		7	0.079438	−0.037675			2	−0.028206	−0.113563
		8	0.098522	−0.012272			3	−0.015049	−0.109206
		9	0.121752	0.022956			4	0.001231	−0.099644
		10	0.503338	0.481555			5	0.020686	−0.085204
12	11	1	0.015982	−0.069798			6	0.043677	−0.065581
		2	0.024997	−0.074285			7	0.070830	−0.039995
		3	0.034156	−0.074131			8	0.944372	0.623896
		4	0.043790	−0.070617	13	9	1	−0.017389	−0.095590
		5	0.054149	−0.063891			2	−0.008934	−0.099109
		6	0.065515	−0.053621			3	0.001863	−0.096521
		7	0.078264	−0.039034			4	0.014684	−0.089554
		8	0.092958	−0.018715			5	0.029637	−0.078490
		9	0.110521	0.009948			6	0.047027	−0.063068
		10	0.132666	0.052280			7	0.067346	−0.042607
		11	0.347003	0.401864			8	0.091328	−0.015928
12	12	1	0.022771	−0.059449			9	0.774437	0.580865
		2	0.031776	−0.063952	13	10	1	−0.002927	−0.083170
		3	0.040408	−0.064601			2	0.005067	−0.087085
		4	0.049122	−0.062489			3	0.014356	−0.085792
		5	0.058175	−0.057754			4	0.024891	−0.080789
		6	0.067800	−0.050137			5	0.036816	−0.072325
		7	0.078281	−0.039010			6	0.050389	−0.060181
		8	0.090017	−0.023199			7	0.065995	−0.043768
		9	0.103664	−0.000505			8	0.084201	−0.022048
		10	0.120475	0.033696			9	0.105863	0.006715
		11	0.143566	0.091751			10	0.615348	0.528441
		12	0.193947	0.295648					
13	2	1	−1.020378	−0.490105	13	11	1	0.007628	−0.072617
		2	2.020377	0.490105			2	0.015408	−0.076746
							3	0.023732	−0.076418
13	3	1	−0.506031	−0.324239			4	0.032743	−0.072938
		2	−0.449735	−0.306454			5	0.042611	−0.066531
		3	1.955765	0.630694			6	0.053556	−0.057014
							7	0.065876	−0.043886
13	4	1	−0.289420	−0.240219			8	0.080005	−0.026244
		2	−0.258687	−0.232349			9	0.096594	−0.002552
		3	−0.205024	−0.207450			10	0.116703	0.029910
		4	1.753131	0.680018			11	0.465143	0.465037
13	5	1	−0.176109	−0.189177					
		2	−0.156637	−0.186381	13	12	1	0.015382	−0.063288
		3	−0.122893	−0.170454			2	0.023100	−0.067492
		4	−0.078337	−0.144971			3	0.030818	−0.067892
		5	1.533976	0.690983			4	0.038824	−0.065622

Table A.7. (continued)

N	M	I	A(N,M,I)	C(N,M,I)	N	M	I	A(N,M,I)	C(N,M,I)
		5	0.047302	−0.060887			7	0.054897	−0.048074
		6	0.056444	−0.053540			8	1.036868	0.646052
		7	0.066482	−0.043158					
		8	0.077739	−0.028970	14	9	1	−0.027030	−0.097117
		9	0.090699	−0.009644			2	−0.019516	−0.100334
		10	0.106166	0.017233			3	−0.009363	−0.097827
		11	0.125627	0.056547			4	0.002928	−0.091298
		12	0.321416	0.386713			5	0.017368	−0.081103
							6	0.034165	−0.067124
13	13	1	0.021005	−0.054436			7	0.053685	−0.048921
		2	0.028757	−0.058585			8	0.076476	−0.025720
		3	0.036127	−0.059535			9	0.871287	0.609445
		4	0.043501	−0.058259					
		5	0.051078	−0.054942	14	10	1	−0.011580	−0.084931
		6	0.059028	−0.049472			2	−0.004548	−0.088528
		7	0.067533	−0.041504			3	0.004100	−0.087207
		8	0.076831	−0.030398			4	0.014144	−0.082451
		9	0.087274	−0.015037			5	0.025647	−0.074573
		10	0.099441	0.006644			6	0.038794	−0.063473
		11	0.114446	0.038943			7	0.053879	−0.048768
		12	0.135068	0.093324			8	0.071335	−0.029776
		13	0.179913	0.283257			9	0.091783	−0.005398
							10	0.716445	0.565105
14	2	1	−1.060461	−0.490831					
		2	2.060461	−0.490831	14	11	1	−0.000170	−0.074686
							2	0.006622	−0.078499
14	3	1	−0.533185	−0.324929			3	0.014283	−0.078064
		2	0.479874	−0.308462			4	0.022800	−0.074680
		3	2.013059	0.633391			5	0.032273	−0.068624
							6	0.042866	−0.059816
14	4	1	−0.310144	−0.240992			7	0.054817	−0.047926
		2	−0.281132	−0.233670			8	0.068463	−0.032355
		3	−0.229990	−0.210735			9	0.084290	−0.012126
		4	1.821266	0.685397			10	0.103025	0.014349
							11	0.570731	0.512429
14	5	1	−0.192947	−0.190068					
		2	−0.174709	−0.187427	14	12	1	0.008361	−0.065816
		3	−0.142478	−0.172710			2	0.015058	−0.069728
		4	−0.099930	−0.149393			3	0.022076	−0.069962
		5	1.610065	0.699598			4	0.029552	−0.067659
							5	0.037615	−0.063070
14	6	1	−0.123352	−0.155736			6	0.046411	−0.056130
		2	−0.110490	−0.155747			7	0.056132	−0.046558
		3	−0.088443	−0.146054			8	0.067039	−0.033834
		4	−0.059523	−0.129460			9	0.079506	−0.017101
		5	−0.024111	−0.106556			10	0.094096	0.005064
		6	1.405919	0.693553			11	0.111723	0.035156
							12	0.432431	0.449638
14	7	1	−0.078656	−0.130915					
		2	−0.068666	−0.132521	14	13	1	0.014760	−0.057849
		3	−0.052554	−0.126123			2	0.021453	−0.061764
		4	−0.031776	−0.114051			3	0.028064	−0.062506
		5	−0.006522	−0.096788			4	0.034842	−0.061074
		6	0.023467	−0.074184			5	0.041933	−0.057693
		7	1.214708	0.674581			6	0.049474	−0.052317
							7	0.057619	−0.044707
14	8	1	−0.048365	−0.112041			8	0.066569	−0.034420
		2	−0.039964	−0.114637			9	0.076605	−0.020713
		3	−0.027495	−0.110509			10	0.088151	−0.002338
		4	−0.011849	−0.101635			11	0.101914	0.022943
		5	0.006905	−0.088422			12	0.119200	0.059643
		6	0.029002	−0.070735			13	0.299416	0.372795

Table A.7. *(continued)*

N	M	I	A(N,M,I)	C(N,M,I)	N	M	I	A(N,M,I)	C(N,M,I)
14	14	1	0.019487	−0.050186			6	0.022403	−0.070544
		2	0.026238	−0.054008			7	0.041203	−0.054142
		3	0.032614	−0.055130			8	0.062969	−0.033595
		4	0.038947	−0.054419			9	0.959920	0.632967
		5	0.045399	−0.052075	15	10	1	−0.019626	−0.086339
		6	0.052097	−0.048066			2	−0.013383	−0.089664
		7	0.059168	−0.042197			3	−0.005271	−0.088341
		8	0.066767	−0.034099			4	0.004351	−0.083828
		9	0.075102	−0.023149			5	0.015475	−0.076474
		10	0.084482	−0.008285			6	0.028227	−0.066261
		11	0.095428	0.012430			7	0.042832	−0.052943
		12	0.108942	0.043015			8	0.059624	−0.036054
		13	0.127523	0.094166			9	0.079072	−0.014863
		14	0.167807	0.272004			10	0.808700	0.594768
15	2	1	−1.097617	−0.491458					
		2	2.097617	0.491458	15	11	1	−0.007450	−0.076297
							2	−0.001467	−0.079835
15	3	1	−0.558336	−0.325521			3	0.005652	−0.079332
		2	−0.507671	−0.310191			4	0.013759	−0.076068
		3	2.066007	0.635712			5	0.022893	−0.070355
							6	0.033174	−0.062181
15	4	1	−0.329324	−0.241651			7	0.044787	−0.051331
		2	−0.301829	−0.234806			8	0.057997	−0.037396
		3	−0.252948	−0.213548			9	0.073180	−0.019723
		4	1.884101	0.690005			10	0.090865	0.002701
							11	0.666610	0.549817
15	5	1	−0.208525	−0.190823					
		2	−0.191357	−0.188323	15	12	1	0.001756	−0.067695
		3	−0.160491	−0.174645			2	0.007624	−0.071342
		4	−0.119748	−0.153153			3	0.014079	−0.071459
		5	1.680121	0.706944			4	0.021133	−0.069178
							5	0.028861	−0.064779
15	6	1	−0.136498	−0.156597			6	0.037374	−0.058256
		2	−0.124518	−0.156563			7	0.046827	−0.049425
		3	−0.103401	−0.147517			8	0.057431	−0.037926
		4	−0.075614	−0.132182			9	0.069479	−0.023180
		5	−0.041680	−0.111215			10	0.083393	−0.004280
		6	1.481712	0.704074			11	0.099799	0.020236
							12	0.532243	0.497284
15	7	1	−0.090036	−0.131891					
		2	−0.080850	−0.133342	15	13	1	0.008779	−0.060130
		3	−0.065446	−0.127335			2	0.014620	−0.063805
		4	−0.045441	−0.116138			3	0.020637	−0.064394
		5	−0.021137	−0.100291			4	0.026961	−0.062900
		6	0.007597	−0.079774			5	0.033693	−0.059574
		7	1.295312	0.688771			6	0.040939	−0.054417
							7	0.048828	−0.047269
15	8	1	−0.058390	−0.113143			8	0.057528	−0.037821
		2	−0.050767	−0.115520			9	0.067265	−0.025565
		3	−0.038897	−0.111607			10	0.078368	−0.009694
		4	−0.023825	−0.103332			11	0.091330	0.011113
		5	−0.005717	−0.091156			12	0.106947	0.039155
		6	0.015565	−0.075053			13	0.404106	0.435302
		7	0.040351	−0.054703					
		8	1.121680	0.664514	15	14	1	0.014143	−0.053241
							2	0.020013	−0.056879
15	9	1	−0.035972	−0.098361			3	0.025750	−0.057827
		2	−0.029235	−0.101322			4	0.031576	−0.056973
		3	−0.019633	−0.098904			5	0.037611	−0.054542
		4	−0.007812	−0.092773			6	0.043958	−0.050539
		5	0.006156	−0.083327			7	0.050725	−0.044833

Table A.7. (continued)

N	M	I	A(N,M,I)	C(N,M,I)	N	M	I	A(N,M,I)	C(N,M,I)
		8	0.058045	−0.037157			7	0.026973	−0.060251
		9	0.066092	−0.027072			8	1.199963	0.680158
		10	0.075114	−0.013872	16	9	1	−0.044303	−0.099396
		11	0.085490	0.003612			2	−0.038218	−0.102138
		12	0.097844	0.027465			3	−0.029094	−0.099811
		13	0.113340	0.061879			4	−0.017697	−0.094037
		14	0.280298	0.359980			5	−0.004166	−0.085242
15	15	1	0.018170	−0.046538			6	0.011570	−0.073467
		2	0.024108	−0.050064			7	0.029712	−0.058535
		3	0.029685	−0.051279			8	0.050576	−0.040084
		4	0.035191	−0.050957			9	1.041619	0.652711
		5	0.040762	−0.049298	16	10	1	−0.027135	−0.087496
		6	0.046496	−0.046315			2	−0.021550	−0.090585
		7	0.052488	−0.041899			3	−0.013895	−0.089277
		8	0.058844	−0.035827			4	−0.004646	−0.084992
		9	0.065696	−0.027731			5	0.006132	−0.078105
		10	0.073230	−0.017008			6	0.018515	−0.068653
		11	0.081725	−0.002653			7	0.032675	−0.056482
		12	0.091651	0.017156			8	0.048869	−0.041268
		13	0.103914	0.046191			9	0.067459	−0.022503
		14	0.120784	0.094483			10	0.893576	0.619360
		15	0.157255	0.261738	16	11	1	−0.014263	−0.077597
16	2	1	−1.132243	−0.492005			2	−0.008950	−0.080895
		2	2.132243	0.492005			3	−0.002286	−0.080349
16	3	1	−0.581757	−0.326035			4	0.005469	−0.077213
		2	−0.533457	−0.311694			5	0.014303	−0.071820
		3	2.115214	0.637730			6	0.024297	−0.064207
16	4	1	−0.347172	−0.242220			7	0.035593	−0.054237
		2	−0.321026	−0.235794			8	0.048404	−0.041625
		3	−0.274186	−0.215984			9	0.063020	−0.025917
		4	1.942384	0.693998			10	0.079847	−0.006432
16	5	1	−0.223015	−0.191470			11	0.754566	0.580293
		2	−0.206788	−0.189099	16	12	1	−0.004450	−0.069172
		3	−0.177158	−0.176323			2	0.000732	−0.072584
		4	−0.138048	−0.156390			3	0.006721	−0.072615
		5	1.745009	0.713282			4	0.013424	−0.070383
16	6	1	−0.148725	−0.157331			5	0.020868	−0.066184
		2	−0.137508	−0.157263			6	0.029134	−0.060054
		3	−0.117232	−0.148785			7	0.038344	−0.051876
		4	−0.090481	−0.134532			8	0.048668	−0.041398
		5	−0.057883	−0.115196			9	0.060342	−0.028216
		6	1.551828	0.713108			10	0.073692	−0.011716
16	7	1	−0.100621	−0.132718			11	0.089173	0.009035
		2	−0.092121	−0.134040			12	0.623351	0.535164
		3	−0.077354	−0.128381	16	13	1	0.003118	−0.061843
		4	−0.058057	−0.117942			2	0.008256	−0.065297
		5	−0.034624	−0.103296			3	0.013789	−0.065770
		6	−0.007020	−0.084506			4	0.019747	−0.064259
		7	1.369798	0.700883			5	0.026189	−0.061031
16	8	1	−0.067719	−0.114069			6	0.033196	−0.056120
		2	−0.060754	−0.116260			7	0.040872	−0.049427
		3	−0.049415	−0.112545			8	0.049357	−0.040731
		4	−0.034868	−0.104798			9	0.058836	−0.029675
		5	−0.017357	−0.093508			10	0.069568	−0.015710
		6	0.003178	−0.078726			11	0.081920	0.002010
							12	0.096438	0.024833
							13	0.498713	0.483018

Table A.7. *(continued)*

N	M	I	A(N,M,I)	C(N,M,I)	N	M	I	A(N,M,I)	C(N,M,I)
16	14	1	0.008992	−0.055309			3	−0.192661	−0.177793
		2	0.014141	−0.058750			4	−0.155037	−0.159206
		3	0.019370	−0.059563			5	1.805419	0.718809
		4	0.024804	−0.058635					
		5	0.030525	−0.056208	17	6	1	−0.160149	−0.157965
		6	0.036615	−0.052317			2	−0.149601	−0.157871
		7	0.043164	−0.046878			3	−0.130090	−0.149896
		8	0.050284	−0.039699			4	−0.104290	−0.136581
		9	0.058124	−0.030467			5	−0.072907	−0.118639
		10	0.066884	−0.018695			6	1.617037	0.720952
		11	0.076854	−0.003625					
		12	0.088469	0.015969	17	7	1	−0.110512	−0.133428
		13	0.102433	0.042224			2	−0.102606	−0.134640
		14	0.379341	0.421953			3	−0.088415	−0.129294
							4	−0.069771	−0.119517
16	15	1	0.013547	−0.049291			5	−0.047139	−0.105901
		2	0.018743	−0.052670			6	−0.020560	−0.088568
		3	0.023778	−0.053739			7	1.439003	0.711349
		4	0.028849	−0.053290					
		5	0.034060	−0.051538	17	8	1	−0.076441	−0.114859
		6	0.039489	−0.048520			2	−0.070039	−0.116891
		7	0.045218	−0.044164			3	−0.059173	−0.113357
		8	0.051338	−0.038307			4	−0.045110	−0.106076
		9	0.057965	−0.030678			5	−0.028154	−0.095554
		10	0.065253	−0.020850			6	−0.008307	−0.081890
		11	0.073425	−0.008156			7	0.014595	−0.064968
		12	0.082818	0.008503			8	1.272628	0.693595
		13	0.093994	0.031075					
		14	0.107995	0.063476	17	9	1	−0.052096	−0.100271
		15	0.263528	0.348149			2	−0.046565	−0.102825
							3	−0.037862	−0.100587
16	16	1	0.017016	−0.043375			4	−0.026851	−0.095136
		2	0.022284	−0.046633			5	−0.013728	−0.086910
		3	0.027208	−0.047890			6	0.001531	−0.075995
		4	0.032046	−0.047839			7	0.019069	−0.062288
		5	0.036912	−0.046675			8	0.039129	−0.045535
		6	0.041887	−0.044432			9	1.117373	0.669546
		7	0.047042	−0.041053					
		8	0.052455	−0.036402	17	10	1	−0.034167	−0.088465
		9	0.058216	−0.030249			2	−0.029139	−0.091350
		10	0.064444	−0.022230			3	−0.021881	−0.090064
		11	0.071304	−0.011772			4	−0.012965	−0.085992
		12	0.079051	0.002079			5	−0.002507	−0.079521
		13	0.088111	0.021044			6	0.009531	−0.070728
		14	0.099315	0.048675			7	0.023273	−0.059520
		15	0.114733	0.094419			8	0.038922	−0.045671
		16	0.147977	0.252333			9	0.056761	−0.028822
							10	0.972172	0.640135
17	2	1	−1.164659	−0.492486					
		2	2.164659	0.492486	17	11	1	−0.020654	−0.078673
							2	−0.015906	−0.081761
17	3	1	−0.603668	−0.326486			3	−0.009632	−0.081188
		2	−0.557497	−0.313014			4	−0.002186	−0.078180
		3	2.161166	0.639500			5	0.006378	−0.073083
							6	0.016104	−0.065964
17	4	1	−0.363861	−0.242716			7	0.027102	−0.056744
		2	−0.338922	−0.236662			8	0.039540	−0.045224
		3	−0.293934	−0.218114			9	0.053648	−0.031078
		4	1.996717	0.697492			10	0.069744	−0.013827
							11	0.835861	0.605723
17	5	1	−0.236557	−0.192031					
		2	−0.221164	−0.189778					

Table A.7. (continued)

N	M	I	A(N,M,I)	C(N,M,I)	N	M	I	A(N,M,I)	C(N,M,I)
17	12	1	−0.010288	−0.070375			5	0.031091	−0.048727
		2	−0.005683	−0.073577			6	0.035799	−0.046430
		3	−0.000086	−0.073546			7	0.040724	−0.043057
		4	0.006316	−0.071375			8	0.045932	−0.038508
		5	0.013511	−0.067372			9	0.051504	−0.032609
		6	0.021553	−0.061602			10	0.057542	−0.025090
		7	0.030535	−0.053996			11	0.064186	−0.015545
		8	0.040597	−0.044377			12	0.071635	−0.003341
		9	0.051928	−0.032455			13	0.080195	0.012556
		10	0.064785	−0.017797			14	0.090373	0.033974
		11	0.079517	0.000228			15	0.103110	0.064588
		12	0.707314	0.566244			16	0.248699	0.337194
17	13	1	−0.002231	−0.063202	17	17	1	0.015998	−0.040607
		2	0.002318	−0.066454			2	0.020706	−0.043624
		3	0.007448	−0.066839			3	0.025089	−0.044891
		4	0.013101	−0.065335			4	0.029378	−0.045031
		5	0.019298	−0.062220			5	0.033671	−0.044229
		6	0.026098	−0.057556			6	0.038035	−0.042531
		7	0.033584	−0.051282			7	0.042527	−0.039913
		8	0.041872	−0.043242			8	0.047204	−0.036289
		9	0.051113	−0.033181			9	0.052133	−0.031512
		10	0.061516	−0.020708			10	0.057392	−0.025352
		11	0.073364	−0.005250			11	0.063089	−0.017458
		12	0.087058	0.014056			12	0.069375	−0.007282
		13	0.585461	0.521211			13	0.076482	0.006082
							14	0.084803	0.024262
17	14	1	0.004088	−0.056878			15	0.095098	0.050618
		2	0.008636	−0.060131			16	0.109270	0.094076
		3	0.013446	−0.060836			17	0.139752	0.243681
		4	0.018560	−0.059871					
		5	0.024028	−0.057487	18	2	1	−1.195128	−0.492912
		6	0.029909	−0.053742			2	2.195128	0.492912
		7	0.036278	−0.048586					
		8	0.043231	−0.041881	18	3	1	−0.624252	−0.326884
		9	0.050892	−0.033402			2	−0.580008	−0.314183
		10	0.059426	−0.022799			3	2.204260	0.641066
		11	0.069060	−0.009558					
		12	0.080119	0.007112	18	4	1	−0.379529	−0.243153
		13	0.093083	0.028459			2	−0.355679	−0.237429
		14	0.469244	0.469601			3	−0.312382	−0.219992
							4	2.047590	0.700574
17	15	1	0.009066	−0.051176					
		2	0.013648	−0.054390	18	5	1	−0.249266	−0.192523
		3	0.018244	−0.055341			2	−0.234618	−0.190376
		4	0.022974	−0.054815			3	−0.207148	−0.179091
		5	0.027908	−0.053042			4	−0.170883	−0.161679
		6	0.033111	−0.050075			5	1.861914	0.723670
		7	0.038648	−0.045871					
		8	0.044600	−0.040314	18	6	1	−0.170868	−0.158518
		9	0.051065	−0.033203			2	−0.160910	−0.158405
		10	0.058176	−0.024231			3	−0.142100	−0.150876
		11	0.066111	−0.012936			4	−0.117175	−0.138383
		12	0.075128	0.001394			5	−0.086906	−0.121647
		13	0.085616	0.019905			6	1.677960	0.727829
		14	0.098200	0.044587					
		15	0.357506	0.409507	18	7	1	−0.119793	−0.134044
							2	−0.112406	−0.135163
17	16	1	0.012979	−0.045870			3	−0.098738	−0.130098
		2	0.017617	−0.049009			4	−0.080698	−0.120904
		3	0.022076	−0.050145			5	−0.058807	−0.108183
		4	0.026538	−0.049982					

Table A.7. *(continued)*

N	M	I	A(N,M,I)	C(N,M,I)	N	M	I	A(N,M,I)	C(N,M,I)
		6	−0.033165	−0.092095			5	0.012925	−0.063222
		7	1.503605	0.720486			6	0.019540	−0.058792
18	8	1	−0.084626	−0.115541			7	0.026851	−0.052898
		2	−0.078711	−0.117434			8	0.034951	−0.045430
		3	−0.068272	−0.114068			9	0.043969	−0.036200
		4	−0.054656	−0.107202			10	0.054072	−0.024926
		5	−0.038217	−0.097349			11	0.065486	−0.011201
		6	−0.019006	−0.084645			12	0.078516	0.005561
		7	0.003084	−0.069031			13	0.665728	0.552731
		8	1.340405	0.705270	18	14	1	−0.000568	−0.058133
							2	0.003471	−0.061213
18	9	1	−0.059414	−0.101022			3	0.007930	−0.061830
		2	−0.054359	−0.103411			4	0.012775	−0.060849
		3	−0.046030	−0.101260			5	0.018027	−0.058527
		4	−0.035375	−0.096099			6	0.023730	−0.054936
		5	−0.022631	−0.088374			7	0.029942	−0.050053
		6	−0.007819	−0.078203			8	0.036744	−0.043781
		7	0.009161	−0.065532			9	0.044239	−0.035952
		8	0.028495	−0.050186			10	0.052564	−0.026314
		9	1.187973	0.684087			11	0.061904	−0.014497
							12	0.072509	0.000034
18	10	1	−0.040776	−0.089291			13	0.084730	0.018080
		2	−0.036223	−0.091997			14	0.552004	0.507970
		3	−0.029314	−0.090739					
		4	−0.020701	−0.086863	18	15	1	0.004780	−0.052617
		5	−0.010540	−0.080764			2	0.008843	−0.055674
		6	0.001172	−0.072544			3	0.013074	−0.056526
		7	0.014523	−0.062157			4	0.017522	−0.055953
		8	0.029671	−0.049445			5	0.022232	−0.054191
		9	0.046841	−0.034147			6	0.027249	−0.051307
		10	1.045347	0.657947			7	0.032630	−0.047281
							8	0.038443	−0.042029
18	11	1	−0.026669	−0.079582			9	0.044772	−0.035403
		2	−0.022402	−0.082484			10	0.051728	−0.027176
		3	−0.016466	−0.081896			11	0.059460	−0.017018
		4	−0.009294	−0.079012			12	0.068169	−0.004442
		5	−0.000979	−0.074183			13	0.078145	0.011289
		6	0.008496	−0.067503			14	0.089813	0.031340
		7	0.019212	−0.058930			15	0.443142	0.456986
		8	0.031300	−0.048324					
		9	0.044947	−0.035451	18	16	1	0.009048	−0.047594
		10	0.060404	−0.019962			2	0.013157	−0.050597
		11	0.911449	0.627325			3	0.017235	−0.051629
							4	0.021397	−0.051393
18	12	1	−0.015793	−0.071378			5	0.025706	−0.050102
		2	−0.011677	−0.074393			6	0.030212	−0.047820
		3	−0.006416	−0.074315			7	0.034966	−0.044532
		4	−0.000278	−0.072211			8	0.040027	−0.040165
		5	0.006695	−0.068395			9	0.045465	−0.034587
		6	0.014529	−0.062952			10	0.051368	−0.027599
		7	0.023297	−0.055848			11	0.057855	−0.018906
		8	0.033110	−0.046959			12	0.065087	−0.008069
		9	0.044122	−0.036073			13	0.073294	0.005581
		10	0.056540	−0.022877			14	0.082827	0.023119
		11	0.070637	−0.006924			15	0.094248	0.046408
		12	0.785235	0.592326			16	0.338109	0.397887
18	13	1	−0.007289	−0.064317	18	17	1	0.012444	−0.042879
		2	−0.003238	−0.067387			2	0.016611	−0.045800
		3	0.001550	−0.067701			3	0.020593	−0.046965
		4	0.006940	−0.066218			4	0.024555	−0.047008

300 ■ Appendix

Table A.7. (continued)

N	M	I	A(N,M,I)	C(N,M,I)	N	M	I	A(N,M,I)	C(N,M,I)
		5	0.028573	−0.046121			5	−0.069730	−0.110197
		6	0.032702	−0.044362			6	−0.044949	−0.095185
		7	0.036990	−0.041722			7	1.564162	0.728535
		8	0.041487	−0.038137	19	8	1	−0.092336	−0.116135
		9	0.046252	−0.033494			2	−0.086846	−0.117908
		10	0.051355	−0.027618			3	−0.076795	−0.114696
		11	0.056889	−0.020248			4	−0.063593	−0.108201
		12	0.062980	−0.010994			5	−0.047637	−0.098937
		13	0.069810	0.000742			6	−0.029016	−0.087065
		14	0.077655	0.015937			7	−0.007670	−0.072570
		15	0.086979	0.036313			8	1.403893	0.715513
		16	0.098636	0.065331	19	9	1	−0.066309	−0.101674
		17	0.235490	0.327023			2	−0.061667	−0.103918
18	18	1	0.015092	−0.038165			3	−0.053675	−0.101850
		2	0.019328	−0.040965			4	−0.043347	−0.096952
		3	0.023258	−0.042221			5	−0.030960	−0.089671
		4	0.027089	−0.042497			6	−0.016565	−0.080147
		5	0.030909	−0.041963			7	−0.000103	−0.068365
		6	0.034773	−0.040676			8	0.018570	−0.054204
		7	0.038728	−0.038627			9	1.254056	0.696782
		8	0.042820	−0.035765	19	10	1	−0.047007	−0.090004
		9	0.047095	−0.031992			2	−0.042865	−0.092551
		10	0.051612	−0.027160			3	−0.036265	−0.091324
		11	0.056443	−0.021041			4	−0.027929	−0.087629
		12	0.061685	−0.013300			5	−0.018045	−0.081863
		13	0.067477	−0.003410			6	−0.006641	−0.074147
		14	0.074032	0.009488			7	0.006343	−0.064468
		15	0.081713	0.026940			8	0.021029	−0.052718
		16	0.091221	0.052132			9	0.037595	−0.038703
		17	0.104314	0.093529			10	1.113786	0.673407
		18	0.132411	0.235693	19	11	1	−0.032345	−0.080360
19	2	1	−1.223869	−0.493292			2	−0.028492	−0.083097
		2	2.223869	0.493292			3	−0.022852	−0.082502
19	3	1	−0.643659	−0.327238			4	−0.015928	−0.079736
		2	−0.601169	−0.315224			5	−0.007843	−0.075152
		3	2.244827	0.642462			6	0.001395	−0.068861
19	4	1	−0.394294	−0.243540			7	0.011842	−0.060851
		2	−0.371431	−0.238113			8	0.023603	−0.051025
		3	−0.329685	−0.221662			9	0.036827	−0.039209
		4	2.095409	0.703314			10	0.051716	−0.025147
19	5	1	−0.261237	−0.192958			11	0.982076	0.645940
		2	−0.247259	−0.190909	19	12	1	−0.020995	−0.072230
		3	−0.220739	−0.180245			2	−0.017298	−0.075078
		4	−0.185724	−0.163869			3	−0.012331	−0.074965
		5	1.914959	0.727980			4	−0.006425	−0.072929
19	6	1	−0.180964	−0.159004			5	0.000344	−0.069288
		2	−0.171530	−0.158877			6	0.007984	−0.064141
		3	−0.153365	−0.151748			7	0.016548	−0.057480
		4	−0.129250	−0.139981			8	0.026124	−0.049218
		5	−0.100003	−0.124297			9	0.036839	−0.039200
		6	1.735111	0.733908			10	0.048859	−0.027194
19	7	1	−0.128533	−0.134583			11	0.062404	−0.012872
		2	−0.121602	−0.135622			12	0.857947	0.614595
		3	−0.108414	−0.130811	19	13	1	−0.012077	−0.065253
		4	−0.090935	−0.122136			2	−0.008453	−0.068158
							3	−0.003960	−0.068416

Table A.7. (continued)

N	M	I	A(N,M,I)	C(N,M,I)	N	M	I	A(N,M,I)	C(N,M,I)
		4	0.001201	−0.066962			4	0.020023	−0.048319
		5	0.006996	−0.064084			5	0.023825	−0.047390
		6	0.013442	−0.059872			6	0.027772	−0.045626
		7	0.020588	−0.054320			7	0.031906	−0.043028
		8	0.028511	−0.047351			8	0.036272	−0.039552
		9	0.037316	−0.038827			9	0.040920	−0.035112
		10	0.047141	−0.028538			10	0.045913	−0.029573
		11	0.058169	−0.016185			11	0.051331	−0.022740
		12	0.070640	−0.001353			12	0.057280	−0.014330
		13	0.740487	0.579318			13	0.063906	−0.003928
							14	0.071419	0.009098
19	14	1	−0.004989	−0.059169			15	0.080136	0.025759
		2	−0.001384	−0.062091			16	0.090563	0.047806
		3	0.002773	−0.062636			17	0.320763	0.387016
		4	0.007387	−0.061652					
		5	0.012453	−0.059399	19	18	1	0.011941	−0.040244
		6	0.017997	−0.055961			2	0.015709	−0.042966
		7	0.024066	−0.051334			3	0.019289	−0.044136
		8	0.030726	−0.045450			4	0.022835	−0.044327
		9	0.038064	−0.038185			5	0.026412	−0.043716
		10	0.046194	−0.029351			6	0.030068	−0.042366
		11	0.055267	−0.018676			7	0.033841	−0.040281
		12	0.065482	−0.005781			8	0.037772	−0.037423
		13	0.077109	0.009882			9	0.041903	−0.033716
		14	0.628854	0.539802			10	0.046286	−0.029044
							11	0.050984	−0.023232
19	15	1	0.000692	−0.053779			12	0.056083	−0.016030
		2	0.004313	−0.056686			13	0.061695	−0.007067
		3	0.008234	−0.057456			14	0.067989	0.004227
		4	0.012444	−0.056855			15	0.075216	0.018774
		5	0.016963	−0.055121			16	0.083802	0.038205
		6	0.021823	−0.052332			17	0.094528	0.065788
		7	0.027068	−0.048486			18	0.223648	0.317554
		8	0.032757	−0.043523					
		9	0.038961	−0.037334	19	19	1	0.014282	−0.035995
		10	0.045774	−0.029749			2	0.018115	−0.038600
		11	0.053320	−0.020523			3	0.021661	−0.039833
		12	0.061762	−0.009310			4	0.025107	−0.040204
		13	0.071323	0.004394			5	0.028531	−0.039873
		14	0.082316	0.021334			6	0.031980	−0.038897
		15	0.522250	0.495426			7	0.035494	−0.037282
							8	0.039109	−0.034997
19	16	1	0.005271	−0.048924			9	0.042861	−0.031977
		2	0.008929	−0.051792			10	0.046794	−0.028121
		3	0.012687	−0.052735			11	0.050958	−0.023280
		4	0.016601	−0.052449			12	0.055419	−0.017234
		5	0.020708	−0.051151			13	0.060267	−0.009660
		6	0.025048	−0.048913			14	0.065629	−0.000055
		7	0.029663	−0.045735			15	0.071704	0.012400
		8	0.034603	−0.041566			16	0.078826	0.029177
		9	0.039929	−0.036308			17	0.087648	0.053305
		10	0.045717	−0.029809			18	0.099799	0.092832
		11	0.052068	−0.021850			19	0.125817	0.228292
		12	0.059114	−0.012118	20	2	1	−1.251068	−0.493634
		13	0.067036	−0.000151			2	2.251068	0.493634
		14	0.076094	0.014738	20	3	1	−0.662014	−0.327555
		15	0.086669	0.033642			2	−0.621129	−0.316157
		16	0.419861	0.445119			3	2.283144	0.643713
19	17	1	0.008968	−0.044464	20	4	1	−0.408252	−0.243885
		2	0.012677	−0.047270			2	−0.386289	−0.238726
		3	0.016326	−0.048344					

Table A.7. (continued)

N	M	I	A(N,M,I)	C(N,M,I)	N	M	I	A(N,M,I)	C(N,M,I)
		3	−0.345972	−0.223154			9	0.029216	−0.042476
		4	2.140513	0.705766			10	0.043593	−0.029594
20	5	1	−0.272551	−0.193344			11	1.048347	0.662168
		2	−0.259179	−0.191385	20	12	1	−0.025922	−0.072964
		3	−0.233536	−0.181278			2	−0.022589	−0.075662
		4	−0.199675	−0.165821			3	−0.017879	−0.075522
		5	1.964941	0.731828			4	−0.012183	−0.073554
20	6	1	−0.190502	−0.159435			5	−0.005600	−0.070076
		2	−0.181539	−0.159298			6	0.001858	−0.065197
		3	−0.163969	−0.152528			7	0.010227	−0.058928
		4	−0.140605	−0.141408			8	0.019578	−0.051211
		5	−0.112303	−0.126651			9	0.030012	−0.041931
		6	1.788917	0.739321			10	0.041668	−0.030912
20	7	1	−0.136790	−0.135060			11	0.054724	−0.017911
		2	−0.130264	−0.136029			12	0.926107	0.633868
		3	−0.117518	−0.131448	20	13	1	−0.016619	−0.066052
		4	−0.100561	−0.123236			2	−0.013364	−0.068809
		5	−0.079995	−0.111990			3	−0.009129	−0.069021
		6	−0.056007	−0.097918			4	−0.004170	−0.067601
		7	1.621135	0.735681			5	0.001453	−0.064835
20	8	1	−0.099621	−0.116659			6	0.007742	−0.060825
		2	−0.094504	−0.118326			7	0.014732	−0.055581
		3	−0.084808	−0.115255			8	0.022485	−0.049051
		4	−0.071993	−0.109093			9	0.031087	−0.041132
		5	−0.056488	−0.100352			10	0.040654	−0.031666
		6	−0.038416	−0.089210			11	0.051333	−0.020429
		7	−0.017755	−0.075681			12	0.063321	−0.007116
		8	1.463585	0.724575			13	0.810474	0.602120
20	9	1	−0.072826	−0.102246	20	14	1	−0.009191	−0.060043
		2	−0.068544	−0.104362			2	−0.005961	−0.062821
		3	−0.060858	−0.102371			3	−0.002065	−0.063307
		4	−0.050834	−0.097711			4	0.002348	−0.062329
		5	−0.038781	−0.090828			5	0.007248	−0.060148
		6	−0.024779	−0.081874			6	0.012649	−0.056856
		7	−0.008798	−0.070863			7	0.018585	−0.052465
		8	0.009270	−0.057714			8	0.025109	−0.046929
		9	1.316151	0.707969			9	0.032297	−0.040154
20	10	1	−0.052900	−0.090626			10	0.040241	−0.031999
		2	−0.049115	−0.093031			11	0.049069	−0.022261
		3	−0.042792	−0.091837			12	0.058941	−0.010659
		4	−0.034710	−0.088309			13	0.070069	0.003196
		5	−0.025087	−0.082842			14	0.700660	0.566775
		6	−0.013973	−0.075573	20	15	1	−0.003203	−0.054744
		7	−0.001335	−0.066511			2	0.000035	−0.057513
		8	0.012921	−0.055584			3	0.003690	−0.058213
		9	0.028939	−0.042651			4	0.007695	−0.057596
		10	1.178052	0.686964			5	0.012048	−0.055899
20	11	1	−0.037716	−0.081036			6	0.016769	−0.053209
		2	−0.034222	−0.083625			7	0.021892	−0.049537
		3	−0.028845	−0.083028			8	0.027467	−0.044842
		4	−0.022146	−0.080373			9	0.033552	−0.039043
		5	−0.014276	−0.076014			10	0.040230	−0.032010
		6	−0.005262	−0.070070			11	0.047602	−0.023560
		7	0.004930	−0.062554			12	0.055804	−0.013436
		8	0.016382	−0.053398			13	0.065012	−0.001280
							14	0.075467	0.013416
							15	0.595940	0.527466

Table A.7. (continued)

N	M	I	A(N,M,I)	C(N,M,I)	N	M	I	A(N,M,I)	C(N,M,I)
20	16	1	0.001656	−0.050002			9	0.038247	−0.033514
		2	0.004926	−0.052742			10	0.042061	−0.029746
		3	0.008410	−0.053608			11	0.046112	−0.025085
		4	0.012111	−0.053287			12	0.050458	−0.019364
		5	0.016048	−0.051997			13	0.055176	−0.012340
		6	0.020247	−0.049816			14	0.060372	−0.003660
		7	0.024742	−0.046757			15	0.066198	0.007217
		8	0.029575	−0.042785			16	0.072887	0.021167
		9	0.034801	−0.037825			17	0.080830	0.039737
		10	0.040482	−0.031763			18	0.090746	0.066024
		11	0.046707	−0.024432			19	0.212971	0.308714
		12	0.053585	−0.015601	20	20	1	0.013553	−0.034055
		13	0.061262	−0.004939			2	0.017039	−0.036484
		14	0.069938	0.008023			3	0.020257	−0.037686
		15	0.079893	0.023984			4	0.023376	−0.038123
		16	0.495616	0.483548			5	0.026464	−0.037945
20	17	1	0.005617	−0.045695			6	0.029565	−0.037211
		2	0.008931	−0.048385			7	0.032711	−0.035932
		3	0.012297	−0.049380			8	0.035932	−0.034091
		4	0.015773	−0.049304			9	0.039258	−0.031646
		5	0.019394	−0.048357			10	0.042720	−0.028527
		6	0.023192	−0.046613			11	0.046357	−0.024632
		7	0.027201	−0.044083			12	0.050215	−0.019814
		8	0.031459	−0.040736			13	0.054354	−0.013860
		9	0.036010	−0.036510			14	0.058856	−0.006460
		10	0.040910	−0.031298			15	0.063842	0.002866
		11	0.046228	−0.024954			16	0.069496	0.014902
		12	0.052055	−0.017265			17	0.076128	0.031052
		13	0.058509	−0.007934			18	0.084346	0.054203
		14	0.065756	0.003474			19	0.095669	0.092028
		15	0.074029	0.017606			20	0.119862	0.221415
		16	0.083671	0.035486	21	2	1	−1.276880	−0.493942
		17	0.398968	0.433947			2	2.276880	0.493942
20	18	1	0.008847	−0.041706	21	3	1	−0.679427	−0.327841
		2	0.012215	−0.044331			2	−0.640016	−0.316999
		3	0.015502	−0.045422			3	2.319443	0.644841
		5	0.022197	−0.044896	21	4	1	−0.421487	−0.244196
		6	0.025690	−0.043529			2	−0.400349	−0.239279
		7	0.029324	−0.041460			3	−0.361352	−0.224498
		8	0.033136	−0.038666			4	2.183188	0.707973
		9	0.037162	−0.035086	21	5	1	−0.283275	−0.193691
		10	0.041450	−0.030632			2	−0.270454	−0.191814
		11	0.046055	−0.025168			3	−0.245624	−0.182209
		12	0.051050	−0.018506			4	−0.212831	−0.167572
		13	0.056532	−0.010374			5	2.012185	0.735285
		14	0.062634	−0.000381	21	6	1	−0.199541	−0.159820
		15	0.069547	0.012071			2	−0.191002	−0.159675
		16	0.077558	0.027938			3	−0.173984	−0.153231
		17	0.087131	0.048871			4	−0.151319	−0.142690
		18	0.305157	0.376826			5	−0.123892	−0.128756
20	19	1	0.011469	−0.037905			6	1.839738	0.744172
		2	0.014895	−0.040446	21	7	1	−0.144615	−0.135485
		3	0.018135	−0.041607			2	−0.138450	−0.136392
		4	0.021329	−0.041903			3	−0.126112	−0.132021
		5	0.024538	−0.041503			4	−0.109643	−0.124225
		6	0.027802	−0.040467					
		7	0.031153	−0.038810					
		8	0.034624	−0.036509					

Appendix ■ 303

Table A.7. (continued)

N	M	I	A(N,M,I)	C(N,M,I)	N	M	I	A(N,M,I)	C(N,M,I)
		5	−0.089672	−0.113595			4	−0.009216	−0.068156
		6	−0.066419	−0.100351			5	−0.003750	−0.065499
		7	1.674911	0.742070			6	0.002392	−0.061675
21	8	1	−0.106524	−0.117124			7	0.009234	−0.056708
		2	−0.101739	−0.118696			8	0.016823	−0.050566
		3	−0.092368	−0.115755			9	0.025231	−0.043172
		4	−0.079914	−0.109895			10	0.034555	−0.034404
		5	−0.064832	−0.101621			11	0.044917	−0.024094
		6	−0.047274	−0.091123			12	0.056476	−0.012009
		7	−0.027245	−0.078439			13	0.876270	0.621934
		8	1.519896	0.732652	21	14	1	−0.013190	−0.060794
21	9	1	−0.079003	−0.102752			2	−0.010287	−0.063440
		2	−0.075039	−0.104752			3	−0.006618	−0.063875
		3	−0.067630	−0.102836			4	−0.002385	−0.062909
		4	−0.057890	−0.098393			5	0.002367	−0.060800
		5	−0.046151	−0.091865			6	0.007636	−0.057647
		6	−0.032519	−0.083417			7	0.013446	−0.053473
		7	−0.016987	−0.073081			8	0.019842	−0.048247
		8	0.000523	−0.060807			9	0.026883	−0.041903
		9	1.374696	0.717905			10	0.034651	−0.034330
21	10	1	−0.058486	−0.091174			11	0.043251	−0.025373
		2	−0.055015	−0.093453			12	0.052814	−0.014822
		3	−0.048942	−0.092291			13	0.063513	−0.002388
		4	−0.041096	−0.088916			14	0.768077	0.590001
		5	−0.031718	−0.083721	21	15	1	−0.006919	−0.055563
		6	−0.020880	−0.076849			2	−0.004012	−0.058205
		7	−0.008567	−0.068330			3	−0.000589	−0.058845
		8	0.005290	−0.058118			4	0.003237	−0.058219
		9	0.020804	−0.046107			5	0.007443	−0.056566
		10	1.238610	0.698958			6	0.012039	−0.053974
21	11	1	−0.042812	−0.081628			7	0.017051	−0.050466
		2	−0.039631	−0.084084			8	0.022516	−0.046017
		3	−0.034490	−0.083489			9	0.028487	−0.040565
		4	−0.027998	−0.080939			10	0.035035	−0.034010
		5	−0.020328	−0.076784			11	0.042242	−0.026215
		6	−0.011527	−0.071153			12	0.050225	−0.016982
		7	−0.001577	−0.064073			13	0.059124	−0.006049
		8	0.009585	−0.055502			14	0.069130	0.006947
		9	0.022055	−0.045346			15	0.664991	0.554729
		10	0.035968	−0.033457	21	16	1	−0.001802	−0.050903
		11	1.110755	0.676455			2	0.001131	−0.053522
21	12	1	−0.030601	−0.073604			3	0.004378	−0.054323
		2	−0.027584	−0.076167			4	0.007896	−0.053977
		3	−0.023103	−0.076006			5	0.011684	−0.052704
		4	−0.017596	−0.074104			6	0.015759	−0.050587
		5	−0.011187	−0.070777			7	0.020147	−0.047646
		6	−0.003901	−0.066142			8	0.024883	−0.043862
		7	0.044282	−0.060223			9	0.030017	−0.039172
		8	0.013418	−0.052983			10	0.035594	−0.033501
		9	0.023589	−0.044337			11	0.041700	−0.026708
		10	0.034909	−0.034153			12	0.048420	−0.018625
		11	0.047525	−0.022243			13	0.055879	−0.009004
		12	0.990248	0.650738			14	0.064233	0.002489
21	13	1	−0.020936	−0.066744			15	0.073697	0.016331
		2	−0.018002	−0.069367			16	0.566382	0.515715
		3	−0.013994	−0.069541	21	17	1	0.002397	−0.046698
							2	0.005370	−0.049278
							3	0.008493	−0.050203

Table A.7. (continued)

N	M	I	A(N,M,I)	C(N,M,I)	N	M	I	A(N,M,I)	C(N,M,I)
		4	0.011778	−0.050089			8	0.031923	−0.035486
		5	0.015245	−0.049137			9	0.035133	−0.033042
		6	0.018916	−0.047426			10	0.038489	−0.029966
		7	0.022818	−0.044972			11	0.042026	−0.026176
		8	0.026981	−0.041760			12	0.045787	−0.021555
		9	0.031445	−0.037741			13	0.049824	−0.015938
		10	0.036263	−0.032831			14	0.054209	−0.009097
		11	0.041487	−0.026922			15	0.059038	−0.000691
		12	0.047201	−0.019846			16	0.064453	0.009794
		13	0.053501	−0.011386			17	0.070670	0.023194
		14	0.060520	−0.001230			18	0.078048	0.040976
		15	0.068439	0.011065			19	0.087254	0.066087
		16	0.077509	0.026154			20	0.203293	0.300441
		17	0.471636	0.472299	21	21	1	0.012894	−0.032310
21	18	1	0.005853	−0.042848			2	0.016080	−0.034581
		2	0.008874	−0.045374			3	0.019015	−0.035746
		3	0.011911	−0.046394			4	0.021851	−0.036228
		4	0.015024	−0.046474			5	0.024653	−0.036168
		5	0.018245	−0.045796			6	0.027458	−0.035622
		6	0.021603	−0.044433			7	0.030295	−0.034606
		7	0.025124	−0.042403			8	0.033186	−0.033114
		8	0.028836	−0.039694			9	0.036159	−0.031116
		9	0.032780	−0.036255			10	0.039235	−0.028564
		10	0.036985	−0.032027			11	0.042446	−0.025388
		11	0.041512	−0.026894			12	0.045824	−0.021484
		12	0.046418	−0.020718			13	0.049414	−0.016709
		13	0.051789	−0.013292			14	0.053270	−0.010860
		14	0.057729	−0.004339			15	0.057470	−0.003637
		15	0.064392	0.006557			16	0.062124	0.005419
		16	0.071987	0.020003			17	0.067404	0.017060
		17	0.080826	0.036967			18	0.073603	0.032626
		18	0.380111	0.423414			19	0.081287	0.054878
21	19	1	0.008698	−0.039258			20	0.091876	0.091146
		2	0.011772	−0.041717			21	0.114456	0.215004
		3	0.014753	−0.042808	22	2	1	−1.301440	−0.494222
		4	0.017739	−0.043048			2	2.301440	0.494222
		5	0.020774	−0.042605	22	3	1	−0.695987	−0.328100
		6	0.023891	−0.041546			2	−0.657936	−0.317763
		7	0.027116	−0.039889			3	2.353923	0.645863
		8	0.030478	−0.037622	22	4	1	−0.434070	−0.244476
		9	0.034006	−0.034708			2	−0.413690	−0.239780
		10	0.037735	−0.031080			3	−0.375918	−0.225713
		11	0.041707	−0.026648			4	2.223678	0.709969
		12	0.045971	−0.021282	22	5	1	−0.293467	−0.194003
		13	0.050597	−0.014797			2	−0.281150	−0.192202
		14	0.055671	−0.006936			3	−0.257076	−0.183051
		15	0.061314	0.002673			4	−0.225276	−0.169152
		16	0.067704	0.014599			5	2.056969	0.738407
		17	0.075101	0.029744	22	6	1	−0.208130	−0.160166
		18	0.083931	0.049671			2	−0.199975	−0.160015
		19	0.291040	0.367257			3	−0.183470	−0.153867
21	20	1	0.011028	−0.035816			4	−0.161458	−0.143848
		2	0.014157	−0.038193			5	−0.134845	−0.130649
		3	0.017105	−0.039333			6	1.887878	0.748545
		4	0.020000	−0.039706					
		5	0.022898	−0.039466					
		6	0.025833	−0.038676					
		7	0.028833	−0.037352					

Table A.7. (continued)

N	M	I	A(N,M,I)	C(N,M,I)	N	M	I	A(N,M,I)	C(N,M,I)
22	7	1	−0.152049	−0.135866	22	13	1	−0.025047	−0.067350
		2	−0.146209	−0.136718			2	−0.022395	−0.069850
		3	−0.134250	−0.132539			3	−0.018588	−0.069993
		4	−0.118236	−0.125119			4	−0.013974	−0.068645
		5	−0.098823	−0.115041			5	−0.008653	−0.066089
		6	−0.076253	−0.102533			6	−0.002649	−0.062438
		7	1.725819	0.747815			7	0.004050	−0.057722
22	8	1	−0.113082	−0.117539			8	0.011483	−0.051924
		2	−0.108593	−0.119026			9	0.019706	−0.044989
		3	−0.099523	−0.116206			10	0.028801	−0.036822
		4	−0.087407	−0.110619			11	0.038870	−0.027293
		5	−0.072723	−0.102765			12	0.050045	−0.016224
		6	−0.055645	−0.092840			13	0.938350	0.639340
		7	−0.036204	−0.080900	22	14	1	−0.017003	−0.061447
		8	1.573176	0.739897			2	−0.014385	−0.063972
22	9	1	−0.084872	−0.103202			3	−0.010918	−0.064365
		2	−0.081190	−0.105100			4	−0.006845	−0.063413
		3	−0.074036	−0.103253			5	−0.00229	−0.061375
		4	−0.064560	−0.099008			6	0.002917	−0.058353
		5	−0.053119	−0.092802			7	0.008610	−0.054376
		6	−0.039835	−0.084805			8	0.014879	−0.049432
		7	−0.024723	−0.075066			9	0.021782	−0.043466
		8	−0.007729	−0.063556			10	0.029381	−0.036397
		9	1.430063	0.726792			11	0.037767	−0.028103
22	10	1	−0.063797	−0.091660			12	0.047049	−0.018422
		2	−0.060601	−0.093825			13	0.057369	−0.007137
		3	−0.054757	−0.092695			14	0.831624	0.610259
		4	−0.047129	−0.089463	22	15	1	−0.010467	−0.056269
		5	−0.037983	−0.084513			2	−0.007849	−0.058795
		6	−0.027405	−0.077998			3	−0.004630	−0.059383
		7	−0.015399	−0.069960			4	−0.000964	−0.058755
		8	−0.001917	−0.060373			5	0.003111	−0.057145
		9	0.013133	−0.049161			6	0.007593	−0.054649
		10	1.295854	0.709648			7	0.012499	−0.051296
22	11	1	−0.047658	−0.082152			8	0.017862	−0.047069
		2	−0.044752	−0.084487			9	0.023724	−0.041928
		3	−0.039824	−0.083898			10	0.030145	−0.035792
		4	−0.033522	−0.081446			11	0.037196	−0.028555
		5	−0.026041	−0.077478			12	0.044975	−0.020066
		6	−0.017442	−0.072129			13	0.053597	−0.010126
		7	−0.007722	−0.065437			14	0.063218	0.001536
		8	0.003165	−0.057379			15	0.729991	0.578292
		9	0.015297	−0.047887	22	16	1	−0.005110	−0.051672
		10	0.028784	−0.036846			2	−0.002471	−0.054179
		11	1.169717	0.689139			3	0.000569	−0.054921
22	12	1	−0.035052	−0.074167			4	0.003924	−0.054559
		2	−0.032313	−0.076608			5	0.007582	−0.053308
		3	−0.028036	−0.076431			6	0.011543	−0.051259
		4	−0.022704	−0.074593			7	0.015838	−0.048429
		5	−0.016456	−0.071406			8	0.020484	−0.044819
		6	−0.009333	−0.066992			9	0.025520	−0.040385
		7	−0.001327	−0.061386			10	0.031004	−0.035054
		8	0.007603	−0.054568			11	0.036998	−0.028725
		9	0.017525	−0.046475			12	0.043567	−0.021274
		10	0.028533	−0.037005			13	0.050828	−0.012503
		11	0.040749	−0.026013			14	0.058901	−0.002170
		12	1.050810	0.665643			15	0.067958	0.010067
							16	0.632864	0.543188

Table A.7. *(continued)*

N	M	I	A(N,M,I)	C(N,M,I)	N	M	I	A(N,M,I)	C(N,M,I)
22	17	1	−0.000692	−0.047541			5	0.019519	−0.040500
		2	0.001982	−0.050015			6	0.022321	−0.039681
		3	0.004893	−0.050879			7	0.025204	−0.038351
		4	0.008014	−0.050736			8	0.028200	−0.036496
		5	0.011345	−0.049789			9	0.031317	−0.034104
		6	0.014906	−0.048115			10	0.034600	−0.031120
		7	0.018711	−0.045743			11	0.038064	−0.027481
		8	0.022792	−0.042661			12	0.041759	−0.023097
		9	0.027167	−0.038846			13	0.045722	−0.017840
		10	0.031923	−0.034195			14	0.050022	−0.011535
		11	0.037042	−0.028684			15	0.054736	−0.003933
		12	0.042658	−0.022124			16	0.059977	0.005317
		13	0.048810	−0.014389			17	0.065907	0.016757
		14	0.055631	−0.005227			18	0.072766	0.031243
		15	0.063253	0.005668			19	0.080947	0.050259
		16	0.071871	0.018749			20	0.278209	0.358253
		17	0.539692	0.504527	22	21	1	0.010614	−0.033940
22	18	1	0.002969	−0.043785			2	0.013486	−0.036167
		2	0.005687	−0.046214			3	0.016181	−0.037280
		3	0.008507	−0.047171			4	0.018819	−0.037707
		4	0.011449	−0.047212			5	0.021451	−0.037592
		5	0.014531	−0.046521			6	0.024107	−0.036993
		6	0.017772	−0.045174			7	0.026810	−0.035935
		7	0.021193	−0.043196			8	0.029588	−0.034409
		8	0.024827	−0.040573			9	0.032449	−0.032401
		9	0.028687	−0.037286			10	0.035435	−0.029863
		10	0.032827	−0.033268			11	0.038552	−0.026747
		11	0.037279	−0.028441			12	0.041849	−0.022960
		12	0.042106	−0.022691			13	0.045351	−0.018398
		13	0.047371	−0.015865			14	0.049118	−0.012895
		14	0.053169	−0.007753			15	0.053206	−0.006235
		15	0.059619	0.001940			16	0.057712	0.001909
		16	0.066884	0.013632			17	0.062764	0.012027
		17	0.075192	0.027940			18	0.068562	0.024916
		18	0.449933	0.461640			19	0.075443	0.041976
22	19	1	0.006009	−0.040322			20	0.084022	0.066013
		2	0.008777	−0.042695			21	0.194480	0.292681
		3	0.011534	−0.043723	22	22	1	0.012295	−0.030733
		4	0.014342	−0.043917			2	0.015219	−0.032859
		5	0.017230	−0.043447			3	0.017907	−0.033986
		6	0.020225	−0.042380			4	0.020501	−0.034499
		7	0.023341	−0.040750			5	0.023056	−0.034529
		8	0.026624	−0.038526			6	0.025607	−0.034130
		9	0.030059	−0.035724			7	0.028178	−0.033325
		10	0.033736	−0.032232			8	0.030794	−0.032106
		11	0.037624	−0.028048			9	0.033464	−0.030463
		12	0.041834	−0.023001			10	0.036226	−0.028354
		13	0.046370	−0.017006			11	0.039082	−0.025735
		14	0.051344	−0.009832			12	0.042076	−0.022528
		15	0.056833	−0.001233			13	0.045226	−0.018638
		16	0.062987	0.009193			14	0.048580	−0.013922
		17	0.069992	0.022017			15	0.052185	−0.008184
		18	0.078134	0.038155			16	0.056116	−0.001137
		19	0.363007	0.413471			17	0.060475	0.007660
22	20	1	0.008533	−0.037073			18	0.065424	0.018928
		2	0.011352	−0.039379			19	0.071237	0.033950
		3	0.014069	−0.040459			20	0.078445	0.055370
		4	0.016778	−0.040780			21	0.088381	0.090209
							22	0.109526	0.209011

Table A.7. (continued)

N	M	I	A(N,M,I)	C(N,M,I)	N	M	I	A(N,M,I)	C(N,M,I)
23	2	1	−1.324862	−0.494478	23	11	1	−0.052275	−0.082619
		2	2.324862	0.494478			2	−0.049614	−0.084845
							3	−0.044879	−0.084262
23	3	1	−0.711775	−0.328336			4	−0.038754	−0.081902
		2	−0.674982	−0.318458			5	−0.031451	−0.078106
		3	2.386757	0.646794			6	−0.023044	−0.073012
							7	−0.013543	−0.066668
23	4	1	−0.446061	−0.244730			8	−0.002916	−0.059066
		2	−0.426381	−0.240237			9	0.008899	−0.050154
		3	−0.389748	−0.226817			10	0.021994	−0.039847
		4	2.262191	0.711785			11	1.225582	0.700481
23	5	1	−0.303177	−0.194285	23	12	1	−0.039296	−0.074667
		2	−0.291322	−0.192554			2	−0.036802	−0.076996
		3	−0.267953	−0.183817			3	−0.032710	−0.076807
		4	−0.237079	−0.170584			4	−0.027537	−0.075031
		5	2.099531	0.741241			5	−0.021441	−0.071973
23	6	1	−0.216311	−0.160479			6	−0.014473	−0.067761
		2	−0.208505	−0.160324			7	−0.006638	−0.062438
		3	−0.192480	−0.154445			8	0.002097	−0.055995
		4	−0.171078	−0.144898			9	0.011784	−0.048387
		5	−0.145224	−0.132362			10	0.022502	−0.039535
		6	1.933598	0.752508			11	0.034352	−0.029328
							12	1.108163	0.678918
23	7	1	−0.159128	−0.136209					
		2	−0.153582	−0.137013	23	13	1	−0.028970	−0.067886
		3	−0.141975	−0.133009			2	−0.026565	−0.070274
		4	−0.126391	−0.125931			3	−0.022939	−0.070390
		5	−0.107499	−0.116350			4	−0.018475	−0.069079
		6	−0.085566	−0.104499			5	−0.013288	−0.066620
		7	1.774142	0.753011			6	−0.007415	−0.063126
							7	−0.000851	−0.058638
23	8	1	−0.119329	−0.117912			8	0.006430	−0.053149
		2	−0.115104	−0.119324			9	0.014477	−0.046619
		3	−0.106313	−0.116615			10	0.023356	−0.038975
		4	−0.094514	−0.111278			11	0.033154	−0.030114
		5	−0.080205	−0.103803			12	0.043981	−0.019898
		6	−0.063578	−0.094391			13	0.997106	0.654769
		7	−0.044683	−0.083112					
		8	1.623725	0.746435	23	14	1	−0.020644	−0.062022
							2	−0.018278	−0.064436
23	9	1	−0.090463	−0.103606			3	−0.014991	−0.064792
		2	−0.087031	−0.105411			4	−0.011061	−0.063857
		3	−0.087031	−0.103629			5	−0.006570	−0.061888
		4	−0.070885	−0.099566			6	−0.001537	−0.058986
		5	−0.059723	−0.093651			7	0.004039	−0.055193
		6	−0.046768	−0.086060			8	0.010191	−0.050500
		7	−0.032050	−0.076851			9	0.016959	−0.044870
		8	−0.015537	−0.066015			10	0.024395	−0.038243
		9	1.482569	0.734790			11	0.032581	−0.030518
							12	0.041605	−0.021572
23	10	1	−0.068856	−0.092095			13	0.051586	−0.011236
		2	−0.065905	−0.094156			14	0.891725	0.628112
		3	−0.060269	−0.093058					
		4	−0.052846	−0.089957	23	15	1	−0.013858	−0.056886
		5	−0.043918	−0.085232			2	−0.011496	−0.059304
		6	−0.033587	−0.079038			3	−0.008457	−0.059458
		7	−0.021873	−0.071429			4	−0.004935	−0.059221
		8	−0.008741	−0.062395			5	−0.000978	−0.057655
		9	0.005880	−0.051880			6	0.003395	−0.055253
		10	1.350115	0.719241			7	0.008206	−0.052040

Table A.7. (continued)

N	M	I	A(N,M,I)	C(N,M,I)	N	M	I	A(N,M,I)	C(N,M,I)
		8	0.013472	−0.048107	23	19	1	0.003411	−0.041199
		9	0.019220	−0.043159			2	0.005909	−0.043487
		10	0.025528	−0.037385			3	0.008471	−0.044457
		11	0.032425	−0.030636			4	0.011123	−0.044615
		12	0.040015	−0.022775			5	0.013887	−0.044125
		13	0.048387	−0.013657			6	0.016780	−0.043060
		14	0.057669	−0.003074			7	0.019798	−0.041470
		15	0.791406	0.598911			8	0.022996	−0.039317
23	16	1	−0.008277	−0.052337			9	0.026380	−0.036600
		2	−0.005898	−0.054740			10	0.030001	−0.033248
		3	−0.003041	−0.055433			11	0.033825	−0.029295
		4	0.000172	−0.055058			12	0.037942	−0.024577
		5	0.003709	−0.053835			13	0.042427	−0.018980
		6	0.007575	−0.051846			14	0.047315	−0.012379
		7	0.011777	−0.049129			15	0.052660	−0.004606
		8	0.016329	−0.045688			16	0.058612	0.004662
		9	0.021298	−0.041465			17	0.065310	0.015808
		10	0.026669	−0.036455			18	0.072957	0.029414
		11	0.032560	−0.030526			19	0.430194	0.451531
		12	0.038985	−0.023615	23	20	1	0.006104	−0.038066
		13	0.046066	−0.015549			2	0.008651	−0.040298
		14	0.053887	−0.006157			3	0.011168	−0.041321
		15	0.062596	−0.004820			4	0.013716	−0.041599
		16	0.695593	0.567014			5	0.016325	−0.041290
23	17	1	−0.003655	−0.048263			6	0.019011	−0.040466
		2	−0.001244	−0.050639			7	0.021809	−0.039131
		3	0.001481	−0.051447			8	0.024700	−0.037336
		4	0.004455	−0.051283			9	0.027785	−0.034966
		5	0.007669	−0.050345			10	0.030958	−0.032134
		6	0.011133	−0.048710			11	0.034420	−0.028603
		7	0.014831	−0.046438			12	0.037984	−0.024527
		8	0.018871	−0.043448			13	0.041940	−0.019546
		9	0.023152	−0.039831			14	0.046153	−0.013730
		10	0.027817	−0.035444			15	0.050775	−0.006799
		11	0.032865	−0.030257			16	0.055861	0.001462
		12	0.038384	−0.024144			17	0.061571	0.011459
		13	0.044389	−0.017027			18	0.068058	0.023716
		14	0.051034	−0.008672			19	0.075590	0.039106
		15	0.058398	0.001120			20	0.347421	0.404070
		16	0.066644	0.012681	23	21	1	0.008356	−0.035111
		17	0.603776	0.532147			2	0.010952	−0.037277
23	18	1	0.000194	−0.044575			3	0.013441	−0.038337
		2	0.002646	−0.046912			4	0.015912	−0.038718
		3	0.005275	−0.047812			5	0.018403	−0.038563
		4	0.008068	−0.047822			6	0.020936	−0.037940
		5	0.011028	−0.047126			7	0.023539	−0.036861
		6	0.014162	−0.045809			8	0.026213	−0.035349
		7	0.017515	−0.043866			9	0.029021	−0.033345
		8	0.021048	−0.041362			10	0.031907	−0.030890
		9	0.024872	−0.038183			11	0.034969	−0.027882
		10	0.028875	−0.034430			12	0.038215	−0.024224
		11	0.033363	−0.029780			13	0.041650	−0.019928
		12	0.038034	−0.024480			14	0.045352	−0.014781
		13	0.043231	−0.018136			15	0.049362	−0.008653
		14	0.048884	−0.010732			16	0.053758	−0.001297
		15	0.055157	−0.001985			17	0.058642	0.007615
		16	0.062143	0.008368			18	0.064166	0.018608
		17	0.070035	0.020767			19	0.070550	0.032491
		18	0.515472	0.493876					

Table A.7. (continued)

N	M	I	A(N,M,I)	C(N,M,I)	N	M	I	A(N,M,I)	C(N,M,I)
		20	0.078160	0.050676	24	5	1	−0.312447	−0.194542
		21	0.266495	0.349764			2	−0.301019	−0.192877
23	22	1	0.010228	−0.032246			3	−0.278309	−0.184517
		2	0.012873	−0.034337			4	−0.248301	−0.171890
		3	0.015348	−0.035418			5	2.140076	0.743825
		4	0.017763	−0.035883	24	6	1	−0.224120	−0.160763
		5	0.020166	−0.035864			2	−0.216634	−0.160605
		6	0.022582	−0.035420			3	−0.201057	−0.154973
		7	0.025037	−0.034566			4	−0.180228	−0.145857
		8	0.027534	−0.033327			5	−0.155085	−0.133918
		9	0.030133	−0.031642			6	1.977123	0.756116
		10	0.032777	−0.029557	24	7	1	−0.165886	−0.136519
		11	0.035580	−0.026947			2	−0.160606	−0.137280
		12	0.038476	−0.023825			3	−0.149328	−0.133438
		13	0.041565	−0.020057			4	−0.134146	−0.126672
		14	0.044837	−0.015569			5	−0.115746	−0.117541
		15	0.048365	−0.010179			6	−0.094409	−0.106282
		16	0.052188	−0.003701			7	1.820121	0.757733
		17	0.056406	0.004192					
		18	0.061135	0.013968	24	8	1	−0.125290	−0.118250
		19	0.066561	0.026384			2	−0.121304	−0.119593
		20	0.073001	0.042777			3	−0.112772	−0.116987
		21	0.081024	0.065832			4	−0.101272	−0.111878
		22	0.186420	0.285386			5	−0.087317	−0.104747
							6	−0.071114	−0.095799
23	23	1	0.011749	−0.029300			7	−0.052731	−0.085111
		2	0.014443	−0.031296			8	1.671800	0.752364
		3	0.016915	−0.032383					
		4	0.019296	−0.032915	24	9	1	−0.095798	−0.103970
		5	0.021637	−0.033014			2	−0.092592	−0.105691
		6	0.023968	−0.032735			3	−0.085890	−0.103970
		7	0.026314	−0.032093			4	−0.076896	−0.100075
		8	0.028683	−0.031103			5	−0.066000	−0.094425
		9	0.031120	−0.029730			6	−0.053356	−0.087200
		10	0.033589	−0.027985			7	−0.039009	−0.078467
		11	0.036174	−0.025797			8	−0.022943	−0.068229
		12	0.038830	−0.023139			9	1.532484	0.742027
		13	0.041635	−0.019922					
		14	0.044584	−0.016060	24	10	1	−0.073685	−0.092486
		15	0.047728	−0.011412			2	−0.070954	−0.094454
		16	0.051110	−0.005790			3	−0.065509	−0.093387
		17	0.054801	0.001083			4	−0.058278	−0.090406
		18	0.058897	0.009633			5	−0.049557	−0.085886
		19	0.063548	0.020550			6	−0.039461	−0.079984
		20	0.069018	0.035063			7	−0.028021	−0.072761
		21	0.075800	0.055713			8	−0.015218	−0.064218
		22	0.085149	0.089235			9	−0.000998	−0.054317
		23	0.105012	0.203397			10	1.401679	0.727899
24	2	1	−1.347247	−0.494711	24	11	1	−0.056685	−0.083038
		2	2.347347	0.494711			2	−0.054241	−0.085164
							3	−0.049681	−0.084589
24	3	1	−0.726858	−0.328551			4	−0.043721	−0.082316
		2	−0.691234	−0.319094			5	−0.036586	−0.078677
		3	2.418093	0.647644			6	−0.028363	−0.073816
							7	−0.019070	−0.067786
24	4	1	−0.457513	−0.244962			8	−0.008690	−0.060569
		2	−0.438482	−0.240654			9	0.002829	−0.052190
		3	−0.402913	−0.227825			10	0.015559	−0.042523
		4	2.298908	0.713442			11	1.278649	0.710688

Table A.7. (continued)

N	M	I	A(N,M,I)	C(N,M,I)
24	12	1	−0.043350	−0.075114
		2	−0.041075	−0.077341
		3	−0.037150	−0.077143
		4	−0.032125	−0.075425
		5	−0.026171	−0.072488
		6	−0.019351	−0.068461
		7	−0.011678	−0.063393
		8	−0.003130	−0.057285
		9	0.006334	−0.050108
		10	0.016780	−0.041798
		11	0.028293	−0.032267
		12	1.162621	0.690824
24	13	1	−0.032719	0.068363
		2	−0.030535	−0.070648
		3	−0.027071	−0.070743
		4	−0.022744	−0.069468
		5	−0.017683	−0.067099
		6	−0.011934	−0.063752
		7	−0.005499	−0.059470
		8	0.001635	−0.054260
		9	0.009515	−0.048088
		10	0.018189	−0.040903
		11	0.027735	−0.032622
		12	0.038242	−0.023135
		13	1.052869	0.668551
24	14	1	−0.024127	−0.062531
		2	−0.021983	−0.064844
		3	−0.018857	−0.065168
		4	−0.015060	−0.053253
		5	−0.010682	−0.062346
		6	−0.005761	−0.059561
		7	−0.000290	−0.055934
		8	0.005750	−0.051466
		9	0.012378	−0.046144
		10	0.019668	−0.039899
		11	0.027663	−0.032671
		12	0.036447	−0.024354
		13	0.046122	−0.014814
		14	0.948731	0.643985
24	15	1	−0.017105	−0.057431
		2	−0.014969	−0.059749
		3	−0.012092	−0.060254
		4	−0.008696	−0.059630
		5	−0.004850	−0.058110
		6	−0.000583	−0.055800
		7	0.004161	−0.052701
		8	0.009290	−0.048895
		9	0.014974	−0.044259
		10	0.021148	−0.038824
		11	0.027900	−0.032498
		12	0.035313	−0.025117
		13	0.043457	−0.016750
		14	0.052436	−0.007059
		15	0.849316	0.617138
24	16	1	−0.011313	−0.052921
		2	−0.009163	−0.055227
		3	−0.006469	−0.055875
		4	−0.003385	−0.055494
		5	0.000046	−0.054297
		6	0.003828	−0.052365
		7	0.007921	−0.049773
		8	0.012411	−0.046465
		9	0.017303	−0.042445
		10	0.022558	−0.037726
		11	0.028368	−0.032133
		12	0.034638	−0.025703
		13	0.041559	−0.018228
		14	0.049152	−0.009617
		15	0.057555	0.000342
		16	0.754990	0.587928
24	17	1	−0.006499	−0.048891
		2	−0.004321	−0.051174
		3	−0.001761	−0.051934
		4	0.001076	−0.051760
		5	0.004211	−0.050811
		6	0.007533	−0.049264
		7	0.011242	−0.046993
		8	0.015072	−0.044237
		9	0.019386	−0.040701
		10	0.023957	−0.036555
		11	0.028939	−0.031656
		12	0.034306	−0.025980
		13	0.040205	−0.019361
		14	0.046703	−0.011667
		15	0.053823	−0.002782
		16	0.061755	0.007593
		17	0.664373	0.556173
24	18	1	−0.002475	−0.045255
		2	−0.000259	−0.047504
		3	0.002201	−0.048354
		4	0.004866	−0.048336
		5	0.007700	−0.047659
		6	0.010794	−0.046318
		7	0.014016	−0.044487
		8	0.017417	−0.042119
		9	0.021386	−0.038895
		10	0.025182	−0.035448
		11	0.029642	−0.031020
		12	0.034054	−0.026209
		13	0.039415	−0.020075
		14	0.044838	−0.013352
		15	0.050958	−0.005370
		16	0.057699	0.003929
		17	0.065252	0.014885
		18	0.577315	0.521587
24	19	1	0.000905	−0.041941
		2	0.003164	−0.044148
		3	0.005555	−0.045066
		4	0.008071	−0.045193
		5	0.010734	−0.044684
		6	0.013521	−0.043644
		7	0.016435	−0.042115
		8	0.019637	−0.039942
		9	0.022803	−0.037505
		10	0.026719	−0.033941

Table A.7. *(continued)*

N	M	I	A(N,M,I)	C(N,M,I)	N	M	I	A(N,M,I)	C(N,M,I)
		11	0.029997	−0.030672			9	0.026977	−0.032554
		12	0.034402	−0.025868			10	0.029650	−0.030414
		13	0.038730	−0.020747			11	0.032257	−0.028021
		14	0.043507	−0.014657			12	0.035139	−0.024936
		15	0.048732	−0.007553			13	0.038322	−0.021171
		16	0.054520	0.000812			14	0.041338	−0.017209
		17	0.060958	0.010675			15	0.044935	−0.011998
		18	0.068216	0.022447			16	0.048628	−0.006114
		19	0.493393	0.483741			17	0.052763	0.001032
							18	0.057327	0.009615
24	20	1	0.003753	−0.038888			19	0.062480	0.020215
		2	0.006057	−0.041046			20	0.068458	0.033512
		3	0.008399	−0.042017			21	0.075549	0.050964
		4	0.010808	−0.042259			22	0.255757	0.341745
		5	0.013294	−0.041940					
		6	0.015900	−0.041093	24	23	1	0.009865	−0.030709
		7	0.018611	−0.039782			2	0.012311	−0.032676
		8	0.021436	−0.038014			3	0.014593	−0.033724
		9	0.024426	−0.035765			4	0.016814	−0.034214
		10	0.027521	−0.033081			5	0.019018	−0.034270
		11	0.031146	−0.029441			6	0.021232	−0.033942
		12	0.034308	−0.025968			7	0.023451	−0.033279
		13	0.038424	−0.021075			8	0.025757	−0.032215
		14	0.042551	−0.015682			9	0.028049	−0.030887
		15	0.047082	−0.009322			10	0.030536	−0.029034
		16	0.052019	−0.001870			11	0.032941	−0.026957
		17	0.057545	0.007016			12	0.035565	−0.024275
		18	0.063743	0.017651			13	0.038329	−0.021160
		19	0.070809	0.030651			14	0.041191	−0.017439
		20	0.412170	0.441924			15	0.044281	−0.013016
							16	0.047572	−0.007758
24	21	1	0.006154	−0.036040			17	0.051172	−0.001443
		2	0.008506	−0.038142			18	0.055132	0.006201
		3	0.010815	−0.039151			19	0.059561	0.015675
		4	0.013140	−0.039493			20	0.064672	0.027621
		5	0.015514	−0.039306			21	0.070703	0.043424
		6	0.017945	−0.038667			22	0.078235	0.065565
		7	0.020429	−0.037625			23	0.179020	0.278512
		8	0.023099	−0.036040					
		9	0.025743	−0.034203	24	24	1	0.011248	−0.027993
		10	0.028697	−0.031687			2	0.013739	−0.029871
		11	0.031580	−0.028925			3	0.016021	−0.030917
		12	0.034836	−0.025342			4	0.018216	−0.031460
		13	0.038156	−0.021392			5	0.020370	−0.031614
		14	0.041933	−0.016415			6	0.022514	−0.031424
		15	0.045799	−0.010844			7	0.024647	−0.030930
		16	0.050108	−0.004137			8	0.026838	−0.030091
		17	0.054842	0.003809			9	0.029015	−0.028989
		18	0.060165	0.013408			10	0.031329	−0.027477
		19	0.066188	0.025170			11	0.033589	−0.025684
		20	0.073197	0.039844			12	0.035989	−0.023440
		21	0.333154	0.395179			13	0.038543	−0.020740
							14	0.041131	−0.017557
24	22	1	0.008175	−0.033339			15	0.043928	−0.013710
		2	0.010574	−0.035378			16	0.046862	−0.009153
		3	0.012864	−0.036413			17	0.050054	−0.003639
		4	0.015129	−0.036835			18	0.053531	0.003055
		5	0.017403	−0.036782			19	0.057380	0.011390
		6	0.019717	−0.036300			20	0.061783	0.021945
		7	0.022047	−0.035464			21	0.066927	0.036007
		8	0.024510	−0.034155			22	0.073331	0.055932

Table A.7. (continued)

N	M	I	A(N,M,I)	C(N,M,I)	N	M	I	A(N,M,I)	C(N,M,I)
		23	0.082153	0.088238			9	−0.007535	−0.056515
		24	0.100863	0.198122			10	1.450793	0.735755
25	2	1	−1.368683	−0.494926	25	11	1	−0.060903	−0.083415
		2	2.368683	0.494926			2	−0.058654	−0.085451
							3	−0.054256	−0.084885
25	3	1	−0.741297	−0.328748			4	−0.048449	−0.082692
		2	−0.706762	−0.319677			5	−0.041474	−0.079199
		3	2.448059	0.648426			6	−0.033425	−0.074550
							7	−0.024332	−0.068804
25	4	1	−0.468472	−0.245175			8	−0.014184	−0.061972
		2	−0.450043	−0.241037			9	−0.002945	−0.054030
		3	−0.415470	−0.228750			10	0.009446	−0.044927
		4	2.333986	0.714962			11	1.329176	0.719926
25	5	1	−0.321316	−0.194777	25	12	1	−0.047230	−0.075517
		2	−0.310282	−0.193172			2	−0.045150	−0.077650
		3	−0.288191	−0.185159			3	−0.041377	−0.077446
		4	−0.258993	−0.173084			4	−0.036489	−0.075783
		5	2.178782	0.746191			5	−0.030670	−0.072958
25	6	1	−0.231589	−0.161021			6	−0.023991	−0.069100
		2	−0.224397	−0.160862			7	−0.016473	−0.064265
		3	−0.209240	−0.155458			8	−0.008104	−0.058459
		4	−0.188950	−0.146734			9	−0.001149	−0.051664
		5	−0.164473	−0.135339			10	−0.011340	−0.043833
		6	2.018650	0.759414			11	−0.022540	−0.034894
							12	−0.214454	−0.701568
25	7	1	−0.172349	−0.136802					
		2	−0.167311	−0.137524	25	13	1	−0.036308	−0.068792
		3	−0.156342	−0.133831			2	−0.034320	−0.070982
		4	−0.141540	−0.127351			3	−0.031004	−0.071058
		5	−0.123602	−0.118629			4	−0.026804	−0.069819
		6	−0.102823	−0.107905			5	−0.021861	−0.067534
		7	1.863966	0.762043			6	−0.016228	−0.064321
							7	−0.009922	−0.060231
25	8	1	−0.130991	−0.118557			8	−0.002924	−0.055270
		2	−0.127221	−0.119837			9	0.004794	−0.049420
		3	−0.118931	−0.117327			10	0.013273	−0.042642
		4	−0.107714	−0.112428			11	0.022584	−0.034867
		5	−0.094093	−0.105611			12	0.032798	−0.026010
		6	−0.078289	−0.097082			13	1.105923	0.680946
		7	−0.060385	−0.086925					
		8	1.717625	0.757767	25	14	1	−0.027463	−0.062987
							2	−0.025518	−0.065206
25	9	1	−0.100902	−0.104300			3	−0.022537	−0.065502
		2	−0.097898	−0.105944			4	−0.018860	−0.064606
		3	−0.091397	−0.104281			5	−0.014588	−0.062761
		4	−0.082624	−0.100540			6	−0.009771	−0.060083
		5	−0.071979	−0.095134			7	−0.004406	−0.056611
		6	−0.059629	−0.088240			8	0.001530	−0.052347
		7	−0.045632	−0.079936			9	0.008022	−0.047302
		8	−0.029985	−0.070233			10	0.015177	−0.041393
		9	1.580044	0.748609			11	0.022982	−0.034605
							12	0.031548	−0.026830
25	10	1	−0.078304	−0.092840			13	0.040941	−0.017971
		2	−0.075769	−0.094722			14	1.002942	0.658203
		3	−0.070501	−0.093684					
		4	−0.063450	−0.090817	25	15	1	−0.020217	−0.057916
		5	−0.054925	−0.086485			2	−0.018283	−0.060142
		6	−0.045053	−0.080848			3	−0.015551	−0.060612
		7	−0.033874	−0.073973			4	−0.012269	−0.059993
		8	−0.021382	−0.065870			5	−0.008537	−0.058526

Table A.7. (continued)

N	M	I	A(N,M,I)	C(N,M,I)	N	M	I	A(N,M,I)	C(N,M,I)
		6	−0.004338	−0.056280			16	0.053506	0.000123
		7	0.000300	−0.053317			17	0.060779	0.009946
		8	0.005319	−0.049695			18	0.635905	0.545761
		9	0.010966	−0.045241					
		10	0.016954	−0.040149	25	19	1	−0.001511	−0.042582
		11	0.023615	−0.034163			2	0.000536	−0.044712
		12	0.030844	−0.027322			3	0.002777	−0.045580
		13	0.038774	−0.019488			4	0.005148	−0.045706
		14	0.047483	−0.010542			5	0.007784	−0.045118
		15	0.904940	0.633385			6	0.010293	−0.044292
							7	0.013648	−0.042297
25	16	1	−0.014226	−0.053437			8	0.016012	−0.040961
		2	−0.012280	−0.055655			9	0.019203	−0.038460
		3	−0.009730	−0.056262			10	0.024225	−0.034053
		4	−0.006767	−0.055881			11	0.026723	−0.031592
		5	−0.003416	−0.054698			12	0.030579	−0.027487
		6	0.000228	−0.052868			13	0.034843	−0.022690
		7	0.004311	−0.050319			14	0.040233	−0.016383
		8	0.008740	−0.047138			15	0.045011	−0.010159
		9	0.013383	−0.043435			16	0.050657	−0.002512
		10	0.018781	−0.038783			17	0.056882	0.006342
		11	0.024316	−0.033639			18	0.063787	0.016770
		12	0.030527	−0.027559			19	0.553170	0.511471
		13	0.037271	−0.020612					
		14	0.044668	−0.012645	25	20	1	0.001479	−0.039587
		15	0.052795	−0.003537			2	0.003569	−0.041673
		16	0.811399	0.606468			3	0.005756	−0.042596
							4	0.008031	−0.042816
25	17	1	−0.009232	−0.049443			5	0.010448	−0.042449
		2	−0.007260	−0.051640			6	0.013025	−0.041554
		3	−0.004848	−0.052357			7	0.015402	−0.040540
		4	−0.002117	−0.052162			8	0.018347	−0.038621
		5	0.000879	−0.051264			9	0.021346	−0.036313
		6	0.004177	−0.049710			10	0.024325	−0.033953
		7	0.007815	−0.047517			11	0.028185	−0.030127
		8	0.011434	−0.044984			12	0.030435	−0.027632
		9	0.015921	−0.041405			13	0.035676	−0.021855
		10	0.020253	−0.037606			14	0.038707	−0.017912
		11	0.025191	−0.032939			15	0.043784	−0.011387
		12	0.030451	−0.027620			16	0.048355	−0.004818
		13	0.036231	−0.021444			17	0.053791	0.003245
		14	0.042593	−0.014304			18	0.059696	0.012672
		15	0.049500	−0.006173			19	0.066471	0.023804
		16	0.057147	0.003248			20	0.473171	0.474112
		17	0.721867	0.577320					
					25	21	1	0.004015	−0.036813
25	18	1	−0.005043	−0.045848			2	0.006150	−0.038850
		2	−0.003036	−0.048015			3	0.008301	−0.039812
		3	−0.000726	−0.048819			4	0.010494	−0.040124
		4	0.001824	−0.048779			5	0.012767	−0.039913
		5	0.004483	−0.048170			6	0.015083	−0.039304
		6	0.007751	−0.046642			7	0.017547	−0.038195
		7	0.010435	−0.045269			8	0.019926	−0.036894
		8	0.014601	−0.042266			9	0.023362	−0.034108
		9	0.01727	−0.040246			10	0.024924	−0.033298
		10	0.022055	−0.036060			11	0.028334	−0.029964
		11	0.026242	−0.032037			12	0.032751	−0.025098
		12	0.030231	−0.027809			13	0.033983	−0.023707
		13	0.035679	−0.021917			14	0.038804	−0.017806
		14	0.041032	−0.015644			15	0.042477	−0.012817
		15	0.047011	−0.008309			16	0.046778	−0.006543

Table A.7. *(continued)*

N	M	I	A(N,M,I)	C(N,M,I)	N	M	I	A(N,M,I)	C(N,M,I)
		17	0.051285	0.000504	25	24	1	0.009526	−0.029309
		18	0.056394	0.009060			2	0.011794	−0.031162
		19	0.062243	0.019180			3	0.013905	−0.032175
		20	0.068693	0.031758			4	0.015953	−0.032683
		21	0.395689	0.432743			5	0.017988	−0.032792
							6	0.020022	−0.032574
25	22	1	0.006168	−0.034211			7	0.022074	−0.032010
		2	0.008348	−0.036193			8	0.024017	−0.031278
		3	0.010476	−0.037183			9	0.026504	−0.029812
		4	0.012607	−0.037570			10	0.028155	−0.028849
		5	0.014776	−0.037485			11	0.031096	−0.026352
		6	0.017026	−0.036955			12	0.032808	−0.024695
		7	0.019221	−0.036171			13	0.035503	−0.021765
		8	0.021340	−0.035185			14	0.038100	−0.018724
		9	0.024679	−0.032515			15	0.040722	−0.015128
		10	0.025895	−0.032125			16	0.043691	−0.010706
		11	0.030229	−0.027673			17	0.046790	−0.005556
		12	0.031117	−0.027074			18	0.050125	0.000586
		13	0.035583	−0.021773			19	0.053932	0.007926
		14	0.038063	−0.018702			20	0.057987	0.017243
		15	0.041779	−0.013661			21	0.062967	0.028582
		16	0.045324	−0.008299			22	0.068493	0.043984
		17	0.049433	−0.001734			23	0.075661	0.065216
		18	0.053748	0.005862			24	0.172188	0.272031
		19	0.058821	0.015044					
		20	0.064336	0.026491	25	25	1	0.010788	−0.026796
		21	0.071016	0.040315			2	0.013098	−0.028566
		22	0.320015	0.386798			3	0.015213	−0.029573
							4	0.017239	−0.030122
25	23	1	0.007991	−0.031733			5	0.019234	−0.030313
		2	0.010217	−0.033654			6	0.021211	−0.030205
		3	0.012332	−0.034661			7	0.023185	−0.029800
		4	0.014413	−0.035114			8	0.025057	−0.029207
		5	0.016507	−0.035132			9	0.027393	−0.028041
		6	0.018615	−0.034796			10	0.029004	−0.027158
		7	0.020791	−0.034037			11	0.031672	−0.025206
		8	0.022797	−0.033206			12	0.033363	−0.023590
		9	0.025487	−0.031417			13	0.035757	−0.021260
		10	0.027226	−0.030315			14	0.038198	−0.018529
		11	0.030451	−0.027371			15	0.040571	−0.015429
		12	0.032439	−0.025277			16	0.043268	−0.011548
		13	0.034975	−0.022600			17	0.046011	−0.007108
		14	0.038427	−0.018207			18	0.048983	−0.001688
		15	0.040862	−0.014907			19	0.052348	0.004773
		16	0.044536	−0.009371			20	0.055864	0.013016
		17	0.047835	−0.003905			21	0.060199	0.023073
		18	0.051722	0.003109			22	0.064907	0.036847
		19	0.056077	0.011315			23	0.071049	0.056035
		20	0.060795	0.021679			24	0.079353	0.087235
		21	0.066561	0.034261			25	0.097036	0.193159
		22	0.073053	0.051189					
		23	0.245889	0.334151					

Table A.8. Confidence bounds for two-parameter Weibull distributions for censored samples of size 3(1)25.

Table A.8a. Percentiles of the distribution of $V_{.90}$.

n	m	0.02	0.05	0.10	0.25	0.40	0.50	0.60	0.75	0.90	0.95	0.98
3	3	0.75	1.10	1.43	2.18	2.88	3.40	4.06	5.50	8.99	13.16	20.93
4	3	0.78	1.16	1.49	2.18	2.82	3.33	3.96	5.38	9.03	13.07	20.23
	4	0.87	1.16	1.46	2.06	2.60	2.99	3.45	4.40	6.47	8.39	11.66
5	3	0.78	1.18	1.51	2.17	2.79	3.27	3.87	5.24	8.78	12.58	20.38
	4	0.97	1.23	1.51	2.09	2.61	2.99	3.44	4.40	6.49	8.48	11.73
	5	0.97	1.23	1.49	2.02	2.49	2.82	3.20	3.93	5.48	6.73	8.66
6	3	0.73	1.18	1.53	2.15	2.73	3.18	3.74	4.98	8.24	11.74	18.65
	4	1.00	1.28	1.55	2.10	2.60	2.98	3.41	4.30	6.33	8.18	11.39
	5	1.02	1.29	1.54	2.05	2.50	2.82	3.21	3.94	5.42	6.73	8.89
	6	1.02	1.27	1.53	2.01	2.42	2.70	3.04	3.67	4.86	5.83	7.31
7	3	0.64	1.18	1.53	2.13	2.66	3.08	3.60	4.79	7.80	11.12	17.54
	4	1.04	1.31	1.58	2.10	2.57	2.91	3.33	4.21	6.16	7.89	10.90
	5	1.08	1.33	1.57	2.06	2.49	2.80	3.15	3.87	5.36	6.68	8.44
	6	1.08	1.32	1.56	2.03	2.42	2.70	3.01	3.63	4.86	5.82	7.23
	7	1.08	1.32	1.55	2.00	2.37	2.62	2.90	3.44	4.46	5.25	6.37
8	3	0.49	1.13	1.52	2.11	2.62	3.01	3.48	4.62	7.51	10.67	16.36
	4	1.04	1.33	1.60	2.10	2.56	2.88	3.27	4.10	5.96	7.79	10.75
	5	1.11	1.36	1.60	2.08	2.49	2.78	3.12	3.82	5.28	6.50	8.62
	6	1.13	1.36	1.59	2.05	2.43	2.71	3.02	3.62	4.83	5.83	7.18
	7	1.12	1.36	1.58	2.03	2.38	2.64	2.93	3.46	4.49	5.31	6.40
	8	1.12	1.36	1.58	2.01	2.34	2.57	2.83	3.32	4.21	4.90	5.84
9	3	0.42	1.12	1.51	2.09	2.57	2.95	3.40	4.43	7.14	10.21	15.61
	4	1.06	1.36	1.61	2.10	2.52	2.84	3.21	4.00	5.77	7.39	10.26
	5	1.17	1.41	1.63	2.08	2.47	2.76	3.08	3.76	5.13	6.34	8.13
	6	1.19	1.41	1.62	2.06	2.43	2.70	2.99	3.59	4.74	5.67	7.06
	7	1.19	1.41	1.62	2.04	2.39	2.64	2.91	3.45	4.48	5.28	6.46
	8	1.19	1.40	1.61	2.02	2.36	2.59	2.84	3.34	4.26	4.95	5.94
	9	1.19	1.40	1.60	2.00	2.33	2.55	2.78	3.22	4.04	4.66	5.50

Nancy R. Mann, K. W. Fertig, and E. M. Scheuer. 1971. Confidence and tolerance bounds and a new goodness-of-fit test for two-parameter Weibull or extreme-value distributions (with tables for censored samples of size 3(1)25). Aerospace Research Laboratories, U.S. Air Force, ARL 71-0077, public domain.

Table A.8a. *(continued)*

n	m	0.02	0.05	0.10	0.25	0.40	0.50	0.60	0.75	0.90	0.95	0.98
10	3	0.09	0.99	1.46	2.05	2.51	2.84	3.27	4.25	6.75	9.36	14.88
	4	0.99	1.34	1.62	2.08	2.48	2.77	3.13	3.90	5.56	7.17	9.60
	5	1.17	1.42	1.64	2.07	2.45	2.71	3.02	3.67	5.00	6.13	8.02
	6	1.20	1.43	1.64	2.05	2.41	2.66	2.94	3.53	4.67	5.59	6.99
	7	1.21	1.43	1.64	2.04	2.38	2.62	2.88	3.41	4.41	5.18	6.29
	8	1.21	1.43	1.63	2.02	2.35	2.58	2.83	3.31	4.22	4.91	5.83
	9	1.21	1.42	1.63	2.01	2.32	2.54	2.77	3.22	4.03	4.63	5.51
	10	1.21	1.42	1.62	1.99	2.30	2.50	2.72	3.13	3.86	4.41	5.16
11	3	−0.09	0.90	1.42	2.01	2.45	2.77	3.17	4.07	6.41	9.11	14.47
	4	0.97	1.35	1.61	2.06	2.44	2.73	3.06	3.79	5.46	7.04	9.98
	5	1.18	1.43	1.64	2.05	2.41	2.68	2.98	3.60	4.90	6.07	7.83
	6	1.24	1.45	1.64	2.04	2.38	2.63	2.91	3.46	4.58	5.52	6.96
	7	1.25	1.45	1.64	2.03	2.35	2.59	2.86	3.36	4.36	5.16	6.34
	8	1.25	1.45	1.64	2.01	2.33	2.56	2.80	3.28	4.15	4.87	5.82
	9	1.25	1.44	1.64	2.00	2.31	2.53	2.76	3.21	4.01	4.63	5.54
	10	1.25	1.44	1.64	1.99	2.29	2.49	2.71	3.14	3.87	4.44	5.23
	11	1.25	1.45	1.63	1.98	2.28	2.46	2.67	3.06	3.76	4.26	4.94
12	3	−0.38	0.75	1.37	1.98	2.41	2.71	3.08	3.89	6.00	8.40	12.96
	4	0.95	1.34	1.60	2.05	2.42	2.69	3.00	3.67	5.17	6.60	9.07
	5	1.20	1.44	1.66	2.05	2.40	2.65	2.93	3.52	4.72	5.79	7.35
	6	1.26	1.46	1.67	2.04	2.38	2.62	2.88	3.39	4.41	5.31	6.61
	7	1.28	1.47	1.67	2.03	2.36	2.58	2.82	3.30	4.21	4.98	6.09
	8	1.28	1.47	1.66	2.02	2.34	2.54	2.78	3.22	4.06	4.75	5.71
	9	1.27	1.46	1.66	2.01	2.31	2.52	2.74	3.16	3.94	4.53	5.40
	10	1.27	1.47	1.65	2.00	2.30	2.49	2.70	3.11	3.82	4.37	5.11
	11	1.27	1.46	1.64	2.00	2.28	2.47	2.67	3.05	3.72	4.23	4.88
	12	1.28	1.47	1.64	1.99	2.27	2.44	2.63	3.00	3.62	4.07	4.68
13	3	−0.45	0.72	1.34	1.99	2.40	2.69	3.04	3.85	5.88	8.16	12.45
	4	0.88	1.31	1.60	2.06	2.42	2.68	2.98	3.64	5.10	6.45	8.82
	5	1.20	1.45	1.67	2.07	2.40	2.64	2.92	3.49	4.71	5.75	7.32
	6	1.27	1.48	1.68	2.07	2.38	2.61	2.86	3.38	4.43	5.30	6.49
	7	1.30	1.49	1.68	2.06	2.36	2.58	2.82	3.30	4.23	4.96	6.02
	8	1.30	1.49	1.68	2.04	2.34	2.55	2.78	3.22	4.06	4.73	5.63
	9	1.30	1.49	1.68	2.03	2.33	2.52	2.75	3.16	3.94	4.55	5.32
	10	1.30	1.49	1.68	2.03	2.31	2.50	2.72	3.12	3.83	4.37	5.11
	11	1.30	1.49	1.68	2.02	2.30	2.48	2.69	3.06	3.74	4.23	4.90
	12	1.30	1.49	1.67	2.01	2.28	2.46	2.65	3.02	3.65	4.09	4.73
	13	1.30	1.49	1.67	2.01	2.27	2.44	2.62	2.97	3.57	3.97	4.51

Table A.8a. (continued)

n	m	0.02	0.05	0.10	0.25	0.40	0.50	0.60	0.75	0.90	0.95	0.98
14	3	−0.81	0.57	1.25	1.93	2.34	2.62	2.95	3.70	5.56	7.69	11.56
	4	0.83	1.29	1.59	2.03	2.37	2.62	2.92	3.54	4.93	6.17	8.28
	5	1.18	1.46	1.67	2.05	2.36	2.60	2.86	3.42	4.58	5.54	6.95
	6	1.28	1.51	1.69	2.05	2.35	2.57	2.82	3.34	4.33	5.12	6.27
	7	1.32	1.52	1.69	2.04	2.34	2.55	2.78	3.27	4.15	4.82	5.75
	8	1.33	1.52	1.69	2.03	2.33	2.52	2.74	3.19	4.03	4.61	5.47
	9	1.33	1.52	1.69	2.03	2.31	2.50	2.71	3.14	3.90	4.45	5.18
	10	1.33	1.51	1.68	2.02	2.30	2.48	2.70	3.09	3.78	4.30	4.94
	11	1.33	1.51	1.68	2.01	2.28	2.47	2.67	3.05	3.71	4.20	4.79
	12	1.33	1.51	1.68	2.01	2.27	2.45	2.65	3.01	3.64	4.09	4.67
	13	1.33	1.51	1.68	2.00	2.26	2.43	2.62	2.97	3.55	3.98	4.51
	14	1.33	1.51	1.68	2.00	2.25	2.42	2.60	2.93	3.46	3.85	4.36
15	3	−1.05	0.43	1.19	1.91	2.33	2.60	2.91	3.64	5.39	7.23	10.78
	4	0.77	1.26	1.59	2.03	2.37	2.61	2.89	3.49	4.78	5.95	7.94
	5	1.15	1.44	1.67	2.06	2.37	2.59	2.85	3.38	4.43	5.36	6.85
	6	1.29	1.50	1.69	2.06	2.36	2.57	2.81	3.30	4.22	4.97	6.19
	7	1.33	1.52	1.70	2.06	2.35	2.55	2.78	3.23	4.08	4.72	5.77
	8	1.34	1.52	1.70	2.05	2.33	2.53	2.74	3.17	3.95	4.57	5.40
	9	1.35	1.52	1.69	2.04	2.32	2.51	2.72	3.13	3.85	4.40	5.16
	10	1.35	1.52	1.69	2.04	2.31	2.49	2.69	3.09	3.76	4.26	4.95
	11	1.35	1.52	1.69	2.03	2.30	2.48	2.67	3.04	3.69	4.15	4.76
	12	1.34	1.52	1.69	2.02	2.28	2.46	2.64	3.00	3.62	4.08	4.62
	13	1.35	1.52	1.68	2.01	2.27	2.44	2.63	2.96	3.55	3.98	4.51
	14	1.35	1.51	1.69	2.01	2.27	2.43	2.60	2.93	3.49	3.89	4.39
	15	1.35	1.52	1.68	2.01	2.25	2.41	2.58	2.89	3.41	3.77	4.23
16	3	−1.38	0.25	1.10	1.90	2.31	2.57	2.88	3.57	5.22	7.07	10.49
	4	0.74	1.23	1.58	2.03	2.37	2.59	2.87	3.45	4.72	5.90	7.94
	5	1.17	1.45	1.68	2.06	2.37	2.59	2.84	3.36	4.42	5.33	6.73
	6	1.30	1.52	1.71	2.07	2.36	2.57	2.81	3.28	4.20	4.98	6.18
	7	1.35	1.53	1.72	2.06	2.35	2.56	2.77	3.21	4.05	4.74	5.81
	8	1.37	1.54	1.72	2.06	2.34	2.53	2.75	3.16	3.94	4.56	5.38
	9	1.37	1.54	1.72	2.05	2.33	2.51	2.72	3.11	3.84	4.38	5.17
	10	1.37	1.54	1.71	2.04	2.31	2.50	2.70	3.08	3.74	4.24	4.97
	11	1.37	1.54	1.71	2.04	2.30	2.48	2.68	3.04	3.67	4.13	4.79
	12	1.37	1.54	1.71	2.03	2.29	2.46	2.65	3.00	3.60	4.05	4.65
	13	1.37	1.54	1.71	2.03	2.28	2.45	2.63	2.97	3.52	3.94	4.49
	14	1.37	1.54	1.71	2.02	2.27	2.43	2.61	2.94	3.48	3.87	4.38
	15	1.37	1.54	1.71	2.02	2.26	2.42	2.59	2.90	3.41	3.79	4.26
	16	1.38	1.54	1.71	2.02	2.26	2.40	2.56	2.87	3.36	3.71	4.16

Table A.8a. *(continued)*

n	m	0.02	0.05	0.10	0.25	0.40	0.50	0.60	0.75	0.90	0.95	0.98
17	3	−1.56	0.10	1.03	1.86	2.26	2.50	2.79	3.41	4.96	6.54	9.76
	4	0.61	1.18	1.56	2.01	2.32	2.54	2.80	3.33	4.48	5.56	7.33
	5	1.08	1.43	1.67	2.04	2.33	2.54	2.78	3.27	4.24	5.09	6.35
	6	1.28	1.52	1.71	2.05	2.33	2.52	2.75	3.20	4.07	4.76	5.81
	7	1.34	1.54	1.72	2.05	2.32	2.51	2.72	3.14	3.95	4.56	5.44
	8	1.36	1.55	1.72	2.04	2.31	2.49	2.69	3.10	3.85	4.41	5.19
	9	1.37	1.56	1.72	2.04	2.30	2.48	2.67	3.05	3.76	4.28	4.97
	10	1.38	1.55	1.72	2.03	2.29	2.47	2.65	3.01	3.69	4.16	4.80
	11	1.38	1.55	1.72	2.03	2.28	2.45	2.64	2.98	3.62	4.05	4.66
	12	1.38	1.55	1.71	2.02	2.27	2.44	2.62	2.95	3.56	3.98	4.53
	13	1.38	1.55	1.71	2.02	2.26	2.42	2.60	2.92	3.50	3.90	4.45
	14	1.38	1.55	1.71	2.01	2.25	2.42	2.58	2.90	3.45	3.83	4.31
	15	1.38	1.55	1.71	2.01	2.25	2.40	2.57	2.88	3.40	3.76	4.19
	16	1.38	1.55	1.71	2.00	2.24	2.39	2.55	2.85	3.35	3.70	4.10
	17	1.38	1.55	1.71	2.00	2.24	2.38	2.53	2.82	3.29	3.62	4.04
18	3	−1.61	0.11	1.01	1.85	2.26	2.50	2.77	3.38	4.81	6.43	9.64
	4	0.66	1.19	1.56	2.01	2.32	2.54	2.79	3.32	4.44	5.52	7.38
	5	1.10	1.45	1.69	2.05	2.34	2.54	2.77	3.26	4.22	5.04	6.40
	6	1.30	1.53	1.73	2.06	2.34	2.53	2.75	3.19	4.06	4.73	5.79
	7	1.37	1.56	1.74	2.06	2.33	2.52	2.73	3.14	3.94	4.55	5.41
	8	1.40	1.57	1.74	2.06	2.32	2.50	2.71	3.09	3.84	4.39	5.15
	9	1.41	1.58	1.74	2.05	2.31	2.49	2.68	3.06	3.75	4.25	4.99
	10	1.41	1.58	1.74	2.05	2.30	2.48	2.67	3.03	3.67	4.14	4.80
	11	1.41	1.58	1.74	2.05	2.29	2.46	2.65	3.00	3.63	4.05	4.63
	12	1.41	1.58	1.74	2.04	2.28	2.45	2.63	2.97	3.56	3.97	4.51
	13	1.41	1.57	1.74	2.04	2.28	2.44	2.61	2.94	3.50	3.91	4.40
	14	1.41	1.57	1.73	2.03	2.27	2.43	2.60	2.92	3.45	3.84	4.30
	15	1.41	1.57	1.73	2.03	2.26	2.42	2.59	2.89	3.40	3.75	4.21
	16	1.41	1.57	1.73	2.02	2.26	2.41	2.57	2.87	3.36	3.71	4.14
	17	1.41	1.58	1.74	2.02	2.25	2.40	2.55	2.84	3.31	3.66	4.04
	18	1.42	1.58	1.74	2.02	2.24	2.39	2.54	2.82	3.27	3.59	3.97
19	3	−2.13	−0.16	0.83	1.81	2.23	2.47	2.72	3.29	4.68	6.09	9.01
	4	0.43	1.11	1.51	1.99	2.30	2.51	2.75	3.25	4.32	5.33	7.04
	5	1.09	1.44	1.67	2.04	2.33	2.52	2.74	3.20	4.13	4.94	6.21
	6	1.30	1.53	1.72	2.05	2.33	2.52	2.72	3.15	3.98	4.67	5.67
	7	1.38	1.56	1.74	2.05	2.32	2.50	2.70	3.11	3.87	4.45	5.33
	8	1.41	1.58	1.74	2.05	2.31	2.49	2.68	3.07	3.79	4.30	5.09
	9	1.42	1.58	1.74	2.05	2.31	2.48	2.67	3.03	3.71	4.23	4.88
	10	1.43	1.58	1.74	2.05	2.30	2.46	2.65	3.00	3.64	4.11	4.73
	11	1.43	1.58	1.74	2.04	2.29	2.45	2.63	2.98	3.57	4.02	4.63
	12	1.43	1.58	1.74	2.04	2.28	2.45	2.62	2.95	3.52	3.94	4.51
	13	1.43	1.58	1.74	2.03	2.27	2.43	2.60	2.93	3.48	3.88	4.41
	14	1.43	1.58	1.73	2.03	2.26	2.42	2.59	2.90	3.43	3.81	4.29

Table A.8a. *(continued)*

n	m	0.02	0.05	0.10	0.25	0.40	0.50	0.60	0.75	0.90	0.95	0.98
	15	1.42	1.58	1.74	2.03	2.26	2.41	2.57	2.88	3.39	3.75	4.22
	16	1.42	1.58	1.73	2.03	2.25	2.40	2.56	2.86	3.36	3.70	4.10
	17	1.42	1.58	1.74	2.02	2.25	2.40	2.55	2.84	3.32	3.65	4.06
	18	1.42	1.58	1.74	2.02	2.24	2.38	2.54	2.81	3.29	3.59	3.99
	19	1.43	1.59	1.74	2.02	2.24	2.38	2.52	2.78	3.24	3.53	3.90
20	3	−2.27	−0.19	0.82	1.79	2.20	2.44	2.69	3.23	4.53	6.04	9.01
	4	0.46	1.10	1.51	1.99	2.29	2.49	2.72	3.21	4.24	5.24	7.00
	5	1.07	1.42	1.67	2.04	2.31	2.50	2.72	3.16	4.04	4.82	6.16
	6	1.30	1.54	1.73	2.06	2.32	2.50	2.71	3.13	3.90	4.55	5.68
	7	1.38	1.58	1.74	2.06	2.31	2.49	2.69	3.08	3.80	4.41	5.33
	8	1.42	1.59	1.75	2.06	2.30	2.48	2.67	3.04	3.71	4.29	5.06
	9	1.43	1.60	1.75	2.06	2.29	2.46	2.65	3.01	3.65	4.15	4.85
	10	1.44	1.60	1.75	2.05	2.29	2.45	2.64	2.98	3.59	4.05	4.71
	11	1.44	1.60	1.75	2.05	2.28	2.44	2.63	2.95	3.54	3.97	4.57
	12	1.44	1.60	1.75	2.04	2.28	2.43	2.61	2.92	3.49	3.92	4.46
	13	1.44	1.60	1.75	2.04	2.27	2.42	2.59	2.90	3.44	3.84	4.39
	14	1.44	1.60	1.75	2.04	2.26	2.41	2.58	2.88	3.40	3.77	4.29
	15	1.43	1.60	1.75	2.03	2.26	2.40	2.56	2.86	3.37	3.72	4.20
	16	1.44	1.60	1.75	2.03	2.25	2.40	2.55	2.84	3.33	3.67	4.12
	17	1.44	1.60	1.75	2.03	2.24	2.39	2.54	2.82	3.29	3.61	4.04
	18	1.44	1.60	1.74	2.02	2.24	2.38	2.53	2.81	3.25	3.57	3.97
	19	1.44	1.60	1.75	2.02	2.24	2.37	2.52	2.78	3.21	3.52	3.92
	20	1.44	1.60	1.75	2.02	2.23	2.36	2.51	2.76	3.17	3.47	3.84
21	3	−2.70	−0.41	0.72	1.74	2.17	2.41	2.66	3.17	4.40	5.76	8.25
	4	0.32	1.04	1.47	1.97	2.28	2.47	2.70	3.16	4.16	5.06	6.51
	5	1.05	1.41	1.66	2.03	2.30	2.49	2.70	3.14	4.01	4.75	5.88
	6	1.29	1.53	1.72	2.05	2.31	2.49	2.69	3.10	3.87	4.53	5.47
	7	1.39	1.58	1.75	2.05	2.31	2.48	2.68	3.06	3.78	4.35	5.13
	8	1.43	1.60	1.76	2.06	2.30	2.47	2.66	3.03	3.71	4.23	4.95
	9	1.45	1.60	1.76	2.05	2.30	2.46	2.64	3.00	3.65	4.13	4.78
	10	1.45	1.60	1.76	2.05	2.29	2.45	2.63	2.97	3.59	4.05	4.64
	11	1.45	1.60	1.76	2.05	2.28	2.44	2.62	2.95	3.53	3.97	4.54
	12	1.45	1.60	1.76	2.04	2.28	2.43	2.61	2.93	3.49	3.90	4.44
	13	1.45	1.60	1.76	2.04	2.27	2.42	2.60	2.90	3.45	3.83	4.33
	14	1.45	1.60	1.75	2.03	2.26	2.41	2.59	2.89	3.41	3.79	4.29
	15	1.45	1.60	1.75	2.03	2.26	2.40	2.57	2.86	3.37	3.72	4.19
	16	1.46	1.60	1.75	2.03	2.25	2.40	2.56	2.85	3.34	3.66	4.11
	17	1.46	1.60	175	2.03	2.25	2.39	2.54	2.83	3.30	3.61	4.04
	18	1.46	1.61	1.75	2.02	2.24	2.38	2.54	2.81	3.27	3.56	3.97
	19	1.46	1.61	1.75	2.02	2.24	2.38	2.52	2.79	3.23	3.53	3.92
	20	1.46	1.61	1.75	2.02	2.23	2.37	2.51	2.77	3.19	3.48	3.85
	21	1.47	1.62	1.76	2.02	2.23	2.36	2.50	2.75	3.15	3.43	3.77

Table A.8a. (continued)

n	m	0.02	0.05	0.10	0.25	0.40	0.50	0.60	0.75	0.90	0.95	0.98
22	3	−3.15	−0.61	0.61	1.71	2.16	2.39	2.63	3.13	4.31	5.49	7.84
	4	0.26	0.98	1.45	1.96	2.25	2.45	2.67	3.13	4.10	4.96	6.51
	5	1.03	1.42	1.66	2.02	2.29	2.47	2.68	3.10	3.95	4.65	5.70
	6	1.27	1.54	1.73	2.04	2.30	2.47	2.67	3.07	3.84	4.43	5.32
	7	1.38	1.59	1.76	2.05	2.30	2.47	2.66	3.05	3.76	4.32	5.09
	8	1.43	1.61	1.76	2.05	2.30	2.46	2.65	3.01	3.68	4.17	4.87
	9	1.46	1.62	1.77	2.05	2.29	2.45	2.64	2.98	3.61	4.07	4.69
	10	1.47	1.62	1.77	2.05	2.29	2.44	2.62	2.95	3.57	3.99	4.55
	11	1.48	1.62	1.77	2.05	2.28	2.43	2.61	2.93	3.51	3.92	4.44
	12	1.48	1.62	1.77	2.04	2.27	2.42	2.60	2.92	3.47	3.86	4.37
	13	1.48	1.62	1.77	2.04	2.27	2.41	2.58	2.89	3.43	3.80	4.27
	14	1.48	1.62	1.77	2.04	2.26	2.41	2.57	2.87	3.38	3.75	4.22
	15	1.47	1.62	1.76	2.04	2.26	2.40	2.56	2.85	3.34	3.69	4.15
	16	1.47	1.62	1.76	2.03	2.25	2.40	2.55	2.84	3.31	3.63	4.07
	17	1.46	1.62	1.77	2.03	2.25	2.39	2.54	2.82	3.28	3.61	4.01
	18	1.47	1.62	1.76	2.03	2.24	2.38	2.53	2.80	3.25	3.57	3.95
	19	1.47	1.62	1.76	2.03	2.24	2.38	2.52	2.78	3.22	3.53	3.90
	20	1.47	1.62	1.76	2.03	2.23	2.37	2.51	2.77	3.20	3.47	3.85
	21	1.48	1.62	1.77	2.03	2.23	2.36	2.50	2.76	3.17	3.44	3.78
	22	1.48	1.63	1.77	2.03	2.23	2.36	2.49	2.74	3.13	3.39	3.71
23	3	−3.04	−0.75	0.49	1.67	2.13	2.37	2.60	3.07	4.19	5.40	7.86
	4	0.14	0.94	1.41	1.94	2.25	2.44	2.65	3.08	4.01	4.90	6.52
	5	0.95	1.37	1.64	2.01	2.29	2.46	2.66	3.06	3.89	4.60	5.82
	6	1.24	1.51	1.72	2.05	2.30	2.47	2.65	3.04	3.77	4.40	5.31
	7	1.37	1.58	1.75	2.06	2.30	2.47	2.64	3.01	3.68	4.26	5.07
	8	1.44	1.60	1.76	2.06	2.29	2.46	2.63	2.98	3.62	4.12	4.86
	9	1.46	1.62	1.77	2.06	2.29	2.45	2.62	2.95	3.57	4.04	4.69
	10	1.47	1.62	1.77	2.06	2.28	2.44	2.61	2.93	3.51	3.95	4.55
	11	1.48	1.63	1.77	2.05	2.28	2.43	2.60	2.91	3.46	3.88	4.42
	12	1.48	1.62	1.77	2.05	2.27	2.42	2.59	2.89	3.42	3.80	4.33
	13	1.48	1.62	1.77	2.05	2.27	2.41	2.57	2.87	3.39	3.74	4.22
	14	1.48	1.62	1.76	2.04	2.26	2.41	2.56	2.86	3.36	3.71	4.16
	15	1.48	1.62	1.76	2.04	2.26	2.40	2.55	2.84	3.33	3.67	4.08
	16	1.48	1.62	1.76	2.04	2.25	2.39	2.54	2.82	3.30	3.62	4.04
	17	1.48	1.62	1.76	2.03	2.25	2.39	2.53	2.81	3.27	3.59	3.99
	18	1.48	1.62	1.76	2.03	2.25	2.38	2.52	2.79	3.25	3.55	3.94
	19	1.47	1.62	1.76	2.03	2.24	2.37	2.52	2.78	3.22	3.52	3.89
	20	1.48	1.62	1.76	2.03	2.24	2.37	2.51	2.77	3.18	3.48	3.84
	21	1.48	1.62	1.76	2.03	2.23	2.36	2.50	2.75	3.16	3.44	3.79
	22	1.48	1.62	1.77	2.03	2.23	2.36	2.49	2.74	3.13	3.41	3.73
	23	1.49	1.63	1.77	2.03	2.22	2.35	2.48	2.72	3.11	3.37	3.68

Table A.8a. *(continued)*

n	m	0.02	0.05	0.10	0.25	0.40	0.50	0.60	0.75	0.90	0.95	0.98
24	3	−3.60	−1.00	0.45	1.62	2.11	2.36	2.59	3.05	4.11	5.20	7.27
	4	0.11	0.91	1.38	1.94	2.24	2.43	2.64	3.06	3.96	4.75	6.16
	5	0.91	1.36	1.64	2.02	2.28	2.45	2.66	3.05	3.85	4.50	5.60
	6	1.24	1.52	1.72	2.05	2.29	2.46	2.66	3.03	3.75	4.34	5.22
	7	1.38	1.58	1.75	2.06	2.29	2.45	2.65	3.00	3.68	4.22	4.95
	8	1.43	1.61	1.77	2.07	2.29	2.45	2.63	2.98	3.60	4.10	4.77
	9	1.46	1.62	1.77	2.07	2.29	2.44	2.62	2.95	3.56	4.01	4.63
	10	1.47	1.63	1.78	2.06	2.28	2.44	2.61	2.93	3.51	3.94	4.51
	11	1.48	1.63	1.78	2.06	2.28	2.43	2.59	2.91	3.46	3.87	4.38
	12	1.48	1.63	1.77	2.06	2.27	2.42	2.58	2.89	3.43	3.81	4.29
	13	1.48	1.63	1.77	2.05	2.27	2.42	2.57	2.87	3.38	3.76	4.23
	14	1.48	1.63	1.77	2.05	2.27	2.41	2.57	2.85	3.34	3.70	4.14
	15	1.48	1.63	1.77	2.05	2.26	2.40	2.56	2.84	3.31	3.65	4.07
	16	1.48	1.63	1.77	2.05	2.26	2.40	2.55	2.82	3.28	3.61	4.02
	17	1.48	1.63	1.77	2.05	2.25	2.39	2.54	2.81	3.26	3.57	3.96
	18	1.48	1.63	1.77	2.04	2.25	2.39	2.53	2.79	3.23	3.54	3.92
	19	1.48	1.63	1.77	2.04	2.25	2.38	2.52	2.78	3.20	3.50	3.89
	20	1.48	1.63	1.77	2.04	2.25	2.38	2.51	2.77	3.18	3.47	3.82
	21	1.48	1.63	1.77	2.04	2.24	2.37	2.51	2.75	3.14	3.42	3.78
	22	1.48	1.63	1.78	2.04	2.23	2.36	2.50	2.74	3.12	3.39	3.75
	23	1.49	1.63	1.78	2.04	2.23	2.36	2.49	2.73	3.10	3.36	3.69
	24	1.49	1.64	1.78	2.04	2.23	2.35	2.48	2.71	3.07	3.32	3.65
25	3	−3.88	−1.13	0.28	1.60	2.10	2.34	2.57	3.00	3.98	4.98	7.02
	4	0.04	0.86	1.35	1.94	2.24	2.43	2.62	3.02	3.87	4.64	6.06
	5	0.92	1.36	1.63	2.02	2.28	2.45	2.64	3.01	3.76	4.41	5.44
	6	1.24	1.51	1.72	2.05	2.30	2.46	2.64	3.00	3.69	4.26	5.13
	7	1.38	1.59	1.76	2.06	2.31	2.46	2.63	2.98	3.63	4.14	4.87
	8	1.44	1.62	1.78	2.06	2.30	2.45	2.63	2.96	3.56	4.04	4.72
	9	1.47	1.64	1.79	2.07	2.30	2.45	2.61	2.93	3.52	3.96	4.61
	10	1.49	1.65	1.79	2.06	2.30	2.44	2.60	2.92	3.47	3.89	4.47
	11	1.50	1.65	1.79	2.06	2.29	2.43	2.59	2.90	3.43	3.82	4.36
	12	1.50	1.65	1.79	2.06	2.28	2.43	2.58	2.88	3.40	3.76	4.31
	13	1.50	1.65	1.79	2.06	2.28	2.42	2.57	2.86	3.36	3.72	4.22
	14	1.50	1.65	1.79	2.05	2.27	2.41	2.57	2.85	3.33	3.69	4.15
	15	1.50	1.65	1.79	2.05	2.26	2.40	2.56	2.84	3.30	3.64	4.06
	16	1.50	1.65	1.79	2.05	2.26	2.40	2.55	2.82	3.27	3.59	4.01
	17	1.50	1.65	1.79	2.05	2.25	2.39	2.54	2.81	3.26	3.57	3.95
	18	1.50	1.65	1.79	2.05	2.25	2.39	2.53	2.79	3.22	3.52	3.91
	19	1.50	1.65	1.78	2.05	2.25	2.38	2.52	2.78	3.21	3.49	3.84
	20	1.50	1.65	1.79	2.04	2.25	2.37	2.51	2.76	3.18	3.46	3.79
	21	1.50	1.65	1.79	2.04	2.24	2.37	2.50	2.75	3.16	3.43	3.76
	22	1.51	1.65	1.79	2.04	2.24	2.36	2.50	2.74	3.14	3.40	3.72
	23	1.51	1.65	1.79	2.04	2.23	2.36	2.49	2.72	3.12	3.38	3.70
	24	1.51	1.66	1.79	2.04	2.23	2.35	2.48	2.71	3.09	3.35	3.66
	25	1.52	1.66	1.80	2.03	2.22	2.35	2.47	2.70	3.07	3.31	3.61

Table A.8b. Percentiles of the distribution of $V_{.95}$.

n	m	0.02	0.05	0.10	0.25	0.40	0.50	0.60	0.75	0.90	0.95	0.98
3	3	1.26	1.64	2.04	2.94	3.83	4.49	5.33	7.20	11.75	17.21	27.32
4	3	1.38	1.73	2.11	2.98	3.82	4.47	5.30	7.19	12.17	17.55	27.59
	4	1.36	1.69	2.04	2.78	3.44	3.95	4.51	5.73	8.40	10.88	15.06
5	3	1.44	1.79	2.16	3.00	3.81	4.46	5.27	7.16	12.07	17.36	28.30
	4	1.45	1.76	2.10	2.82	3.49	3.98	4.56	5.80	8.56	11.14	15.51
	5	1.44	1.74	2.06	2.72	3.29	3.70	4.17	5.11	7.06	8.68	11.14
6	3	1.48	1.83	2.20	2.99	3.77	4.37	5.15	6.93	11.53	16.66	26.85
	4	1.52	1.83	2.15	2.84	3.49	3.98	4.55	5.75	8.47	10.95	15.32
	5	1.51	1.81	2.12	2.76	3.33	3.73	4.21	5.17	7.08	8.82	11.58
	6	1.50	1.80	2.10	2.70	3.20	3.56	3.97	4.77	6.27	7.53	9.39
7	3	1.52	1.87	2.22	2.97	3.71	4.30	5.04	6.80	11.20	16.07	25.31
	4	1.59	1.88	2.19	2.86	3.47	3.94	4.49	5.69	8.39	10.80	14.80
	5	1.59	1.86	2.16	2.78	3.32	3.72	4.18	5.13	7.12	8.84	11.18
	6	1.57	1.84	2.14	2.72	3.21	3.57	3.96	4.76	6.33	7.61	9.40
	7	1.58	1.85	2.14	2.68	3.14	3.45	3.79	4.47	5.76	6.73	8.19
8	3	1.53	1.90	2.24	2.96	3.67	4.22	4.95	6.61	11.02	15.76	24.57
	4	1.63	1.91	2.22	2.87	3.46	3.91	4.44	5.60	8.19	10.74	15.22
	5	1.63	1.90	2.20	2.79	3.32	3.71	4.18	5.09	7.07	8.78	11.57
	6	1.62	1.89	2.18	2.75	3.24	3.58	3.99	4.77	6.35	7.67	9.43
	7	1.62	1.89	2.17	2.71	3.16	3.48	3.83	4.52	5.83	6.91	8.38
	8	1.62	1.89	2.17	2.69	3.10	3.38	3.71	4.31	5.44	6.29	7.50
9	3	1.55	1.93	2.26	2.95	3.63	4.18	4.85	6.45	10.71	15.33	23.80
	4	1.67	1.96	2.25	2.88	3.44	3.86	4.40	5.50	8.02	10.40	14.41
	5	1.69	1.95	2.23	2.80	3.33	3.70	4.13	5.07	6.90	8.59	11.05
	6	1.68	1.95	2.22	2.76	3.24	3.58	3.98	4.77	6.27	7.51	9.46
	7	1.67	1.94	2.21	2.73	3.18	3.50	3.85	4.53	5.86	6.91	8.40
	8	1.68	1.94	2.20	2.70	3.12	3.41	3.73	4.36	5.53	6.39	7.73
	9	1.69	1.95	2.19	2.68	3.08	3.35	3.63	4.19	5.22	6.00	7.09

Nancy R. Mann, K. W. Fertig, and E. M. Scheuer. 1971. Confidence and tolerance bounds and a new goodness-of-fit test for two-parameter Weibull or extreme-value distributions (with tables for censored samples of size 3(1)25). Aerospace Research Laboratories, U.S. Air Force, ARL 71-0077, public domain.

Table A.8b. *(continued)*

n	m	0.02	0.05	0.10	0.25	0.40	0.50	0.60	0.75	0.90	0.95	0.98
10	3	1.51	1.91	2.26	2.91	3.56	4.06	4.70	6.23	10.24	14.50	23.00
	4	1.70	1.98	2.27	2.86	3.39	3.81	4.29	5.42	7.81	10.12	13.69
	5	1.72	1.97	2.24	2.80	3.31	3.65	4.07	4.97	6.82	8.39	11.00
	6	1.71	1.97	2.23	2.76	3.22	3.55	3.93	4.70	6.24	7.50	9.42
	7	1.71	1.96	2.22	2.73	3.18	3.48	3.80	4.50	5.79	6.83	8.29
	8	1.70	1.96	2.21	2.71	3.13	3.41	3.72	4.33	5.52	6.40	7.61
	9	1.71	1.96	2.21	2.69	3.08	3.35	3.64	4.20	5.23	6.01	7.12
	10	1.72	1.96	2.21	2.67	3.04	3.29	3.56	4.08	4.98	5.67	6.65
11	3	1.48	1.93	2.25	2.88	3.49	3.98	4.59	6.07	9.89	14.11	22.60
	4	1.73	2.01	2.27	2.83	3.37	3.76	4.24	5.31	7.71	10.03	14.44
	5	1.75	2.01	2.26	2.78	3.27	3.62	4.03	4.89	6.72	8.34	10.93
	6	1.75	2.00	2.24	2.74	3.19	3.52	3.90	4.63	6.16	7.42	9.39
	7	1.75	1.99	2.23	2.72	3.14	3.45	3.79	4.45	5.79	6.83	8.42
	8	1.75	1.99	2.22	2.70	3.10	3.39	3.70	4.31	5.46	6.38	7.65
	9	1.75	1.98	2.22	2.68	3.07	3.34	3.63	4.21	5.23	6.04	7.23
	10	1.76	2.00	2.22	2.67	3.04	3.28	3.56	4.09	5.03	5.75	6.73
	11	1.77	2.00	2.22	2.66	3.01	3.25	3.50	3.98	4.85	5.49	6.35
12	3	1.42	1.91	2.26	2.87	3.47	3.91	4.50	5.85	9.41	13.40	21.39
	4	1.74	2.02	2.28	2.84	3.34	3.72	4.18	5.16	7.42	9.56	13.27
	5	1.79	2.03	2.28	2.78	3.26	3.61	3.99	4.82	6.54	8.08	10.40
	6	1.79	2.02	2.27	2.75	3.20	3.52	3.86	4.56	5.97	7.22	9.00
	7	1.79	2.02	2.26	2.73	3.14	3.43	3.76	4.40	5.63	6.66	8.08
	8	1.79	2.01	2.25	2.71	3.11	3.38	3.68	4.26	5.36	6.27	7.49
	9	1.78	2.00	2.24	2.69	3.07	3.33	3.62	4.16	5.16	5.95	7.06
	10	1.77	2.01	2.23	2.67	3.04	3.29	3.56	4.07	4.99	5.67	6.63
	11	1.78	2.01	2.24	2.67	3.02	3.25	3.51	3.98	4.84	5.47	6.29
	12	1.80	2.02	2.24	2.66	3.00	3.22	3.46	3.91	4.68	5.26	6.00
13	3	1.44	1.92	2.27	2.88	3.45	3.90	4.47	5.84	9.23	13.11	20.76
	4	1.78	2.05	2.31	2.84	3.35	3.73	4.18	5.18	7.38	9.47	13.09
	5	1.82	2.06	2.30	2.81	3.26	3.60	3.98	4.80	6.57	8.04	10.25
	6	1.82	2.05	2.29	2.78	3.21	3.50	3.86	4.58	6.03	7.24	8.89
	7	1.82	2.05	2.28	2.75	3.15	3.44	3.75	4.41	5.65	6.68	8.10
	8	1.81	2.04	2.27	2.74	3.12	3.39	3.69	4.28	5.40	6.29	7.43
	9	1.81	2.04	2.27	2.72	3.10	3.35	3.63	4.16	5.17	6.00	6.99
	10	1.81	2.04	2.27	2.70	3.07	3.31	3.58	4.09	5.01	5.70	6.66
	11	1.82	2.04	2.27	2.70	3.05	3.27	3.53	4.02	4.87	5.50	6.36
	12	1.82	2.04	2.27	2.69	3.02	3.24	3.48	3.95	4.73	5.30	6.12
	13	1.83	2.05	2.27	2.68	3.00	3.21	3.43	3.87	4.61	5.12	5.80

Table A.8b. *(continued)*

n	m	0.02	0.05	0.10	0.25	0.40	0.50	0.60	0.75	0.90	0.95	0.98
14	3	1.39	1.92	2.26	2.84	3.39	3.83	4.36	5.64	8.84	12.73	19.14
	4	1.77	2.06	2.31	2.82	3.30	3.67	4.09	5.06	7.18	9.10	12.45
	5	1.84	2.08	2.31	2.79	3.22	3.55	3.93	4.72	6.38	7.82	9.93
	6	1.85	2.08	2.30	2.76	3.17	3.46	3.81	4.52	5.91	7.07	8.71
	7	1.85	2.07	2.29	2.73	3.12	3.40	3.72	4.38	5.58	6.53	7.78
	8	1.85	2.07	2.28	2.72	3.09	3.36	3.66	4.25	5.35	6.16	7.30
	9	1.84	2.07	2.28	2.71	3.07	3.32	3.59	4.15	5.14	5.88	6.92
	10	1.85	2.06	2.27	2.70	3.05	3.29	3.56	4.07	4.98	5.65	6.49
	11	1.85	2.07	2.28	2.69	3.03	3.26	3.51	4.00	4.86	5.48	6.26
	12	1.85	2.07	2.28	2.68	3.01	3.23	3.47	3.93	4.73	5.31	6.06
	13	1.85	2.07	2.29	2.68	2.99	3.21	3.44	3.87	4.61	5.14	5.80
	14	1.86	2.08	2.28	2.66	2.98	3.18	3.41	3.81	4.48	4.97	5.60
15	3	1.24	1.88	2.24	2.84	3.38	3.80	4.35	5.56	8.75	12.22	18.38
	4	1.77	2.06	2.31	2.83	3.30	3.65	4.07	4.99	7.00	8.90	11.93
	5	1.84	2.09	2.32	2.80	3.23	3.55	3.92	4.69	6.25	7.64	9.79
	6	1.87	2.09	2.31	2.79	3.18	3.47	3.80	4.48	5.79	6.91	8.55
	7	1.87	2.08	2.30	2.76	3.14	3.41	3.73	4.34	5.50	6.41	7.90
	8	1.86	2.08	2.29	2.74	3.11	3.36	3.66	4.23	5.29	6.10	7.26
	9	1.86	2.07	2.29	2.73	3.09	3.33	3.60	4.14	5.11	5.81	6.87
	10	1.85	2.07	2.28	2.72	3.06	3.30	3.56	4.08	4.96	5.61	6.50
	11	1.85	2.07	2.28	2.70	3.04	3.27	3.52	3.99	4.84	5.43	6.22
	12	1.86	2.07	2.28	2.70	3.02	3.25	3.48	3.94	4.73	5.31	6.00
	13	1.87	2.07	2.28	2.69	3.01	3.22	3.45	3.88	4.63	5.17	5.88
	14	1.87	2.07	2.28	2.68	3.00	3.20	3.41	3.83	4.53	5.02	5.66
	15	1.89	2.08	2.28	2.68	2.98	3.18	3.39	3.77	4.43	4.88	5.46
16	3	1.13	1.85	2.25	2.84	3.38	3.78	4.30	5.47	8.49	11.98	18.76
	4	1.77	2.06	2.33	2.83	3.30	3.64	4.06	4.95	7.00	8.88	12.30
	5	1.87	2.11	2.34	2.81	3.24	3.54	3.89	4.67	6.28	7.67	9.72
	6	1.89	2.11	2.33	2.79	3.19	3.48	3.81	4.47	5.80	6.92	8.63
	7	1.89	2.10	2.33	2.77	3.15	3.42	3.72	4.33	5.51	6.47	7.98
	8	1.89	2.10	2.32	2.75	3.12	3.38	3.66	4.22	5.28	6.12	7.27
	9	1.88	2.09	2.31	2.74	3.10	3.34	3.62	4.13	5.11	5.84	6.92
	10	1.88	2.09	2.31	2.72	3.07	3.32	3.57	4.07	4.95	5.60	6.60
	11	1.89	2.09	2.30	2.71	3.05	3.28	3.54	4.00	4.85	5.44	6.30
	12	1.88	2.09	2.31	2.71	3.03	3.25	3.50	3.94	4.71	5.30	6.06
	13	1.89	2.10	2.31	2.70	3.02	3.23	3.47	3.89	4.60	5.14	5.87
	14	1.90	2.10	2.31	2.69	3.00	3.21	3.43	3.84	4.52	5.02	5.68
	15	1.90	2.11	2.31	2.69	2.99	3.18	3.40	3.79	4.44	4.91	5.53
	16	1.92	2.12	2.33	2.69	2.98	3.16	3.36	3.74	4.35	4.80	5.35

Table A.8b. *(continued)*

n	m	0.02	0.05	0.10	0.25	0.40	0.50	0.60	0.75	0.90	0.95	0.98
17	3	1.07	1.84	2.22	2.80	3.29	3.67	4.15	5.26	8.19	11.24	17.44
	4	1.76	2.08	2.32	2.80	3.24	3.57	3.95	4.81	6.65	8.45	11.48
	5	1.86	2.11	2.33	2.79	3.19	3.48	3.83	4.57	5.98	7.34	9.21
	6	1.90	2.11	2.33	2.77	3.15	3.42	3.75	4.40	5.61	6.67	8.19
	7	1.90	2.11	2.32	2.76	3.11	3.38	3.67	4.26	5.39	6.26	7.56
	8	1.90	2.11	2.32	2.74	3.09	3.34	3.61	4.15	5.17	5.95	7.05
	9	1.89	2.10	2.31	2.73	3.07	3.30	3.56	4.07	5.01	5.72	6.70
	10	1.89	2.10	2.31	2.71	3.05	3.28	3.52	4.00	4.89	5.53	6.37
	11	1.89	2.10	2.31	2.71	3.03	3.25	3.48	3.94	4.77	5.35	6.16
	12	1.89	2.11	2.31	2.70	3.01	3.23	3.45	3.89	4.67	5.23	5.92
	13	1.89	2.11	2.31	2.69	3.00	3.20	3.42	3.84	4.58	5.10	5.79
	14	1.90	2.12	2.31	2.69	2.99	3.19	3.40	3.80	4.51	5.00	5.61
	15	1.90	2.11	2.31	2.68	2.98	3.17	3.38	3.77	4.43	4.88	5.44
	16	1.90	2.12	2.31	2.67	2.97	3.15	3.35	3.72	4.34	4.79	5.30
	17	1.92	2.12	2.32	2.67	2.96	3.14	3.33	3.68	4.27	4.68	5.19
18	3	1.11	1.83	2.23	2.80	3.29	3.66	4.12	5.23	7.96	11.18	17.89
	4	1.81	2.09	2.34	2.82	3.25	3.56	3.95	4.79	6.66	8.46	11.66
	5	1.90	2.14	2.36	2.80	3.20	3.49	3.83	4.56	6.00	7.32	9.43
	6	1.93	2.14	2.36	2.78	3.16	3.43	3.75	4.40	5.63	6.61	8.27
	7	1.94	2.14	2.35	2.77	3.13	3.38	3.68	4.25	5.39	6.23	7.45
	8	1.94	2.14	2.35	2.75	3.10	3.35	3.63	4.15	5.17	5.95	6.97
	9	1.94	2.14	2.34	2.74	3.08	3.32	3.58	4.09	5.03	5.72	6.71
	10	1.93	2.13	2.34	2.73	3.06	3.29	3.54	4.03	4.88	5.51	6.38
	11	1.93	2.13	2.34	2.73	3.04	3.26	3.51	3.98	4.79	5.35	6.15
	12	1.93	2.14	2.34	2.72	3.03	3.24	3.48	3.92	4.68	5.23	5.96
	13	1.93	2.14	2.33	2.71	3.01	3.22	3.45	3.86	4.60	5.12	5.75
	14	1.93	2.13	2.33	2.70	3.01	3.21	3.42	3.83	4.52	5.01	5.60
	15	1.94	2.14	2.34	2.70	3.00	3.19	3.40	3.79	4.44	4.88	5.46
	16	1.94	2.14	2.34	2.70	2.99	3.18	3.38	3.75	4.37	4.83	5.36
	17	1.95	2.15	2.34	2.69	2.97	3.16	3.35	3.71	4.31	4.74	5.22
	18	1.97	2.17	2.35	2.69	2.97	3.15	3.33	3.68	4.24	4.65	5.11
19	3	0.96	1.73	2.19	2.79	3.27	3.62	4.08	5.16	7.80	10.74	16.87
	4	1.76	2.08	2.34	2.81	3.24	3.54	3.92	4.72	6.52	8.23	11.18
	5	1.90	2.14	2.36	2.80	3.19	3.47	3.79	4.50	5.95	7.23	9.20
	6	1.94	2.15	2.36	2.78	3.15	3.42	3.72	4.35	5.56	6.60	8.07
	7	1.95	2.16	2.35	2.76	3.13	3.37	3.65	4.23	5.31	6.16	7.45
	8	1.95	2.15	2.35	2.75	3.10	3.34	3.60	4.14	5.14	5.87	6.94
	9	1.95	2.15	2.35	2.74	3.07	3.31	3.56	4.05	4.98	5.70	6.58
	10	1.95	2.14	2.34	2.73	3.05	3.27	3.52	4.00	4.84	5.49	6.35
	11	1.95	2.14	2.34	2.72	3.03	3.26	3.50	3.95	4.73	5.33	6.17
	12	1.95	2.14	2.34	2.72	3.02	3.24	3.47	3.89	4.65	5.22	5.97
	13	1.95	2.14	2.34	2.71	3.01	3.22	3.44	3.86	4.55	5.10	5.80
	14	1.95	2.14	2.33	2.70	3.00	3.20	3.41	3.81	4.49	4.98	5.61

Table A.8b. *(continued)*

n	m	0.02	0.05	0.10	0.25	0.40	0.50	0.60	0.75	0.90	0.95	0.98
	15	1.95	2.14	2.33	2.70	2.99	3.18	3.39	3.78	4.43	4.88	5.51
	16	1.95	2.15	2.33	2.70	2.98	3.17	3.37	3.74	4.38	4.81	5.35
	17	1.95	2.15	2.34	2.70	2.98	3.16	3.35	3.71	4.31	4.74	5.25
	18	1.96	2.16	2.35	2.70	2.97	3.14	3.33	3.68	4.26	4.65	5.15
	19	1.98	2.17	2.35	2.69	2.96	3.13	3.31	3.64	4.20	4.57	5.02
20	3	0.94	1.75	2.19	2.78	3.24	3.60	4.02	5.00	7.72	10.78	16.96
	4	1.79	2.09	2.34	2.81	3.21	3.52	3.88	4.66	6.49	8.16	11.03
	5	1.92	2.15	2.36	2.80	3.18	3.46	3.79	4.45	5.85	7.09	9.12
	6	1.94	2.17	2.37	2.79	3.15	3.40	3.71	4.31	5.48	6.49	8.11
	7	1.95	2.17	2.36	2.77	3.12	3.37	3.64	4.19	5.25	6.11	7.42
	8	1.96	2.17	2.36	2.76	3.09	3.32	3.59	4.09	5.06	5.88	6.97
	9	1.96	2.16	2.35	2.75	3.07	3.29	3.55	4.02	4.92	5.62	6.57
	10	1.95	2.16	2.35	2.74	3.05	3.27	3.51	3.97	4.79	5.43	6.30
	11	1.95	2.16	2.35	2.73	3.04	3.24	3.49	3.91	4.70	5.29	6.09
	12	1.95	2.16	2.35	2.72	3.02	3.22	3.45	3.87	4.61	5.18	5.92
	13	1.96	2.16	2.35	2.72	3.01	3.21	3.42	3.83	4.53	5.07	5.79
	14	1.96	2.16	2.34	2.71	3.00	3.19	3.40	3.79	4.46	4.96	5.63
	15	1.96	2.16	2.35	2.70	2.99	3.17	3.38	3.76	4.41	4.86	5.49
	16	1.97	2.16	2.35	2.70	2.98	3.16	3.36	3.73	4.35	4.78	5.37
	17	1.98	2.17	2.35	2.70	2.97	3.15	3.34	3.70	4.29	4.70	5.23
	18	1.98	2.18	2.35	2.69	2.96	3.14	3.33	3.67	4.23	4.64	5.14
	19	1.99	2.18	2.36	2.69	2.96	3.13	3.31	3.63	4.18	4.57	5.06
	20	2.00	2.19	2.36	2.70	2.95	3.12	3.29	3.60	4.12	4.49	4.95
21	3	0.74	1.67	2.16	2.75	3.21	3.56	3.97	4.97	7.52	10.37	15.72
	4	1.72	2.08	2.33	2.79	3.20	3.49	3.85	4.62	6.33	7.91	10.58
	5	1.92	2.15	2.37	2.79	3.17	3.44	3.76	4.41	5.82	7.02	8.86
	6	1.97	2.16	2.37	2.77	3.13	3.39	3.69	4.27	5.46	6.44	7.87
	7	1.98	2.17	2.37	2.76	3.11	3.35	3.63	4.17	5.22	6.05	7.22
	8	1.98	2.17	2.36	2.75	3.08	3.32	3.58	4.09	5.08	5.80	6.79
	9	1.98	2.16	2.36	2.75	3.07	3.29	3.54	4.02	4.93	5.58	6.51
	10	1.98	2.16	2.36	2.74	3.05	3.27	3.51	3.96	4.81	5.43	6.26
	11	1.98	2.16	2.36	2.73	3.04	3.25	3.48	3.92	4.71	5.28	6.07
	12	1.98	2.16	2.35	2.72	3.02	3.23	3.46	3.88	4.63	5.17	5.89
	13	1.98	2.16	2.35	2.72	3.01	3.21	3.43	3.84	4.56	5.05	5.73
	14	1.98	2.16	2.35	2.71	3.00	3.19	3.41	3.81	4.49	4.98	5.64
	15	1.98	2.16	2.35	2.70	2.99	3.17	3.39	3.77	4.42	4.89	5.49
	16	1.99	2.17	2.35	2.70	2.98	3.16	3.36	3.74	4.37	4.78	5.34
	17	2.00	2.17	2.35	2.70	2.98	3.15	3.35	3.71	4.30	4.71	5.24
	18	2.00	2.18	2.36	2.69	2.97	3.14	3.33	3.68	4.26	4.64	5.15
	19	2.01	2.18	2.36	2.69	2.96	3.13	3.32	3.65	4.20	4.59	5.08
	20	2.01	2.19	2.36	2.69	2.95	3.12	3.30	3.62	4.15	4.52	4.98
	21	2.03	2.20	2.37	2.69	2.95	3.11	3.28	3.59	4.09	4.44	4.87

Table A.8b. *(continued)*

n	m	0.02	0.05	0.10	0.25	0.40	0.50	0.60	0.75	0.90	0.95	0.98
22	3	0.58	1.60	2.15	2.74	3.18	3.52	3.93	4.86	7.32	9.93	15.15
	4	1.71	2.07	2.33	2.78	3.18	3.47	3.82	4.55	6.26	7.80	10.66
	5	1.92	2.17	2.37	2.78	3.15	3.42	3.73	4.38	5.75	6.85	8.68
	6	1.98	2.19	2.38	2.77	3.12	3.37	3.66	4.25	5.41	6.34	7.71
	7	1.99	2.19	2.38	2.76	3.10	3.34	3.61	4.16	5.22	6.03	7.20
	8	1.99	2.19	2.37	2.75	3.08	3.31	3.57	4.08	5.03	5.75	6.77
	9	1.99	2.19	2.37	2.74	3.06	3.29	3.53	4.01	4.88	5.53	6.44
	10	1.99	2.19	2.37	2.73	3.04	3.26	3.50	3.94	4.78	5.38	6.16
	11	1.99	2.18	2.36	2.72	3.03	3.24	3.47	3.91	4.67	5.24	5.98
	12	1.99	2.19	2.36	2.72	3.02	3.21	3.45	3.87	4.58	5.14	5.81
	13	2.00	2.19	2.36	2.71	3.00	3.20	3.42	3.82	4.52	5.04	5.66
	14	2.00	2.18	2.37	2.71	3.00	3.19	3.40	3.79	4.46	4.94	5.56
	15	2.00	2.19	2.36	2.71	2.99	3.17	3.38	3.75	4.39	4.84	5.44
	16	2.00	2.19	2.37	2.71	2.98	3.16	3.36	3.73	4.34	4.76	5.30
	17	2.00	2.19	2.36	2.71	2.97	3.15	3.34	3.70	4.29	4.69	5.23
	18	2.00	2.19	2.37	2.70	2.97	3.14	3.33	3.67	4.24	4.64	5.15
	19	2.02	2.19	2.37	2.70	2.96	3.14	3.32	3.65	4.20	4.59	5.06
	20	2.02	2.20	2.38	2.70	2.96	3.12	3.30	3.63	4.16	4.51	4.98
	21	2.02	2.21	2.38	2.70	2.95	3.11	3.29	3.60	4.11	4.45	4.89
	22	2.04	2.22	2.38	2.70	2.95	3.11	3.27	3.57	4.06	4.38	4.79
23	3	0.50	1.56	2.13	2.74	3.18	3.49	3.88	4.84	7.21	9.91	15.32
	4	1.69	2.07	2.33	2.79	3.18	3.46	3.80	4.54	6.20	7.84	10.66
	5	1.92	2.16	2.37	2.79	3.16	3.41	3.71	4.35	5.70	6.93	8.89
	6	1.99	2.19	2.38	2.79	3.13	3.37	3.64	4.22	5.33	6.34	7.78
	7	2.00	2.19	2.38	2.77	3.10	3.33	3.59	4.12	5.14	5.97	7.17
	8	2.00	2.19	2.38	2.76	3.08	3.31	3.55	4.04	4.98	5.68	6.76
	9	2.00	2.19	2.38	2.75	3.07	3.28	3.51	3.98	4.85	5.52	6.37
	10	2.00	2.19	2.37	2.74	3.05	3.26	3.48	3.92	4.73	5.33	6.16
	11	2.00	2.19	2.37	2.74	3.03	3.24	3.46	3.88	4.63	5.21	5.93
	12	2.00	2.18	2.37	2.73	3.02	3.22	3.43	3.84	4.55	5.08	5.78
	13	2.00	2.18	2.37	2.73	3.01	3.20	3.41	3.80	4.48	4.98	5.59
	14	2.00	2.18	2.37	2.72	3.00	3.18	3.39	3.77	4.45	4.91	5.52
	15	2.01	2.19	2.36	2.71	3.00	3.17	3.37	3.74	4.39	4.81	5.35
	16	2.01	2.19	2.36	2.71	2.99	3.16	3.35	3.72	4.33	4.75	5.31
	17	2.02	2.19	2.36	2.71	2.98	3.15	3.34	3.69	4.29	4.70	5.20
	18	2.02	2.19	2.37	2.71	2.97	3.14	3.33	3.67	4.24	4.63	5.14
	19	2.01	2.20	2.37	2.71	2.97	3.13	3.31	3.65	4.20	4.58	5.07
	20	2.02	2.20	2.37	2.70	2.96	3.12	3.30	3.62	4.15	4.53	4.99
	21	2.03	2.20	2.37	2.70	2.95	3.12	3.29	3.60	4.11	4.47	4.92
	22	2.04	2.21	2.38	2.70	2.95	3.10	3.28	3.58	4.07	4.42	4.82
	23	2.05	2.21	2.38	2.70	2.94	3.10	3.26	3.55	4.03	4.36	4.74

Table A.8b. *(continued)*

n	m	0.02	0.05	0.10	0.25	0.40	0.50	0.60	0.75	0.90	0.95	0.98
24	3	0.33	1.49	2.08	2.72	3.15	3.47	3.86	4.77	7.09	9.55	14.35
	4	1.66	2.05	2.33	2.79	3.15	3.44	3.78	4.50	6.13	7.65	10.28
	5	1.90	2.17	2.38	2.79	3.14	3.40	3.71	4.35	5.64	6.72	8.52
	6	1.98	2.20	2.39	2.79	3.12	3.37	3.66	4.22	5.34	6.24	7.68
	7	2.00	2.21	2.39	2.78	3.10	3.33	3.61	4.13	5.14	5.92	7.06
	8	2.00	2.21	2.38	2.77	3.08	3.30	3.56	4.05	4.95	5.69	6.67
	9	2.01	2.21	2.38	2.76	3.06	3.28	3.52	3.99	4.84	5.47	6.34
	10	2.01	2.20	2.38	2.75	3.05	3.25	3.48	3.93	4.73	5.33	6.13
	11	2.00	2.20	2.38	2.75	3.04	3.24	3.46	3.89	4.64	5.21	5.94
	12	2.00	2.19	2.37	2.74	3.03	3.22	3.44	3.84	4.57	5.10	5.74
	13	2.01	2.19	2.37	2.73	3.02	3.21	3.41	3.81	4.50	5.00	5.62
	14	2.00	2.19	2.37	2.73	3.01	3.20	3.40	3.77	4.43	4.89	5.48
	15	2.01	2.19	2.37	2.73	3.00	3.19	3.38	3.75	4.37	4.81	5.33
	16	2.01	2.19	2.37	2.73	2.99	3.17	3.37	3.72	4.32	4.74	5.27
	17	2.01	2.19	2.38	2.72	2.99	3.16	3.35	3.70	4.28	4.70	5.18
	18	2.01	2.20	2.38	2.72	2.98	3.15	3.34	3.68	4.23	4.63	5.10
	19	2.02	2.20	2.38	2.71	2.98	3.14	3.32	3.65	4.19	4.56	5.05
	20	2.03	2.21	2.38	2.71	2.97	3.13	3.31	3.63	4.15	4.52	4.96
	21	2.03	2.21	2.38	2.72	2.96	3.13	3.30	3.61	4.10	4.46	4.90
	22	2.03	2.22	2.39	2.71	2.95	3.11	3.28	3.59	4.07	4.41	4.84
	23	2.05	2.22	2.39	2.71	2.95	3.11	3.26	3.57	4.04	4.35	4.77
	24	2.05	2.23	2.40	2.71	2.95	3.10	3.25	3.54	3.99	4.30	4.70
25	3	0.19	1.44	2.06	2.72	3.15	3.45	3.82	4.68	6.85	9.30	14.21
	4	1.64	2.06	2.35	2.79	3.16	3.42	3.75	4.45	5.96	7.49	9.99
	5	1.94	2.18	2.40	2.80	3.14	3.39	3.69	4.29	5.49	6.65	8.40
	6	2.01	2.21	2.41	2.80	3.13	3.35	3.63	4.19	5.25	6.14	7.48
	7	2.03	2.22	2.41	2.78	3.11	3.33	3.59	4.11	5.07	5.85	6.98
	8	2.04	2.22	2.41	2.77	3.09	3.30	3.55	4.03	4.90	5.60	6.60
	9	2.04	2.22	2.41	2.76	3.08	3.28	3.51	3.96	4.79	5.45	6.39
	10	2.04	2.22	2.40	2.76	3.06	3.27	3.48	3.91	4.69	5.27	6.10
	11	2.03	2.22	2.40	2.75	3.05	3.25	3.46	3.87	4.60	5.16	5.88
	12	2.04	2.22	2.40	2.74	3.04	3.23	3.44	3.83	4.53	5.05	5.79
	13	2.03	2.21	2.39	2.73	3.02	3.21	3.42	3.81	4.48	4.96	5.64
	14	2.03	2.22	2.39	2.73	3.01	3.20	3.40	3.77	4.42	4.88	5.52
	15	2.03	2.22	2.39	2.73	3.00	3.18	3.39	3.75	4.35	4.80	5.40
	16	2.03	2.22	2.39	2.73	3.00	3.17	3.37	3.71	4.31	4.73	5.30
	17	2.03	2.22	2.39	2.73	2.99	3.16	3.35	3.69	4.29	4.68	5.17
	18	2.04	2.22	2.39	2.72	2.98	3.15	3.33	3.67	4.23	4.62	5.11
	19	2.05	2.22	2.39	2.72	2.97	3.14	3.32	3.65	4.19	4.56	5.01
	20	2.05	2.22	2.40	2.72	2.97	3.13	3.31	3.62	4.16	4.51	4.93
	21	2.05	2.23	2.40	2.72	2.96	3.12	3.29	3.61	4.12	4.45	4.88
	22	2.05	2.23	2.40	2.72	2.96	3.11	3.28	3.59	4.09	4.42	4.83
	23	2.07	2.24	2.41	2.71	2.95	3.11	3.27	3.57	4.05	4.39	4.77
	24	2.07	2.25	2.41	2.71	2.95	3.10	3.26	3.55	4.02	4.34	4.74
	25	2.08	2.26	2.42	2.71	2.94	3.09	3.25	3.53	3.99	4.28	4.67

Table A.8c. Percentiles of the distribution of $V_{.99}$.

n	m	0.02	0.05	0.10	0.25	0.40	0.50	0.60	0.75	0.90	0.95	0.98
3	3	2.24	2.75	3.32	4.63	5.93	6.96	8.19	11.05	18.15	26.71	42.07
4	3	2.38	2.87	3.44	4.74	6.02	7.07	8.38	11.31	19.38	27.97	43.72
	4	2.37	2.82	3.31	4.38	5.34	6.09	6.91	8.74	12.79	16.62	23.12
5	3	2.45	2.95	3.47	4.80	6.09	7.14	8.48	11.55	19.73	28.71	46.68
	4	2.43	2.86	3.37	4.44	5.44	6.20	7.09	9.00	13.31	17.41	24.31
	5	2.45	2.87	3.32	4.28	5.11	5.71	6.39	7.81	10.75	13.23	16.87
6	3	2.55	3.02	3.55	4.80	6.08	7.11	8.39	11.42	19.28	28.02	45.73
	4	2.52	2.96	3.45	4.49	5.50	6.25	7.16	9.01	13.49	17.54	24.27
	5	2.52	2.94	3.38	4.33	5.18	5.79	6.52	7.95	10.96	13.67	17.96
	6	2.56	2.94	3.39	4.25	4.96	5.49	6.08	7.26	9.49	11.41	14.37
7	3	2.64	3.11	3.61	4.81	6.07	7.04	8.37	11.40	19.27	27.76	44.93
	4	2.61	3.04	3.49	4.52	5.48	6.25	7.14	9.08	13.53	17.47	23.96
	5	2.59	3.00	3.44	4.37	5.19	5.80	6.51	8.02	11.20	13.91	17.78
	6	2.61	3.01	3.41	4.26	5.00	5.53	6.14	7.32	9.75	11.74	14.36
	7	2.65	3.02	3.43	4.21	4.87	5.32	5.82	6.83	8.75	10.20	12.38
8	3	2.67	3.12	3.65	4.82	6.01	7.03	8.27	11.28	19.24	27.78	44.29
	4	2.66	3.07	3.53	4.54	5.51	6.23	7.12	9.04	13.42	17.64	25.03
	5	2.64	3.05	3.47	4.39	5.20	5.83	6.55	8.03	11.12	13.92	18.42
	6	2.64	3.04	3.45	4.30	5.04	5.57	6.17	7.41	9.82	11.87	14.78
	7	2.67	3.06	3.47	4.24	4.90	5.38	5.90	6.95	8.92	10.52	12.89
	8	2.71	3.08	3.47	4.20	4.79	5.21	5.69	6.58	8.27	9.52	11.35
9	3	2.75	3.20	3.66	4.81	6.02	6.97	8.17	11.14	19.00	27.91	43.02
	4	2.72	3.12	3.56	4.56	5.49	6.21	7.10	8.99	13.28	17.57	24.34
	5	2.71	3.10	3.52	4.40	5.23	5.83	6.53	8.05	10.98	13.78	17.99
	6	2.72	3.10	3.52	4.32	5.06	5.59	6.20	7.43	9.82	11.78	14.84
	7	2.71	3.10	3.51	4.27	4.96	5.42	5.95	6.99	9.06	10.66	12.96
	8	2.75	3.12	3.50	4.24	4.84	5.27	5.74	6.68	8.44	9.75	11.85
	9	2.79	3.17	3.50	4.20	4.78	5.16	5.59	6.41	7.90	9.04	10.71

Nancy R. Mann, K. W. Fertig, and E. M. Scheuer. 1971. Confidence and tolerance bounds and a new goodness-of-fit test for two-parameter Weibull or extreme-value distributions (with tables for censored samples of size 3(1)25). Aerospace Research Laboratories, U.S. Air Force, ARL 71-0077, public domain.

Table A.8c. *(continued)*

n	m	0.02	0.05	0.10	0.25	0.40	0.50	0.60	0.75	0.90	0.95	0.98
10	3	2.75	3.20	3.68	4.77	5.91	6.86	8.10	10.93	18.61	27.05	43.21
	4	2.74	3.16	3.60	4.53	5.45	6.15	7.00	8.87	13.12	17.17	23.48
	5	2.75	3.13	3.54	4.41	5.22	5.79	6.47	7.94	11.05	13.70	17.82
	6	2.75	3.11	3.51	4.33	5.04	5.56	6.15	7.38	9.81	11.86	15.01
	7	2.77	3.12	3.50	4.28	4.95	5.42	5.92	6.99	8.99	10.66	13.03
	8	2.77	3.15	3.50	4.23	4.86	5.29	5.75	6.68	8.49	9.88	11.69
	9	2.81	3.16	3.51	4.21	4.78	5.17	5.60	6.43	7.97	9.17	10.82
	10	2.84	3.19	3.53	4.18	4.72	5.07	5.48	6.23	7.57	8.57	10.03
11	3	2.79	3.22	3.67	4.75	5.87	6.79	7.98	10.78	18.19	26.45	43.82
	4	2.79	3.16	3.58	4.53	5.43	6.13	6.96	8.81	13.02	17.25	24.97
	5	2.79	3.15	3.55	4.39	5.19	5.77	6.44	7.88	10.99	13.66	18.13
	6	2.77	3.14	3.52	4.30	5.02	5.53	6.13	7.33	9.79	11.86	15.02
	7	2.78	3.16	3.51	4.26	4.92	5.39	5.92	6.95	9.03	10.72	13.30
	8	2.81	3.16	3.51	4.23	4.83	5.27	5.74	6.69	8.43	9.86	11.88
	9	2.84	3.17	3.51	4.20	4.77	5.16	5.61	6.49	8.02	9.24	11.08
	10	2.86	3.20	3.53	4.18	4.71	5.09	5.49	6.27	7.67	8.73	10.20
	11	2.88	3.24	3.55	4.17	4.66	5.01	5.38	6.09	7.36	8.31	9.57
12	3	2.88	3.28	3.72	4.77	5.85	6.76	7.88	10.59	17.59	25.73	41.44
	4	2.85	3.22	3.62	4.52	5.41	6.08	6.87	8.67	12.72	16.60	23.41
	5	2.82	3.19	3.58	4.39	5.17	5.75	6.41	7.79	10.68	13.37	17.35
	6	2.82	3.19	3.55	4.31	5.03	5.54	6.12	7.26	9.61	11.60	14.59
	7	2.81	3.18	3.54	4.27	4.92	5.39	5.89	6.89	8.85	10.55	12.65
	8	2.84	3.18	3.54	4.24	4.83	5.25	5.73	6.63	8.34	9.75	11.64
	9	2.86	3.18	3.53	4.21	4.77	5.16	5.60	6.44	7.98	9.18	10.90
	10	2.86	3.20	3.54	4.18	4.73	5.10	5.50	6.26	7.66	8.68	10.17
	11	2.90	3.23	3.55	4.17	4.69	5.03	5.41	6.12	7.37	8.31	9.53
	12	2.94	3.27	3.57	4.17	4.64	4.97	5.32	5.97	7.09	7.96	9.03
13	3	2.88	3.30	3.74	4.78	5.87	6.76	7.89	10.68	17.58	25.37	40.28
	4	2.89	3.25	3.66	4.56	5.46	6.11	6.94	8.75	12.81	16.56	23.73
	5	2.87	3.22	3.60	4.43	5.20	5.78	6.43	7.81	10.89	13.47	17.15
	6	2.89	3.21	3.57	4.35	5.05	5.54	6.14	7.32	9.71	11.72	14.47
	7	2.89	3.21	3.57	4.30	4.94	5.38	5.92	6.94	8.97	10.61	13.04
	8	2.89	3.22	3.57	4.28	4.86	5.29	5.75	6.69	8.46	9.88	11.74
	9	2.91	3.23	3.56	4.25	4.81	5.21	5.64	6.47	8.02	9.27	10.88
	10	2.91	3.24	3.57	4.23	4.78	5.14	5.54	6.31	7.72	8.80	10.25
	11	2.94	3.25	3.58	4.22	4.73	5.07	5.44	6.18	7.46	8.42	9.65
	12	2.97	3.27	3.60	4.21	4.68	5.00	5.37	6.05	7.21	8.03	9.27
	13	3.00	3.30	3.61	4.20	4.66	4.95	5.28	5.91	6.99	7.75	8.76

Table A.8c. *(continued)*

n	m	0.02	0.05	0.10	0.25	0.40	0.50	0.60	0.75	0.90	0.95	0.98
14	3	2.90	3.30	3.75	4.75	5.80	6.63	7.76	10.37	17.23	25.09	39.16
	4	2.91	3.28	3.68	4.52	5.39	6.05	6.83	8.61	12.48	16.19	22.41
	5	2.91	3.25	3.62	4.40	5.14	5.71	6.38	7.77	10.64	13.22	16.89
	6	2.88	3.24	3.59	4.33	5.00	5.50	6.07	7.25	9.59	11.55	14.29
	7	2.91	3.23	3.58	4.28	4.90	5.35	5.86	6.92	8.88	10.47	12.51
	8	2.91	3.24	3.58	4.25	4.83	5.25	5.72	6.66	8.40	9.71	11.54
	9	2.93	3.25	3.58	4.23	4.79	5.16	5.59	6.45	8.00	9.17	10.81
	10	2.95	3.26	3.58	4.22	4.74	5.11	5.52	6.31	7.71	8.72	10.06
	11	2.95	3.28	3.60	4.20	4.71	5.06	5.44	6.17	7.46	8.42	9.61
	12	2.98	3.31	3.61	4.20	4.67	5.00	5.36	6.04	7.25	8.11	9.25
	13	3.00	3.31	3.62	4.19	4.65	4.96	5.30	5.93	7.03	7.82	8.81
	14	3.02	3.34	3.63	4.17	4.62	4.92	5.24	5.82	6.81	7.49	8.45
15	3	2.96	3.34	3.77	4.78	5.83	6.72	7.82	10.45	17.15	24.54	38.14
	4	2.94	3.29	3.69	4.55	5.40	6.05	6.82	8.54	12.41	15.96	21.82
	5	2.93	3.28	3.64	4.42	5.18	5.72	6.36	7.72	10.56	12.94	16.99
	6	2.93	3.26	3.61	4.37	5.03	5.52	6.09	7.23	9.45	11.38	14.24
	7	2.93	3.25	3.59	4.31	4.92	5.37	5.90	6.89	8.81	10.31	12.84
	8	2.93	3.25	3.59	4.27	4.86	5.27	5.73	6.64	8.36	9.64	11.58
	9	2.95	3.26	3.58	4.26	4.81	5.19	5.62	6.48	8.00	9.11	10.79
	10	2.94	3.27	3.59	4.24	4.77	5.13	5.54	6.33	7.71	8.72	10.06
	11	2.95	3.26	3.59	4.21	4.73	5.08	5.46	6.18	7.48	8.37	9.61
	12	2.98	3.29	3.61	4.21	4.70	5.04	5.39	6.07	7.29	8.17	9.24
	13	3.01	3.30	3.61	4.20	4.68	4.99	5.33	5.96	7.09	7.89	8.92
	14	3.03	3.32	3.62	4.19	4.65	4.95	5.26	5.86	6.92	7.63	8.59
	15	3.07	3.35	3.62	4.19	4.63	4.91	5.21	5.77	6.70	7.37	8.24
16	3	2.98	3.36	3.78	4.79	5.86	6.70	7.78	10.30	17.06	24.06	38.50
	4	2.96	3.32	3.70	4.57	5.41	6.04	6.82	8.52	12.49	16.24	23.03
	5	2.94	3.30	3.65	4.46	5.18	5.73	6.37	7.71	10.64	13.23	16.90
	6	2.93	3.29	3.63	4.37	5.05	5.53	6.11	7.23	9.53	11.53	14.38
	7	2.94	3.28	3.62	4.32	4.96	5.40	5.89	6.89	8.84	10.54	12.97
	8	2.95	3.28	3.62	4.30	4.88	5.30	5.75	6.65	8.39	9.75	11.59
	9	2.96	3.28	3.61	4.27	4.84	5.22	5.66	6.47	8.03	9.18	10.96
	10	2.97	3.28	3.62	4.25	4.79	5.16	5.55	6.32	7.71	8.75	10.29
	11	2.99	3.29	3.62	4.23	4.74	5.10	5.48	6.19	7.50	8.43	9.76
	12	2.99	3.31	3.63	4.23	4.71	5.04	5.41	6.09	7.27	8.17	9.32
	13	3.01	3.34	3.63	4.21	4.68	5.01	5.36	5.99	7.07	7.86	9.04
	14	3.05	3.35	3.65	4.21	4.66	4.96	5.30	5.90	6.93	7.68	8.67
	15	3.07	3.37	3.66	4.21	4.64	4.92	5.24	5.80	6.78	7.46	8.39
	16	3.12	3.39	3.69	4.20	4.62	4.88	5.17	5.72	6.60	7.26	8.10

Table A.8c. *(continued)*

n	m	0.02	0.05	0.10	0.25	0.40	0.50	0.60	0.75	0.90	0.95	0.98
17	3	2.97	3.37	3.76	4.71	5.69	6.50	7.56	10.06	16.55	23.57	37.77
	4	2.97	3.32	3.70	4.51	5.32	5.94	6.71	8.36	11.93	15.54	21.68
	5	2.96	3.30	3.66	4.42	5.12	5.65	6.27	7.62	10.23	12.58	16.34
	6	2.95	3.28	3.63	4.34	5.01	5.47	6.03	7.14	9.32	11.13	13.79
	7	2.96	3.29	3.63	4.30	4.90	5.34	5.83	6.83	8.70	10.18	12.33
	8	2.97	3.28	3.61	4.27	4.83	5.23	5.69	6.58	8.25	9.56	11.40
	9	2.97	3.29	3.61	4.25	4.79	5.17	5.58	6.39	7.91	9.07	10.65
	10	2.98	3.30	3.61	4.23	4.76	5.12	5.50	6.25	7.66	8.70	10.01
	11	2.99	3.31	3.62	4.22	4.71	5.06	5.42	6.13	7.42	8.33	9.63
	12	3.01	3.33	3.63	4.21	4.69	5.01	5.36	6.02	7.22	8.08	9.21
	13	3.03	3.34	3.63	4.21	4.66	4.97	5.30	5.93	7.05	7.86	8.92
	14	3.04	3.35	3.64	4.19	4.64	4.93	5.25	5.85	6.91	7.67	8.63
	15	3.05	3.37	3.66	4.19	4.62	4.91	5.21	5.79	6.77	7.45	8.29
	16	3.09	3.39	3.67	4.19	4.60	4.87	5.17	5.71	6.62	7.29	8.04
	17	3.12	3.41	3.68	4.18	4.58	4.84	5.12	5.63	6.48	7.11	7.84
18	3	3.03	3.38	3.78	4.72	5.69	6.50	7.53	9.97	16.20	23.42	38.75
	4	3.03	3.35	3.72	4.55	5.33	5.94	6.72	8.32	12.10	15.75	21.93
	5	3.01	3.35	3.68	4.44	5.14	5.67	6.29	7.61	10.29	12.75	16.76
	6	3.00	3.33	3.66	4.37	5.01	5.47	6.04	7.15	9.34	11.10	14.07
	7	3.00	3.33	3.65	4.33	4.93	5.35	5.86	6.84	8.80	10.17	12.25
	8	3.01	3.33	3.65	4.30	4.87	5.28	5.73	6.60	8.28	9.62	11.30
	9	3.03	3.33	3.64	4.27	4.82	5.19	5.61	6.46	7.97	9.12	10.71
	10	3.04	3.34	3.64	4.26	4.77	5.15	5.54	6.31	7.63	8.69	10.04
	11	3.04	3.34	3.64	4.25	4.74	5.08	5.47	6.20	7.45	8.34	9.69
	12	3.05	3.35	3.66	4.25	4.71	5.03	5.39	6.09	7.26	8.11	9.24
	13	3.06	3.37	3.67	4.23	4.68	4.99	5.34	5.99	7.10	7.92	8.84
	14	3.08	3.38	3.67	4.21	4.67	4.97	5.29	5.90	6.95	7.70	8.60
	15	3.11	3.39	3.68	4.21	4.65	4.94	5.25	5.82	6.82	7.50	8.32
	16	3.13	3.41	3.69	4.21	4.64	4.91	5.22	5.76	6.68	7.37	8.15
	17	3.16	3.43	3.70	4.20	4.61	4.88	5.17	5.69	6.57	7.19	7.94
	18	3.19	3.47	3.73	4.20	4.60	4.86	5.13	5.63	6.45	7.04	7.75
19	3	3.03	3.40	3.79	4.73	5.72	6.52	7.53	9.96	16.17	23.13	36.55
	4	3.05	3.37	3.73	4.55	5.35	5.95	6.70	8.27	11.98	15.31	21.70
	5	3.03	3.36	3.69	4.43	5.14	5.65	6.27	7.54	10.29	12.70	16.31
	6	3.02	3.34	3.66	4.36	5.01	5.47	6.01	7.10	9.28	11.15	13.95
	7	3.02	3.33	3.66	4.33	4.93	5.36	5.83	6.80	8.65	10.16	12.41
	8	3.02	3.33	3.64	4.29	4.87	5.25	5.69	6.58	8.25	9.52	11.35
	9	3.03	3.33	3.65	4.26	4.80	5.18	5.59	6.40	7.93	9.09	10.61
	10	3.03	3.34	3.64	4.25	4.76	5.11	5.52	6.26	7.63	8.67	10.08
	11	3.06	3.35	3.65	4.24	4.72	5.07	5.44	6.15	7.42	8.34	9.70
	12	3.07	3.37	3.65	4.23	4.70	5.03	5.39	6.05	7.22	8.13	9.34
	13	3.08	3.38	3.65	4.22	4.68	4.99	5.33	5.97	7.06	7.88	8.99
	14	3.09	3.38	3.66	4.21	4.66	4.96	5.28	5.89	6.93	7.69	8.63

Table A.8c. *(continued)*

n	m	0.02	0.05	0.10	0.25	0.40	0.50	0.60	0.75	0.90	0.95	0.98
	15	3.10	3.40	3.67	4.22	4.64	4.93	5.24	5.81	6.79	7.49	8.46
	16	3.13	3.42	3.68	4.21	4.63	4.90	5.20	5.76	6.70	7.35	8.19
	17	3.14	3.43	3.69	4.21	4.62	4.88	5.16	5.70	6.59	7.23	8.01
	18	3.17	3.44	3.71	4.21	4.60	4.86	5.14	5.64	6.49	7.08	7.84
	19	3.20	3.46	3.72	4.21	4.59	4.83	5.10	5.57	6.38	6.91	7.60
20	3	3.01	3.39	3.79	4.72	5.67	6.42	7.38	9.80	16.16	23.55	37.26
	4	3.02	3.37	3.73	4.54	5.32	5.93	6.66	8.21	11.97	15.38	21.61
	5	3.03	3.36	3.70	4.44	5.13	5.64	6.26	7.53	10.15	12.57	16.40
	6	3.02	3.34	3.67	4.38	5.01	5.46	6.01	7.08	9.17	11.05	14.08
	7	3.02	3.34	3.66	4.33	4.93	5.35	5.82	6.78	8.63	10.10	12.40
	8	3.03	3.35	3.66	4.30	4.85	5.25	5.68	6.53	8.18	9.55	11.47
	9	3.04	3.34	3.65	4.28	4.80	5.17	5.58	6.36	7.86	9.02	10.65
	10	3.04	3.35	3.65	4.26	4.76	5.12	5.50	6.24	7.59	8.61	10.02
	11	3.06	3.36	3.66	4.25	4.73	5.06	5.45	6.12	7.37	8.35	9.57
	12	3.08	3.38	3.66	4.24	4.70	5.03	5.38	6.03	7.20	8.08	9.28
	13	3.09	3.38	3.67	4.23	4.68	4.98	5.33	5.95	7.01	7.86	9.02
	14	3.11	3.40	3.67	4.22	4.65	4.95	5.26	5.87	6.89	7.66	8.68
	15	3.12	3.41	3.69	4.21	4.63	4.92	5.23	5.80	6.79	7.49	8.44
	16	3.14	3.42	3.70	4.21	4.62	4.89	5.20	5.74	6.67	7.34	8.23
	17	3.16	3.44	3.71	4.20	4.61	4.87	5.16	5.69	6.57	7.22	7.98
	18	3.18	3.46	3.72	4.20	4.60	4.86	5.13	5.63	6.47	7.07	7.82
	19	3.21	3.47	3.73	4.21	4.58	4.84	5.11	5.57	6.38	6.96	7.64
	20	3.24	3.49	3.74	4.21	4.57	4.81	5.06	5.52	6.25	6.80	7.48
21	3	3.05	3.40	3.79	4.70	5.66	6.42	7.38	9.68	16.03	22.96	36.33
	4	3.08	3.39	3.74	4.53	5.31	5.91	6.61	8.15	11.66	14.96	20.76
	5	3.05	3.37	3.69	4.44	5.13	5.64	6.23	7.46	10.16	12.46	16.03
	6	3.05	3.35	3.67	4.36	4.99	5.45	5.97	7.03	9.22	10.99	13.71
	7	3.05	3.35	3.66	4.32	4.92	5.33	5.82	6.75	8.57	10.02	12.10
	8	3.05	3.35	3.66	4.29	4.85	5.25	5.68	6.54	8.17	9.45	11.25
	9	3.06	3.35	3.65	4.28	4.80	5.16	5.58	6.38	7.91	8.97	10.54
	10	3.07	3.35	3.66	4.26	4.77	5.11	5.50	6.25	7.62	8.62	10.00
	11	3.08	3.36	3.66	4.25	4.74	5.07	5.44	6.13	7.44	8.34	9.64
	12	3.09	3.36	3.67	4.24	4.71	5.03	5.38	6.06	7.27	8.11	9.23
	13	3.10	3.38	3.68	4.23	4.68	4.99	5.33	5.98	7.10	7.88	8.94
	14	3.12	3.39	3.67	4.22	4.66	4.96	5.29	5.90	6.96	7.72	8.70
	15	3.14	3.40	3.68	4.21	4.64	4.93	5.25	5.83	6.83	7.55	8.49
	16	3.16	3.42	3.69	4.21	4.63	4.90	5.20	5.78	6.72	7.37	8.23
	17	3.17	3.44	3.69	4.20	4.62	4.89	5.17	5.72	6.61	7.24	8.04
	18	3.19	3.45	3.71	4.20	4.60	4.86	5.15	5.66	6.52	7.10	7.89
	19	3.21	3.47	3.72	4.21	4.60	4.85	5.12	5.60	6.42	7.00	7.72
	20	3.24	3.49	3.74	4.20	4.59	4.83	5.09	5.56	6.32	6.87	7.58
	21	3.27	3.51	3.75	4.21	4.57	4.81	5.06	5.51	6.22	6.73	7.42

Table A.8c. *(continued)*

n	m	0.02	0.05	0.10	0.25	0.40	0.50	0.60	0.75	0.90	0.95	0.98
22	3	3.06	3.42	3.80	4.67	5.60	6.38	7.33	9.61	15.80	22.00	34.81
	4	3.10	3.41	3.76	4.52	5.29	5.87	6.56	8.12	11.67	15.09	20.79
	5	3.09	3.38	3.71	4.42	5.09	5.60	6.19	7.42	10.06	12.25	16.00
	6	3.08	3.36	3.69	4.36	4.97	5.43	5.96	6.98	9.18	10.91	13.39
	7	3.08	3.36	3.67	4.32	4.90	5.31	5.79	6.74	8.65	10.07	12.18
	8	3.08	3.38	3.67	4.28	4.83	5.24	5.67	6.54	8.18	9.40	11.20
	9	3.09	3.38	3.68	4.26	4.79	5.17	5.57	6.37	7.85	8.97	10.51
	10	3.09	3.38	3.67	4.24	4.75	5.11	5.50	6.23	7.58	8.61	9.89
	11	3.10	3.38	3.67	4.23	4.72	5.07	5.44	6.14	7.36	8.29	9.53
	12	3.12	3.40	3.68	4.23	4.69	5.01	5.37	6.05	7.19	8.06	9.18
	13	3.14	3.42	3.68	4.22	4.67	4.98	5.32	5.95	7.05	7.85	8.87
	14	3.15	3.42	3.69	4.22	4.66	4.96	5.28	5.88	6.91	7.66	8.67
	15	3.16	3.43	3.70	4.22	4.64	4.94	5.25	5.81	6.79	7.48	8.43
	16	3.17	3.44	3.71	4.21	4.63	4.90	5.20	5.76	6.68	7.35	8.21
	17	3.18	3.45	3.71	4.22	4.61	4.88	5.17	5.70	6.60	7.19	8.03
	18	3.19	3.47	3.72	4.21	4.60	4.85	5.15	5.65	6.51	7.11	7.87
	19	3.21	3.47	3.73	4.21	4.60	4.85	5.12	5.61	6.44	7.00	7.72
	20	3.24	3.49	3.74	4.22	4.59	4.83	5.09	5.57	6.36	6.87	7.58
	21	3.25	3.51	3.75	4.22	4.58	4.82	5.07	5.53	6.27	6.77	7.43
	22	3.29	3.53	3.77	4.22	4.57	4.81	5.04	5.48	6.18	6.65	7.25
23	3	3.10	3.45	3.82	4.71	5.61	6.36	7.33	9.63	15.57	21.97	35.77
	4	3.12	3.43	3.77	4.53	5.31	5.87	6.57	8.18	11.82	15.36	21.34
	5	3.11	3.41	3.72	4.46	5.11	5.62	6.21	7.45	10.11	12.46	16.45
	6	3.10	3.39	3.69	4.39	5.00	5.44	5.96	7.00	9.07	10.93	13.64
	7	3.09	3.38	3.69	4.34	4.91	5.30	5.78	6.70	8.56	10.01	12.34
	8	3.09	3.38	3.68	4.31	4.85	5.23	5.64	6.49	8.13	9.40	11.19
	9	3.09	3.38	3.68	4.29	4.81	5.16	5.55	6.33	7.83	8.93	10.45
	10	3.09	3.38	3.68	4.27	4.77	5.11	5.47	6.20	7.56	8.55	9.95
	11	3.10	3.40	3.69	4.26	4.74	5.06	5.41	6.10	7.36	8.27	9.48
	12	3.11	3.40	3.69	4.25	4.72	5.02	5.36	6.01	7.16	8.02	9.08
	13	3.13	3.41	3.70	4.25	4.69	4.99	5.31	5.92	7.04	7.83	8.80
	14	3.14	3.42	3.70	4.23	4.67	4.96	5.27	5.86	6.92	7.66	8.63
	15	3.16	3.44	3.70	4.22	4.66	4.92	5.24	5.80	6.80	7.47	8.36
	16	3.16	3.44	3.70	4.23	4.64	4.91	5.20	5.75	6.68	7.35	8.16
	17	3.19	3.45	3.71	4.22	4.62	4.89	5.17	5.70	6.62	7.23	8.00
	18	3.20	3.47	3.72	4.22	4.61	4.87	5.15	5.65	6.53	7.10	7.90
	19	3.21	3.47	3.72	4.22	4.60	4.85	5.12	5.61	6.43	7.01	7.73
	20	3.23	3.48	3.73	4.22	4.59	4.83	5.09	5.57	6.34	6.92	7.57
	21	3.24	3.49	3.74	4.22	4.58	4.82	5.07	5.52	6.28	6.80	7.46
	22	3.26	3.51	3.76	4.22	4.57	4.80	5.05	5.50	6.22	6.72	7.32
	23	3.30	3.53	3.76	4.22	4.56	4.79	5.02	5.45	6.14	6.61	7.17

Table A.8c. *(continued)*

n	m	0.02	0.05	0.10	0.25	0.40	0.50	0.60	0.75	0.90	0.95	0.98
24	3	3.10	3.46	3.83	4.67	5.57	6.32	7.27	9.56	15.36	22.10	34.39
	4	3.13	3.44	3.78	4.52	5.24	5.82	6.53	8.10	11.62	14.93	20.76
	5	3.12	3.41	3.73	4.44	5.09	5.59	6.20	7.45	10.09	12.23	15.75
	6	3.11	3.40	3.71	4.38	4.99	5.43	5.96	7.03	9.16	10.80	13.44
	7	3.11	3.39	3.70	4.34	4.92	5.33	5.79	6.72	8.53	9.97	12.10
	8	3.11	3.39	3.69	4.31	4.85	5.24	5.68	6.52	8.12	9.34	11.18
	9	3.11	3.40	3.69	4.29	4.81	5.17	5.57	6.36	7.80	8.90	10.37
	10	3.11	3.39	3.69	4.27	4.77	5.10	5.49	6.23	7.55	8.52	9.94
	11	3.10	3.40	3.68	4.26	4.74	5.07	5.43	6.11	7.35	8.26	9.49
	12	3.11	3.40	3.69	4.25	4.72	5.03	5.37	6.03	7.19	8.04	9.11
	13	3.13	3.41	3.69	4.25	4.70	4.99	5.32	5.95	7.07	7.86	8.84
	14	3.14	3.43	3.69	4.24	4.68	4.97	5.29	5.89	6.91	7.66	8.55
	15	3.16	3.43	3.71	4.25	4.66	4.95	5.26	5.82	6.79	7.47	8.30
	16	3.17	3.44	3.71	4.24	4.65	4.93	5.23	5.77	6.70	7.34	8.16
	17	3.19	3.44	3.72	4.24	4.64	4.91	5.20	5.74	6.60	7.25	7.99
	18	3.19	3.46	3.73	4.23	4.62	4.88	5.17	5.69	6.53	7.12	7.83
	19	3.22	3.48	3.74	4.23	4.62	4.87	5.14	5.64	6.44	7.00	7.72
	20	3.23	3.49	3.74	4.23	4.61	4.85	5.12	5.59	6.38	6.91	7.53
	21	3.25	3.50	3.75	4.23	4.60	4.83	5.10	5.54	6.30	6.81	7.47
	22	3.27	3.52	3.76	4.23	4.58	4.81	5.06	5.52	6.22	6.74	7.36
	23	3.28	3.53	3.77	4.24	4.58	4.80	5.03	5.48	6.15	6.63	7.24
	24	3.31	3.55	3.79	4.24	4.57	4.78	5.02	5.43	6.09	6.54	7.11
25	3	3.09	3.47	3.87	4.71	5.59	6.31	7.24	9.40	15.23	21.80	34.84
	4	3.14	3.45	3.80	4.56	5.27	5.82	6.50	8.00	11.51	14.84	20.21
	5	3.12	3.44	3.77	4.47	5.11	5.58	6.15	7.34	9.81	12.21	15.64
	6	3.12	3.44	3.74	4.43	5.01	5.44	5.94	6.97	9.00	10.75	13.41
	7	3.11	3.43	3.73	4.37	4.92	5.32	5.80	6.70	8.45	9.92	11.90
	8	3.13	3.42	3.72	4.34	4.87	5.24	5.67	6.48	8.03	9.28	11.01
	9	3.13	3.42	3.71	4.32	4.83	5.17	5.57	6.33	7.75	8.88	10.46
	10	3.13	3.43	3.72	4.30	4.79	5.12	5.50	6.22	7.51	8.51	9.95
	11	3.13	3.43	3.71	4.29	4.76	5.07	5.43	6.11	7.32	8.26	9.52
	12	3.15	3.43	3.71	4.26	4.72	5.05	5.38	6.03	7.16	8.04	9.15
	13	3.16	3.44	3.72	4.25	4.70	5.00	5.34	5.94	7.03	7.82	8.90
	14	3.17	3.45	3.72	4.25	4.69	4.97	5.30	5.88	6.90	7.64	8.68
	15	3.18	3.46	3.72	4.24	4.67	4.95	5.26	5.83	6.78	7.47	8.42
	16	3.18	3.47	3.73	4.24	4.66	4.93	5.22	5.76	6.69	7.35	8.23
	17	3.20	3.48	3.74	4.24	4.65	4.91	5.19	5.72	6.63	7.26	8.03
	18	3.23	3.48	3.75	4.23	4.63	4.90	5.17	5.68	6.54	7.14	7.88
	19	3.24	3.49	3.75	4.24	4.62	4.87	5.14	5.63	6.45	7.03	7.70
	20	3.25	3.50	3.75	4.24	4.61	4.85	5.11	5.58	6.38	6.91	7.57
	21	3.26	3.52	3.77	4.24	4.60	4.84	5.09	5.55	6.32	6.82	7.45
	22	3.27	3.54	3.78	4.24	4.58	4.82	5.07	5.51	6.25	6.75	7.36
	23	3.30	3.55	3.79	4.24	4.58	4.81	5.05	5.47	6.18	6.68	7.26
	24	3.31	3.57	3.80	4.23	4.58	4.79	5.02	5.45	6.13	6.58	7.19
	25	3.33	3.58	3.82	4.23	4.57	4.78	5.00	5.41	6.07	6.51	7.07

Index

Abraham, J. A., 230
Absorption. *See* ALR algorithm
Acceptable assignment of probabilities, 9–10
Acceptable quality level (AQL), 52
Acyclic graph. *See* Mathematical graph theory
Additive probabilities of disjoint events, 10, 231
Additive sufficient statistics, 126, 202
Adult volume ventilator (AVV), 157
ALR (Abraham-Locks revised) algorithm, 230, 232–38
American Embassy, attempted rescue of hostages from, 52
Ancestor. *See* Mathematical graph theory
Apple Macintosh™, 204
Approximate reliability-confidence, 67–69
Approximating polynomial, 209
Approximations, 67–69, 215. *See also* Esary-Proschan lower bound, Subset
Arc, 217–18
Assessment, xxi–xxv
Attributes, xxii, 3–6, 23, 39–52, 67–69, 112–14, 130–33. *See also* Binomial distribution, Negative binomial distribution, Beta distribution

Availability, xxv–xxvi, 191–96, 201–10
inherent, 195, 201–5
observed, 195, 205–10

Backtracking, 247–48
Baseball, 15–16
Basu, A. P., 118
Bathtub curve, 162
Bayes, Thomas, 13, 137, 138
Bayesian methods, xxiii, 125–45, 196, 202–6
attributes data, 130–35
conjugate distributions. *See* prior distribution
discrete model, 125–30
lower confidence limits in, 125–26, 132, 136, 141–45
Poisson process, 127, 135–45
posterior distribution. *See* Prior distribution
Bayesian reliability assessment for attributes data, 130–33
for Poisson process, 135–45
Bayes theorem, 3, 12–15, 125, 258
Bernoulli process, 39, 45, 109, 112–16, 125, 126, 139
Best linear invariant (BLI) estimates, 176–78
Best linear unbiased estimates (BLUE), 119, 176
Beta distribution, 39, 48–52, 113, 126
table of, 281–82

Beta function, 131
Beta scale, 168, 173
Bias, 29–30
Binary distribution, 23, 28, 34, 39
Binary element, 217
Binary set, 217. *See also* Set theory
Binary trial, 3, 10–11, 39–40
Binomial coefficient, 40–41, 51
Binomial distribution, xxii, 22–23, 42–43, 113, 127, 131–32. *See also* Attributes, Negative binomial distribution, Beta distribution, Binary distribution, Binary trials
  distribution function, 42
  mean of, 43–45
  symmetry of, 43
  tables of, 43, 50–51
  variance of, 43–45
Birnbaum, Z. W., 214
Boole, G. W., 214
Boolean algebra, xxvii, 215–23, 233–38
  inversion, 215, 222, 230
  minimization, 233–34
Boolean set, 221. *See also* Set theory
Bowker, A. H., 155
Bridge circuit, 219–23, 230–32, 249–52
Brother. *See* Mathematical graph theory
Burst pressure, 154

Calculus, 20, 49, 100–103, 112, 116, 131–36, 151, 171, 239–40

Cardinal number, 239
Castings, 157–58
Censoring of data, 88, 110, 113–14, 119, 134, 169, 176. *See also* Complete data sets
Central limit theorem, 55
Change-of-variables technique, 46, 73
Chevalier de Méré, 32
Child. *See* Mathematical graph theory
Chi-square (χ^2) distribution, 74–76, 83, 95–97, 102–3, 116–17, 135, 140, 142–45. *See also* Gamma distribution
  table of, xxiv, 74, 83, 286
Coherent ST system, 252
Cohort, 182
Coin tossing, 23
Combinatorial mathematics, 40–41
Communication network, 235
Complete data sets, 110, 113, 119, 133–34. *See also* Censoring of data
Component, xxi, 84, 214, 220, 229, 257. *See also* System reliability
Conditional probability, 10–11, 101, 125. *See also* Bayes theorem
Confidence assessment, xxi, 48–51, 67–69, 112–18, 149, 108, 194, 201, 281. *See also* Reliability, Lower confidence limit
Conjugate prior distribution, 127. *See also* Prior distribution

Convolution, 201
Coolant loops, 257–58
Correction for bias, 30
Counting events, 6
Cramer-von Mises test, 63
Craps, 8, 12, 21, 129
Cumulative hazard. *See* Hazard
Curse of dimensionality, 225
Cut. *See* Minimal cut
Cycle, 221, 247
Cyclic graph, 243, 247

Data sets
  censored. *See* Censoring of data
  complete. *See* Complete data sets
Davis, D. J., xxiii
Degrees of freedom, 22, 23, 31, 56–57, 75, 152
DeMoivre, A., 55
DeMorgan, A., 214
DeMorgan's theorems, 215, 221–23, 232
Dependence, 3, 10. *See also* Independence
Depth-first search, 247–48. *See also* Mathematical graph theory
Derivative. *See* Calculus
Descendant. *See* Mathematical graph theory
Dice. *See* Craps
Diffuse prior distribution, 138
Discrete Bayesian model, 125
Discrete distribution, 48
Discrete hazard analysis, 163–64
Discrete random variable, 20
Discrete-valued variable, 19

Disjoint events, 7–10, 215, 229, 257
Disjoint products, xxvii, 229–40, 259–62
Distribution
  beta, 48–52
  binary, 23
  binomial. *See* Binomial distribution
  chi-square. *See* Chi-square distribution
  discrete, 48
  empirical, 206
  exponential. *See* Exponential distribution
  extreme-value, 176–77
  gamma. *See* Gamma distribution
  lognormal. *See* Lognormal distribution
  negative binomial, 45–47, 132
  negative-log gamma, 136–40
  Poisson. *See* Poisson distribution
  Rayleigh, 118
  two-parameter exponential, 118–21
  uniform, 22
  Weibull. *See* Weibull distribution
Distribution function, 20–23, 42, 45, 49, 58, 84, 166
Dixon, W. J., 178
Domination, 244–46, 250. *See also* Mathematical graph theory
Downtime, 195
D-statistic. *See* Lilliefors K–S statistic
Duality, 46, 49, 225

Efficiency, 111
Eisenhart, C., 150
Elder brother. *See* Mathematical graph theory
Electrical capacitance, 218–19
Electrical inductance, 218–19
Empty set, 5. *See also* Set theory
Epstein-Sobel model, 114–17, 135, 141–45, 194
  assessment of Poisson process for reliability, 114–17, 141–45
  Bayesian modification of, xxiv
  confidence limits, 116–17
  Esary-Proschan lower bound, 225–26
Event, 3, 4–5, 19
  compound, 6–9
  counting, 6
  disjoint, 7
  elementary, 5
  joint, 10
  mutually exclusive. *See* Disjoint events
Expected value, 25. *See also* Mean
Exponential distribution, xxiii, 24, 33, 84, 127, 163, 164
  derivation of, 99–103
  Epstein-Sobel technique of maximum-likelihood assessment, 114–17, 141–45
  goodness-of-fit, 93–95
  graph paper, 85–93
  probability plotting, 85–93
  simulation of, 201–208
  two-parameter, 118–21
Exponential families of distributions, 127

Exponentiality, 225
Exponential repair, 205–8
Exponential time to failure, 83–108, 196, 201–8
Exponential time to repair, 194, 196–98, 201–8
Extreme-value distribution, 178

Factorial, 40
Failure rate, 110, 126, 144, 163, 196
Families of distributions, xxii–xxiii, 21–24, 37–38
Father. *See* Mathematical graph theory
Fault tree, xxvii
Fertig, K. W., 180
Finite population correction. *See* Yates correction factor
Fisher, R. A., 111, 162
Fisher-Tippett type I distribution. *See* Weibull distribution
Formation, 244–46. *See also* Mathematical graph theory
Frechet, M., 214
Friedman, H., 155

Gamma distribution, 95–97, 101–2, 114, 127, 134–35, 202
Gamma function, 101
Gauss, K., 55
Gaussian distribution. *See* Normal distribution
Geiger counter, 99
Go–no go processes. *See* Attributes
Goodness-of-fit test, xxiii, 56, 63–66, 72, 93–95, 196
  exponential distribution, 93–95
  normal distribution, 63–66

Graph. *See* Mathematical graph theory, Probability graphing
Graph theory. *See* Mathematical graph theory
Greenberg, B. G., 119
Grouped data, 63
Guidance system, 260
Gumbel, E. J., 162

Hald, A., 155–56
Halley's method, 161, 163–64
Harter, H. L., 49, 102
Hastay, M., 150
Hastings, C., 208
Hazard, xxiv, 161, 163, 169–75
 functions, 169–75
 Weibull hazard plotting, 169–75
Historical data, 195
Hydraulic pump, 121

Importance. *See* Reliability importance
Inclusion-exclusion (IE), xxvii, 214, 218–21, 229, 258, 261
 bridge network reliability by, 218–21
 relation to topological reliability, 241–42, 249–52
 reliability formula, 220
 with minimal cuts, 220–21
Inclusive-or union, 6, 230–31
Incomplete beta function, 49. *See also* Beta distribution
Incomplete data set. *See* Censoring of data
Independence, 3, 10, 28, 31, 39, 100–1, 215

Inductance, 218–19
Inherent availability, 195, 201–205
Inner loop, 233. *See also* ALR algorithm
Instantaneous hazard, 163
Integration. *See* Calculus
Intelligence quotients (IQs), distribution of, 55
Internode, 243. *See also* Mathematical graph theory
Intersection of sets, 6
Interval estimation, 141–45
Iterative formula, 41, 214

Jet aircraft, 257

Kernel, 49, 112, 126–27
k-graph, 253. *See also* Mathematical graph theory
Kolmogorov–Smirnov (K–S) test, 63, 93
Koopman-Pitman-Darmois class, 127
k-terminal network, extensions of, 252–56

LaPlace, P. S., 55
Leaf, 243–44. *See also* Mathematical graph theory
Lexicographical order, 247
Life tests, 161
 and Bayesian reliability analysis, 133–45
 and censored data sets, 88
Likelihood, 111, 127, 131
Lilliefors K–S goodness-of-fit test, 63–66, 72, 93–95, 121, 156, 199
 for unspecified exponential population, 93–95

for unspecified lognormal
population, 72
for unspecified normal
population, 65–66
Linear estimation
of extreme-value location and
scale parameters, 176–77
of two-parameter exponential
location and scale
parameters, 109, 119
Local area network (LAN), 122,
260
Location parameter, 119, 121,
162, 165, 176. *See also*
Shift parameter
Logarithmic scale, 85, 90, 171
Logic, 229. *See also* Boolean
algebra
Lognormal distribution, xxiv,
69–72, 118, 156, 193,
196–201, 208–10
Log-Weibull. *See* Distribution,
extreme value
Lottery, 34
Lower-bound formula, 225
Lower confidence limit (LCL),
xxii, 48, 69, 113, 121,
125–26, 132–44

Maintainability, xxvi, 191–201
similarity to queuing, 193–94
Mann best linear invariant
estimation, xxv, 176,
184–86
Mann, N. R., 176, 180
Marginal probability, 128, 129,
131, 134–41
Massey, F. M., 178
Mathematical graph theory,
xxvii, 214–15, 241

Mathematical induction, 74–75,
101–2
Maximum likelihood, xxiii, 27,
107, 111–17, 121, 139
Mean, xxii, 25, 28
of the binomial distribution,
43–45
of the chi-square distribution,
75
of the exponential distribution,
26
of the logarithms, 70
of the negative binomial
distribution, 46–47
of the normal distribution, 58,
61, 63, 74
sample, 28–29
of the uniform distribution, 26
variance of the sample mean,
29–31
Mean-time-between-failures
(MTBF), 95, 104, 126,
141–42, 196
Mean-time-to-failure (MTTF),
141–42, 144, 196
Median, 25 27, 70
Mica washer, 156
Microsoft Quick-Basic™, 204
MIL-STD 105E, 52
MIL-STD 414, 153
Mini-keno, 16, 34
Minimal cuts, xxvii, 214,
218–21
Minimal path, xxvii, 214,
218–21, 229, 241
Minimization. *See* Boolean
algebra
Minimized inversion, 215,
221–23. *See also* Boolean
algebra

Minimum variance unbiased estimation, 117–18
Mode, 27–28, 113
Monomials, 214, 239
Monte Carlo simulation, xxvi, 180, 191, 193, 195–96, 202–9
Mortality data for hazard analysis, 161
MTBF. *See* Mean-time-between failures (MTBF)
MTTF. *See* Mean-time-to-failure (MTTF)
Multiple failures and redundancy, 95–98
Multiply censored data, 169
Mutually exclusive events. *See* Disjoint events

Natural logarithm, 34, 85, 137, 165, 176–81, 206
Naval Ordnance Test Station (NOTS) tables, 50–51
Negative binomial distribution, 39, 45–47, 103, 127, 132
Negative-log gamma distribution, 136–40
Nelson, Wayne, 90
Nested formula, 214–15, 242, 250, 252
Neutral sequences, 243, 244, 247. *See also* Mathematical graph theory
Newton, Isaac, 204
Node, 218, 243. *See also* Mathematical graph theory
Noncentral-t distribution, 152. *See also* Student-t
Nonparametric techniques, xxv, 118, 150, 161, 195, 196

Normal distribution, xxii, 20, 55–56, 118. *See also* Lognormal distribution
  approximate confidence limits for attributes, 67–69
  goodness of fit, 63–67
  mean and variance of, 74
  one-sided tolerance analysis, 150–54, 287
  probability plotting and graphing, 59–63, 156
  reliability-confidence assessment, xxiv, 56, 150–54, 155–56
  standardized, 56–59
  table of standard normal distribution, 283–84
  two-sided tolerance analysis, 154–56, 288–89
Normalized probability, 13, 49, 72–73
Normal-theory tolerancing, 198
Nuclear reactor, coolant loop for, 257–58
Null hypothesis, 63–64
Null set, 5. *See also* Set theory
Numerically controlled milling machine, 78

Observed availability, 195, 205–10
One-sided analysis, xxiv, 150–54
  lower specification limit, 153–54
  relationship to noncentral-t tolerance-limit factors, 151–53
  upper specification limit, 150–51
One-sided tolerance limit, 56

Ordered failure times, maximum-likelihood estimate for, 114–16
Orderings. *See* Permutations
Order statistics, 61, 119, 177
Ordinal number, 239
Outcomes, 3–6, 40
Outer loop, 232. *See also* ALR algorithm
Overall reliability, 252
Owen, D. B., 150, 155

p-acyclic graph, 243–48
Parallel reductions, 223–25
Parallel system, 217
Parameter(s), xxii–xxiv, 22, 25–28, 55, 109–10, 119–21, 162–67, 176–177. *See also* Mean, Median, Mode, Variance, Location parameter, Scale parameter, Shape parameter, Shift parameter
Parametric maintainability analysis, 194–96
Partial derivative. *See* calculus
Partition, 3, 7–9, 12–13, 19, 125, 127
Pascal, Blaise, 32, 41
Pascal distribution. *See* Negative binomial distribution
Pascal's triangle, 41
Path. *See* Minimal path
Pearson, Karl, 49, 102
Percentage point, 26–27, 49, 75–76, 135
    table of χ^2 distributions, 286
Permutations, 40
p-graph. *See* Mathematical graph theory, p-cyclical graph

Plotting points, 61–63, 119, 121. *See also* Probability graphing
    table of, 285
Poincare, H., 214
Poisson distribution, xxiii, 83, 97–98, 103, 203
Poisson process, 83, 95, 99, 104, 117–18, 125, 135–36, 193
    Bayesian and Epstein-Sobel models for reliability assessment of, 141–45
    mean-time-to-failure for, 144
    in statistical assessment, 109–22
Polynomial, 214, 222, 229, 237
Posterior probability, 13, 126–28, 134
Power supply, 158
Precise Weibull estimation and confidence assessment, 176–81
Prior data, 125
Prior distribution, 126–40, 196, 202–6
Prior probabilities, 13
Prior sufficient statistics, 126, 130–31
Probability, 3–10, 12–15. *See also* Random variable, Distribution, Bayes theorem, Event, Conditional probability
Probability distributions, families of. *See* Families of distributions
Probability density function, 22–24, 56–57, 97, 101, 131–35, 163, 167. *See also* Distribution, Distribution function

Probability function, 19–24
Probability graphing
  exponential distribution, larger sample sizes, 90–93
  exponential distribution, small sample sizes, 85–90
  normal distribution, 59–63
  plotting points, 61, 171
  table of plotting points, 285
  Weibull distribution, 165–69
  Weibull hazard plotting, 169–75
Probability plotting. *See* Probability graphing
Probability polynomial, 229
Pseudorandom number generator, 203, 207
Pugh, E. L., 118

Quality level, xxi. *See also* Acceptable quality level
Queuing, 24, 193–94
Quine, W. V., 214

Radioactivity, measurement of, 99
Raiffa, H. W., 130
Random number. *See* Pseudorandom number generator
Random sample, 28–31
Random variable, xxii, 19, 39, 69, 84. *See also* Distribution
Rank-order statistic, 65, 93. *See also* Lilliefors K–S test, Weibull hazard
Redundancy, xxvi, 95, 217–18
Reliability, xxi–xxv, 27. *See also* Distribution, Confidence, Bayesian methods, System reliability

Reliability assessment, xxiii, 19, 39, 43, 50–51, 107–8, 139–40, 149
Reliability importance, 238–40, 262
Reliability prediction, xxvi
Repairable system, 101, 191
Repair rate, 196
Repair time, 69, 78, 194
Right-skewed dispersion, 69
Ring circuit, 260
Robotic calibration system, 79
Rocket motor, 77
Root, 243. *See also* Mathematical graph theory
Root finder, 203–4

Safety margin, xxii
Sample. *See* Random sample
Sarhan, A. E., 119
Scale, 57. *See also* Scale parameter
Scale parameter, 119, 165, 167, 176, 177
Scheuer, E. M., 180
Schlaifer, R., 130
Search methods, xxvii, 214–15, 232. *See also* Topological reliability
Seed. *See* Monte Carlo simulation
Sensor, 78
Sequential analysis, 150
Serial system, 217
Series reductions, 223–25
Services, xxi
Service time, 69
Set theory, 3–6, 20, 214–21, 230–31
Shannon, C., 214

Shape, 57
Shape parameter, xxiv, 162, 165, 176
Shewhart, W. A., 150
Shift parameter, 57. *See also* Location parameter, Scale parameter
Simulation, 22. *See also* Monte Carlo simulation
Single censoring, 110. *See also* Multiply censored data
Software, xxi
Solar collector, 259
Solar energy module, 259
Source-to-k-terminal (SKT), 253–56
Source-to-terminal systems, 215, 241
Spacecraft failure, 16
Specification limit, 55
Standard deviation, xxii, 25–26, 93, 177
Standardized binary distribution, 33–34
Standardized form, 32–34, 56–59, 67, 177
  table of standardized normal distribution, 283–84
Standby operation, 95, 98, 258
Statistical assessment
  Bernoulli processes in, 109, 112–14, 130–33
  Poisson processes in, 110, 114–17, 133–45
Statistical hypothesis, 63
Statistical Research Group (SRG), 150
Statistical test, xxii, xxiii
Stopping, 248. *See also* Mathematical graph theory

Stress corrosion, 104, 121
Student-t distribution, 74, 76–77. *See also* Noncentral-t
Subgraph. *See* Mathematical graph theory
Subset. *See* Set theory
Subsystems, 257
Success-or-failure process, 23. *See also* Bernoulli process
Sufficiency, 111
Sufficient statistics, 130, 177
Sum-of-disjoint products (SDP), xxvii, 214, 223, 229–40
Supergraph, 253
Survival probability, 99. *See also* Reliability
Symmetry, 43, 57–59, 69, 209
System formula, 233
  deriving, 248–52
  factored, 215
System reliability, xxvi–xxvii, 213–64
System reliability formula, 214, 248, 250, 252, 261
  factoring of, 242
System-reliability graph, 246

t-Distribution. *See* Student-t distribution, Noncentral-t distribution
Terminal, 218, 252
Time-dependent data. *See* Time-to-failure data, Time-to-repair
Time-to-failure data, xxiv, xxvi, 69, 83–108, 135–45, 161
Time-to-repair (TTR), xxvi, 194
Tippett, L. H. C., 77, 162

Tolerance factor, 153–54, 199
  table of one-sided, for normal distribution, 287
  table of two-sided, for normal distribution, 288–89
Tolerance-limit factors, 150, 154–55
Tolerance limits, 55, 150
Topological reliability (TR), xxvii, 214–15, 218, 241–56
  concepts of, 242–46
  extensions of, 252–56
  four processing rules of, 246–47
  principle differences between inclusion-exclusion and, 241–42
  relationship between inclusion-exclusion and, 249–52
Torque, 156
Torsion spring, 156
Transformer, 52
Transistors, 104, 184
Transmission lines, 260
Tree search, 243–48
Triangular distribution, 33
Trimming, 77
Truncation, 110
t-Statistic, 76–77. *See also* Student-t distribution
Two-out-of-three system, 262
Two-parameter exponential distribution, 109
  best linear unbiased estimates, 119–21
Two-parameter Weibull distribution, 167
Two-sided tolerance analysis, xxiv, 56, 154–56
  Hald's approximation, 155–56
  table of, for normal distribution, 288–89
Two-way edges, 246

Unbiased estimation, 109, 121
Unbiased sampling procedure, 28
Unified approach, extensions of, 252–56
Uniform distribution, 22, 26, 33, 203
Union, 6. *See also* Set theory
Unit event, 5. *See also* Set theory
Universal set, 4, 20–21. *See also* Set theory
Upper-bound technique, 226
Upper confidence limit (UCL), 76, 135
Upper specification limit, 150–51
Utility poles, 183

Valves, 16, 58
Variables, standardized, 31–32
Variance, 25–26, 29–33, 43–45, 56, 74
Venn diagram, 5–6, 15, 230–31
Venn, J., 214
Vital statistics, 161
von Bortkewicz, 104
von Neumann, J., 214
Voting circuit, 101

Wald, A., 155
Wald-Wolfowitz factors, 155
Wallis, W. A., 150
Warranty, 121
Webb, S. R., 155
Weibull, Waloddi, xxiv, 162

Weibull cumulative hazard plotting, 169
Weibull distribution, xxiii–xxv, 118, 150, 161–89, 208–10
  hazard functions, 162, 170–71
  hazard graph paper, xxv, 171–73
  precise estimation and confidence assessment, 176–81
  probability density function, 167
  probability graph paper, 166, 167–69
  probability plotting, elements of, 164–65
  tables of, for confidence and tolerance bounds, 316–36
  tables of weights for best linear invariant estimates of, 290–315
time between failures (TBF), 196
time-to-failure (TTF), 208
Weight, 119, 179. *See also* Mathematical graph theory
Weighted average of order statistics, 177
Weight restriction (WR), 246. *See also* Weight
Wolfowitz, J., 155
Worst case, 225

Yates correction factor, 67–69, 79

Zero-one process. *See* attributes
Z-tables, 32